Scientific Advancements for an Innovative Agri-Food Supply Chain towards the 2030 Sustainable Development Goals III

Scientific Advancements for an Innovative Agri-Food Supply Chain towards the 2030 Sustainable Development Goals III

Editor

Dimitris Skalkos

Basel • Beijing • Wuhan • Barcelona • Belgrade • Novi Sad • Cluj • Manchester

Dimitris Skalkos
Department of Chemistry
University of Ioannina
Ioannina
Greece

Editorial Office
MDPI AG
Grosspeteranlage 5
4052 Basel, Switzerland

This is a reprint of the Special Issue, published open access by the journal *Sustainability* (ISSN 2071-1050), freely accessible at: www.mdpi.com/journal/sustainability/special_issues/Q13695J7I6.

For citation purposes, cite each article independently as indicated on the article page online and using the guide below:

Lastname, A.A.; Lastname, B.B. Article Title. *Journal Name* **Year**, *Volume Number*, Page Range.

ISBN 978-3-7258-1658-3 (Hbk)
ISBN 978-3-7258-1657-6 (PDF)
https://doi.org/10.3390/books978-3-7258-1657-6

Contents

Preface

Two years ago, when we announced the first edition of this Special Issue, the world was changing rapidly due to the Coronavirus effect. The first edition was entitled "Innovative Agrifood Supply Chain in the Post-COVID 19 Era". This Special Issue focused on innovative scientific insights and technological advances in natural resources, organic pollutants identification, new food product development, traceability, packaging, chain management, consumer attitudes, and eating motivations, aiming to tackle the foreseen changes in the global economy and society. A year after the second edition was completed, entitled "Prospects, Challenges and Sustainability of the Agri-Food Supply Chain in the New Global Economy II Era", on the Agri-food Supply Chain (AFSC), the world was still changing in an unpredictable and unprecedented way with unforeseen consequences. The AFSC is at the center of these changes and has been the subject of worldwide studies, due to global economic change, necessitating drastic changes to adjust to and accept the new conditions. The process "from farm to fork" is, no doubt, a key factor in the sustainability and progress of the food produced for the consumers worldwide.

A year after (two years from the beginning), with the completion of this third Special Issue on the Innovative Agri-food Supply Chain (IAFSC), the world is still changing, with serious regional troubles and disputes drastically affecting the global economy and growth. Innovation plays a vital role in modernizing the AFSC towards the 2030 sustainable development goals (SDGs) agenda set by United Nations in 2015 to achieve a sustainable future. Among other things, SDGs provide a basis to end poverty, eradicate hunger, protect the planet, and improve quality of life in the word, ensuring a balance between social, economic, and environmental sustainability.

In this special edition, selected papers on the prospects, challenges, and sustainability of the Innovative AFSC in the new global economy are presented. The driving force of the chain remains, no doubt, the end users of the food, namely, the consumers. The topic of this Special Issue includes modern, state-of-the-art research topics on the subjects of the "primary sector" (five papers), "innovative foods" (one paper), food waste management (three papers), "consumers' and stakeholders" perceptions of foods (four papers), and food diets (one paper). Overall, a total of 14 papers are presented in this issue.

Dimitris Skalkos
Editor

sustainability

Editorial

Scientific Advancements for an Innovative Agri-Food Supply Chain towards the 2030 Sustainable Development Goals III

Dimitris Skalkos

Laboratory of Food Chemistry, Department of Chemistry, University of Ioannina, 45110 Ioannina, Greece; dskalkos@uoi.gr; Tel.: +30-2651008345

check for updates

Citation: Skalkos, D. Scientific Advancements for an Innovative Agri-Food Supply Chain towards the 2030 Sustainable Development Goals III. *Sustainability* **2024**, *16*, 5693. https://doi.org/10.3390/su16135693

Received: 28 June 2024
Accepted: 2 July 2024
Published: 3 July 2024

The world has been changing at an unprecedented speed in the last five years, with unforeseen consequences globally. First it was the COVID-19 pandemic, which changed the social and economic conditions, and now are the regional disputes that are affecting the global economy, generating crises worldwide [1]. In this uncertain environment, as mentioned by the Lancet Commission in 2019, "Food is the single strongest lever to optimize human health and environmental sustainability on the planet" [2]. Food is currently threatening both people and the planet [3]. Food production in a sustainable manner is becoming an urgent issue today for the sustainability of human life and the planet. In 2000, at the Millenium Summit, the Millennium Development Goals (MDGs) were set. At the World Summit on Sustainable Development in 2002, in Johannesburg, terms such as "sustainable diet" [4], "sustainable food consumption" [5], "sustainable nutrition" [6], and "nutritional sustainability" [7] were considered as efforts to produce and consume food in a more sustainable manner. In 2006, the connection between biodiversity, food, and nutrition was established at the Convention of Biological Diversity [8], followed by the World Food Summit on Food Security, where the declaration of the World Summit on Food Security was adopted in 2009 with the commitment for the end of hunger by all the participating countries. In 2012, in Rio de Janeiro, Brazil, at the conference on Sustainable Development with the outcome document, "The Future we Want", the open working group for the Sustainable Development Goals (SDGs) was established. This led to the 2030 Agenda for Sustainable Development, adopted by the United Nations in 2015, and the 17 Sustainable Development Goals (SDGs) [9]. These goals are as follows [10]:

1. *No Poverty*: end poverty in all its forms everywhere.
2. *Zero Hunger*: End hunger, achieve food security, improve nutrition, and promote sustainable agriculture.
3. *Good Health and Well-Being*: Ensure healthy lives and promote well-being for all at all ages.
4. *Quality Education*: Ensure inclusive and equitable quality education and promote lifelong learning opportunities for all.
5. *Gender Equality*: Achieve gender equality and empower all women and girls.
6. *Cleaning and Sanitation*: Ensure availability and sustainable management of water and sanitation for all.
7. *Affordable and Clean Energy*: Ensure access to affordable, reliable, sustainable, and modern energy for all.
8. *Decent Work and Economic Growth*: Promote sustained, inclusive, and sustainable economic growth, full and productive employment, and decent work for all.
9. *Industry, Innovation, and Infrastructure*: Build resilient infrastructure, promote inclusive and sustainable industrialization, and foster innovation.
10. *Reduce inequalities*: Reduce inequality within and among countries.
11. *Sustainable Cities and Communities*: Make cities and human settlements inclusive, safe, resilient, and sustainable.

12. *Responsible Production and Consumption*: Ensure sustainable consumption and production patterns.
13. *Climate Actions*: Take urgent action to combat climate change and its impacts.
14. *Life below Water*: Conserve and sustainably use the oceans, seas, and marine resources for sustainable development.
15. *Life on Land*: Protect, restore, and promote sustainable use of terrestrial ecosystems; sustainably manage forests; combat desertification; and halt and reverse land degradation and biodiversity loss.
16. *Peace, Justice, and Strong Institutions*: Promote peaceful and inclusive societies for sustainable development, provide access to justice for all, and build effective, accountable, and inclusive institutions at all levels.
17. *Partnership for the Goals*: Strengthen the means of implementation and revitalize the global partnership for sustainable development.

The SDGs have been designed to create a vision for achieving a sustainable future. More specifically, they were designed to, among other things, provide a basis to end poverty, eradicate hunger, protect the planet, and improve the quality of life in the world, ensuring a balance between social, economic, and environmental sustainability [11]. Out of the 17 goals of SDG presented above, two of them aim to "end hunger, achieve food security, improve nutrition, and promote sustainable agriculture". This goal is characterized by multiple dimensions (social, economic, and environmental) that go much further than food security [12]. All SDGs were meant to ensure homogeneity between different policy sectors, especially those related to Agri-Food Supply Chain (AFSC), which is at the core of the SDGs' objectives [13]. The following are a few examples of these interrelation links:

1. The promotion of nutrition as poverty makes it difficult to meet nutritional recommendations and restricts access to proper food production included in SDG1.
2. The promotion of sustainable food production to cope with the undernourishment included in SDG2.
3. The promotion of healthy and sustainable nutrition supporting good health is included in SDG3.
4. The promotion of food systems within the sustainable consumption and production part.

The global food system plays a central role in the achievement of the SDGs [14]. The world's major staple crops, i.e., maize, wheat, and rice, must be sustainably produced and contribute to human health and wellbeing to achieve these goals. The intense involvement of the private sector in developing the SDGs is ubiquitous [15]. Almost each of the 169 targets listed under the SDGs is, to a greater or lesser extent, related to food and farming. Indeed, there is overwhelming historical evidence from the developed and the newly emerging economies of the developing world that indicates that agricultural growth has been the primary engine of overall economic growth [16].

From a realistic point of view, it should be mentioned that the implementation of the SDGs has deep economic implications and difficulties, especially for the least developed countries [11]. The urge for achieving the SDGs by 2030 implies that more attention should be placed on the analysis and assessment of the costs of implementing the SDGs and the (intemporal) costs of not pursuing them [17]. The implementation of the SDGs is a global priority, and the role of economic instruments and sustainable financing options has become essential in overcoming the costs associated with implementing the 2030 agenda through bridging the funding gaps for socioeconomic and environmental challenges [18].

In this third special edition, selected subjects of the Agri-Food Supply Chain (AFSC) towards the 2030 sustainable development are presented in 14 published papers. These include modern, state-of-the-art research topics from the subjects of primary sector (five papers), innovative foods (one paper), food waste management (three papers), consumers' and stakeholders' perceptions of foods (four papers), and food diet (one paper).

The five papers that are covering topics of the primary sector are as follows:
POPULATION DYNAMICS OF THE OLIVE FLY *BACTROCERA OLEAE*:

Katsikogiannis et al. investigated the population dynamics of the olive fly, *Bactrocera oleae*, one of the most studied pests of its kind, due to its economic impact on the olive trees' production worldwide. Their findings indicate that ply populations are influenced mostly by climate and attitude over longer periods in the season and from bait sprays for shorter periods of time, which appear to be less effective in autumn.

A NATURE-BASED SOLUTION DRIVEN BY MEDITERRANEAN LEAVING LABS:

Yahya et al. investigated insights on an open innovation ecosystem of Mediterranean Living Labs for the synergetic development and participatory assessment of decentralized wetland-aquaponics in disadvantaged rural areas. This study addressed the knowledge gap and the limited research on the subject while revealing the role of public participation in ascertaining the solution and evaluating its feasibility.

INNOVATIVE IV AND CSD CONDITIONS FOR IMPROVED GRAPE PRODUCTION:

Alba et al. investigated optimizing combined effects of irrigation volumes (IV) and effective cold storage duration (CSD) practices on the quality of grapes produced. Their findings indicate that the proposed IV and CSD conditions gave superior organoleptic and dietary characteristics to the grapes produced, meeting consumers' demands year-round in a better way.

RESEARCH ON WINERY VINEYARD AGRICULTURAL PRODUCTION:

Merkouropoulos et al. investigated an innovative holistic approach in viticulture towards wine production, applying a multidisciplinary methodology for the wine "Asproudi" of the Monemvasia vineyard of Greece. Their findings revealed the genetic relationship of the winery genotypes to the varieties maintained in the reference collection, whereas some other genotypes remained unknown.

EFFECTIVE STRATEGIES FOR RURAL POVERTY REDUCTION:

Ruben investigated the analytical linkages between key problems that cause smallholder poverty, the constraints that limit the effectiveness of ongoing rural development initiatives, and the prospects for alternative strategies to support behavioral change. He concludes that coordinated structural reforms in farms and community organizations, value chain integration, and more effective public–private cooperation are needed to improve poverty conditions.

The paper, which is covering the topic of new foods, specifically new packaging, is as follows:

AN INNOVATIVE INTELLIGENT CHEESE PACKAGING WITH SPIRULINA:

Kontogianni et al. proposed the use of whey protein-based edible films containing spirulina as an innovative cheese packaging alternative, and they found that the proposed intelligent packaging was appropriate for "kefolotyri" yellow Greek cheese.

The three papers covering topics of food waste management are as follows:

INNOVATIVE FEED ADDITIVE FOR IMPROVED NUTRITIONAL PRODUCT:

Kotsou et al. investigated the use of spent coffee grounds (SCG), a by-product, as a feed additive of alternative protein sources with a reduced environmental impact for insect species *Tendbrio molitor* larvae, commonly used for fish, poultry, and pig feeding. They found increased protein composition in the final product.

POTENTIAL SUSTAINABLE USES OF BREWERY BY-PRODUCTS:

Soceanu et al. investigated sustainable strategies for the recovery and valorization of brewery by-products, determining the chemical characteristics of different types of brewery waste, such as moisture content, ash, pH, total content of phenolic compounds, and total protein content. The experimental values obtained have shown that brewery waste is a valuable by-product indeed.

REDUCTION OF FOOD WASTE THROUGH ORGANIZATIONAL THEORIES:

Ramanathan et al. investigated food waste from an operational and supply chains of view, especially from the lens of existing theories in the operations management literature and newer sustainability theories. They proved that existing theories could help explain the motivations of firms engaging in food waste reduction but also call for more research that could help explain interesting observations, not apparent at first glance.

The four papers covering topics of the consumers' and stakeholders' perceptions on foods are as follows:

THE USE OF EDIBLE INSECTS AS A VALUE ANIMAL FEED SOURCE:

Gomes et al. investigated the value chain of the use of edible insects in animal feed in Brazil through the framework of SWOT, the business model, and the multiple case study of two companies, highlighting the sustainability characteristics, identifying the actors in the chain, and how value is generated. They found that the value chain can become a more significant aspect of sustainable agriculture by closing nutrient and energy loops, promoting food security, and minimizing climate change and biodiversity losses.

THE REVALORIZATION OF SURPLUS MATERIAL FROM FRUIT AND VEGETABLE SECTOR:

Fox et al. investigated how industry stakeholders in Ireland manage surplus fruit and vegetable material remaining as a way to reduce food waste through valorization after their main processing. They found that joined-up thinking is required among all stakeholders, including consumers and policymakers, to create positive sustainable changes in view of achieving food waste reduction targets and encourage revalorization.

WINE TOURISM OPPORTUNITIES FOR GROWTH IN THE POST-COVID-19 ERA:

Santorinaios et al. investigated consumers' perceptions and attitudes for wine tourism opportunities in Greece using a formatted questionnaire promoted through the Google platform. Based on the participants' answers, they found three distinctive types of support for the successful development of wine tourism in wine-producing countries such as Greece.

REDUCED-SALT GREEN TABLE OLIVES OPPORTUNITIES FOR DEVELOPMENT:

Paltaki et al. investigated consumers' behavior, attitude, and expectation for the development of a new reduced-salt table olive product from Chalkidiki, an area of Greece. They found that such an innovative food is promising as the interests of consumers and industry have turned to foods that add nutritional value and meet updated food expectations.

Finally, the paper covering a topic of food diet is as follows:

THE GLOBAL GROWTH OF SUSTAINABLE DIET:

Gialeli et al. investigated trends and turning points over time, based on literature evaluation, covering the research on sustainable diets (SD), though a comprehensive bibliometric analysis of publications during the period 1986–2022. Among several dietary patterns, Mediterranean diet (MD) was identified as the most popular among the local SDs, with synergies among scientists in the Mediterranean region.

Acknowledgments: As the Guest Editor of the Special issue titled "Scientific advancements and pathways for an innovative Agri-Food Supply Chain towards the 2030 Sustainable Development Goals III", I express my deep appreciation to all authors whose valuable work was published under this issue and thus contributed to the success of this edition.

Conflicts of Interest: The author declares no conflicts of interest.

List of Contributions:

1. Gomes, J.G.C.; Okano, M.T.; Ursini, E.L.; Lodo do Santos, H.C. Insect Production for Animal Feed: A multiple Case Study in Brazil. *Sustainability* **2023**, *15*, 11419. https://doi.org/10.3390/su151411419.

2. Gialeli, M.; Troumbis, A.Y.; Giaginis, C.; Papadopoulou, S.K.; Antomiadis, I.; Vasios, G.K. The Global Growth of "Sustainable Diet" during Recent Dacades, a Bibliometric Analysis. *Sustainability* **2023**, *15*, 11957. https://doi.org/10.3390/su151511957.

3. Kontogianni, V.G.; Kosma, I.; Mataragas, M.; Pappa, E.; Badeka, A.V.; Bosnea, L. Innovative Intelligent Cheese Packaging with Whey Protein-Based Edible Films Containing Spirulina. *Sustainability* **2023**, *15*, 13909. https://doi.org/10.3390/su151813909.

4. Katsikogiannis, G.; Karvoudakis, D.; Tscheulin, T.; Kizos, T. Population Dynamics of the Olive Fly, *Bactrocera oleae* (Diptera: Tephritidae), Are Influenced by Different Climates, Seasons, and Pest Management. *Sustainability* **2023**, *15*, 14466. https://doi.org/10.3390/su151914466.

5. Yahya, F.; Samrani, A.E.; Khalil, M.; Abdin, A.E.D.; El-Kholy, R.; Embady, M.; Negm, M.; Ketelaere, D.D.; Spiteri, A.; Pana, E.; et al. Decentralized Wetland-Aquaponics Addressing Environmental Degradation and Food Security Challenges in Disadvantaged Rural Areas: A

6. Nature-Based Solution Driven by Mediterranean Living Labs. *Sustainability* **2023**, *15*, 15024. https://doi.org/10.3390/su152015024.
6. Markopoulos, G.; Miliordos, D.E.; Tsimbidis, G.; Hatzopoulos, P.; Kotseridis, Y. How to Improve a Successful Product? The Case of "Asproudi" of the Monemvasia Winey Vineyard *Sustainability* **2023**, *15*, 15597. https://doi.org/10.3390/su152115597.
7. Fox, S.; Kenny, O.; Noci, F.; Dermiki, M. A Pilot Study on Industry Stakeholders' Views towards Revalorization of Surplus Material from the Fruit and Vegetable Sector as a Way to Reduce Food Waste. *Sustainability* **2023**, *15*, 16147. https://doi.org/10.3390/su152316147.
8. Kotsou, K.; Chatzimitakos, T.; Athanasiadis, V.; Bozinou, E.; Athanassiou, C.G.; Lalas, S.I. Utilization of Spent Coffee Grounds as a Feed Additive for Enhancing the Nutritional Value of *Tenebrio molitor* Larvae. *Sustainability* **2023**, *15*, 16224. https://doi.org/10.3390/su152316224.
9. Santorinaios, A.; Kosma. I.S.; Skalkos, D. Consumers' Motives on Wine Tourism in Greece in the Post-COVID-19 Era. *Sustainability* **2023**, *15*, 16225. https://doi.org/10.3390/su152316225.
10. Soceanu, A.; Dobrinas, S.; Popescu, V.; Buzatu, A.; Sirbu, A. Sustainable Strategies for the Recovery and Valorization of Brewery By-Products-A Multidisciplinary Approach. *Sustainability* **2024**, *16*, 220. https://doi.org/10.3390/su16010220.
11. Ramanathan, R.; Ramanathan U.; Pelc, K.; Hermens I. How Do Existing Organizational Theories Help in Understanding the Responses of Food Companies for Reducing Food Waste?. *Sustainability* **2024**, *16*, 1534. https://doi.org/10.3390/su16041534.
12. Paltaki, A.; Mantzouridou F.T.; Loizou, E.; Chatzitheodoridis, F.; Alvanoudi, P.; Choutas, S.; Michailidis, A. Consumers' Attitudes towards Differentiated Agricultural Products: The Case of Reduced-Salt Green Table Olives. *Sustainability* **2024**, *16*, 2392. https://doi.org/10.3390/su16062392.
13. Alba, V.; Russi, A.; Forte, G.; Milella, R.A.; Roccotelli, S.; Campi, P.; Modugno, A.F.; Pipoli, V.; Gentilesco, G.; Tarricone, L.; Caputo, A.R. From Farm to Fork: Irrigation Management and Cold Storage Strategies for the Shelf Life of Seedless Sugrathirtyfive Table Grape Variety. *Sustainability* **2024**, *16*, 3543. https://doi.org/10.3390/su16093543.
14. Ruben, R. What smallholders want: Effective strategies for rural poverty reduction. *Sustainability* **2024**, *16*, 5525. https://doi.org/10.3390/su16135525.

References

1. Hussein, H.; Knol, M. The Ukraine War, Food Trade and the Network of Global Crises. *Int. Spect.* **2023**, *58*, 2211894. [CrossRef]
2. EAT-Lancet Commission. The Eat-Lancet Commission Brief for Everyone. *Lancet.* 2019. Available online: https://eatforum.org/lancet-commission/eatinghealthyandsustainable/ (accessed on 28 June 2024).
3. Willett, W.; Rockström, J.; Loken, B.; Springmann, M.; Lang, T.; Vermeulen, S.; Garnett, T.; Tilman, D.; DeClerck, F.; Wood, A.; et al. Food in the Anthropocene: The EAT–Lancet Commission on Healthy Diets from Sustainable Food Systems. *Lancet* **2019**, *393*, 447–492. [CrossRef] [PubMed]
4. Gussow, J.D.; Clancy, K.L. Dietary Guidelines for Sustainability. *J. Nutr. Educ.* **1986**, *18*, 1–5. [CrossRef]
5. Friel, S.; Barosh, L.J.; Lawrence, M. Towards Healthy and Sustainable Food Consumption: An Australian Case Study. *Public. Health Nutr.* **2014**, *17*, 1156–1166. [CrossRef] [PubMed]
6. Smetana, S.M.; Bornkessel, S.; Heinz, V. A Path from Sustainable Nutrition to Nutritional Sustainability of Complex Food Systems. *Front. Nutr.* **2019**, *6*, 39. [CrossRef] [PubMed]
7. Portugal-Nunes, C.; Nunes, F.M.; Fraga, I.; Saraiva, C.; Gonçalves, C. Assessment of the Methodology That Is Used to Determine the Nutritional Sustainability of the Mediterranean Diet—A Scoping Review. *Front. Nutr.* **2021**, *8*, 772133. [CrossRef] [PubMed]
8. Burlingame, B.; Lawrence, M.; Macdiarmid, J.; Dernini, S.; Oenema, S. IUNS Task Force on Sustainable Diets—Linking Nutrition and Food Systems. *Trends Food Sci. Technol.* **2022**, *130*, 42–50. [CrossRef]
9. Kumar, S.; Kumar, N.; Vivekadhish, S. Millennium Development Goals (MDGS) to Sustainable Development Goals (SDGS): Addressing Unfinished Agenda and Strengthening Sustainable Development and Partnership. *Indian. J. Community Med.* **2016**, *41*, 1–4. [CrossRef] [PubMed]
10. United Nations Transforming Our World: The 2030 Agenda for Sustainable Development United Nations United Nations Transforming Our World: The 2030 Agenda for Sustainable Development. United Nations 2015. Available online: https://sdgs.un.org/2030agenda (accessed on 28 June 2024).
11. Filho, W.L.; Dinis, M.A.P.; Ruiz-De-maya, S.; Doni, F.; Eustachio, J.H.; Swart, J.; Paço, A. The Economics of the UN Sustainable Development Goals: Does Sustainability Make Financial Sense? *Discov. Sustain.* **2022**, *3*, 20. [CrossRef] [PubMed]
12. Galabada, J.K. Towards the Sustainable Development Goal of Zero Hunger: What Role Do Institutions Play? *Sustainability* **2022**, *14*, 4598. [CrossRef]
13. Lencucha, R.; Kulenova, A.; Thow, A.M. Framing Policy Objectives in the Sustainable Development Goals: Hierarchy, Balance, or Transformation? *Glob. Health* **2023**, *19*, 5. [CrossRef] [PubMed]

14. Tanumihardjo, S.A.; McCulley, L.; Roh, R.; Lopez-Ridaura, S.; Palacios-Rojas, N.; Gunaratna, N.S. Maize Agro-Food Systems to Ensure Food and Nutrition Security in Reference to the Sustainable Development Goals. *Glob. Food Sec.* **2020**, *25*. [CrossRef]
15. Terlau, W.; Hirsch, D.; Blanke, M. Smallholder Farmers as a Backbone for the Implementation of the Sustainable Development Goals. *Sustain. Dev.* **2019**, *27*, 523–529. [CrossRef]
16. Abraham, M.; Pingali, P. Transforming Smallholder Agriculture to Achieve the SDGs. In *The Role of Smallholder Farms in Food and Nutrition Security*; Springer: Berlin/Heidelberg, Germany, 2020.
17. Zdzienicka, A.; Prihardini, D. *Meeting the Sustainable Development Goals in Small Developing States with Climate Vulnerabilities: Cost and Financing*; IMF Working Papers; IMF Publications: Washington, DC, USA, 2021; Volume 2021. [CrossRef]
18. Clark, R.; Reed, J.; Sunderland, T. Bridging Funding Gaps for Climate and Sustainable Development: Pitfalls, Progress and Potential of Private Finance. *Land Use Policy* **2018**, *71*, 335–346. [CrossRef]

sustainability

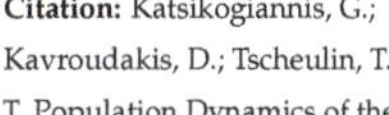

Article

Population Dynamics of the Olive Fly, *Bactrocera oleae* (Diptera: Tephritidae), Are Influenced by Different Climates, Seasons, and Pest Management

Georgios Katsikogiannis *, Dimitris Kavroudakis, Thomas Tscheulin and Thanasis Kizos *

Department of Geography, University of the Aegean, University Hill, 81132 Mytilene, Greece; dimitrisk@aegean.gr (D.K.); t.tscheulin@aegean.gr (T.T.)
* Correspondence: geod16001@aegean.gr (G.K.); akizos@aegean.gr (T.K.)

Abstract: Pest management practices interact with many species and have an impact on the ecology and the economy of the area. In this paper, we examine the population dynamics of the olive fly, *Bactrocera oleae* (Rossi), Diptera: Tephritidae, on Samos Island, Greece, observing the spatial and temporal changes of the pest along an altitude associated with area-wide pest management. More specifically, we analyze data from an extended McPhail trap network and focus on experimental sites, where we monitor the pest population in relation to sprays, temperature, and relative humidity inside the tree canopy during the season for a three-year period. Our findings indicate that fly populations are influenced mostly by climate and altitude over longer periods in the season and from bait sprays for shorter periods of time, which appeared to be less effective in autumn, probably due to population movements and overlapping generations. Apart from the factors that were taken into account, such as the weather conditions and pest management regimes that were proven important, more factors will have to be considered for infestation level, such as fruit availability, inhibition factors (natural enemies, symbiotic agents, food supplies), and cultivation practices. Site microclimate conditions and the landscape can be used to explain changes at the plot level.

Keywords: olive fruit fly; *Bactrocera*; spatial analysis; population dynamics; bait sprays

Citation: Katsikogiannis, G.; Kavroudakis, D.; Tscheulin, T.; Kizos, T. Population Dynamics of the Olive Fly, *Bactrocera oleae* (Diptera: Tephritidae), Are Influenced by Different Climates, Seasons, and Pest Management. *Sustainability* **2023**, *15*, 14466. https://doi.org/10.3390/su151914466

Academic Editors: Marko Vinceković and Dimitris Skalkos

Received: 8 September 2023
Revised: 28 September 2023
Accepted: 30 September 2023
Published: 4 October 2023

1. Introduction

Bactrocera oleae (Rossi), Diptera: Tephritidae, known as the olive fruit fly (OFF), is perhaps one of the most studied pests due to its economic impact [1]. There are, however, still considerable knowledge gaps regarding its biology, like the diapause season and reproductive quiescence [1,2]. It is widely known that, OFFs reportedly overwinter as pupae on the ground or inside infested fruits, but in mild weather conditions, adults can overwinter in a facultative reproductive dormancy and remain active all year round in the canopy [3–6]. Adults can feed on many organic sources, like plant nectar, pollen, and insect honeydew [7]. Each fertilized female lays about 12 eggs a day and about 200–250 eggs in a lifetime [8].

Larvae feed inside fruits, causing premature fruit drop [3,9], and/or, if fruits are harvested, lower the quality of the pressed olive oil due to increased acidity [1,8,10]. Populations and the number of generations per year depend on many different factors, including microclimate (temperature and humidity), fruit availability, and quality [11–14]. Spring reproductive dormancy drives the synchronization of the first ('base generation') and the following generations [2]. Laboratory studies at constant temperatures show that temperatures of 35 °C and above are lethal to pupae [15], and older studies determined that the lower temperature threshold for larval development ranges from 10 to 12.5 °C and the upper temperature from 30 to 32 °C (summarized by Tsitsipis 1980 and Fletcher 1987) [16,17]. In the field, larval development occurs at 12–35 °C [18]. Higher temperature is a major mortality factor for early stages of OFFs, especially for eggs and young larvae.

Also, pupal mortality in the soil during winter plays a pivotal role in population dynamics [19]. To minimize production loss of quantity and quality, OFFs are usually managed with chemical attractants and/or pheromone traps, bait sprays, and cover sprays with insecticides as well as sanitation management, such as the removal of fallen fruits. Long-term use of certain insecticides causes gradual loss of their effectiveness in OFF population control. Examples in Greece include resistance to pyrethroids [20], spinosad [21], and organophosphates [22,23]. Results after a 9-year survey from various regions of Greece, including our study area, suggest increasing resistance for all above substances [24], with a few exceptions. A few developments, including Geographical Information System (G.I.S.) applications and more environmentally friendly spraying substances, but also the extensive abandonment of plantations and the decrease in funds for the practiced control program, brought forward the need for integrated pest management that considers landscape and pest ecology.

Population dynamics of most arthropod species are influenced at the landscape scale [25] by environmental (e.g., relief, microclimate, interactions between organisms, such as parasitoids) and anthropogenic factors (e.g., rural, urban, and industrial activities) [26,27]. Olive plantations are cultivated along a landscape that exhibits different microclimates, especially when these landscapes are continuous [6,13,28,29]. Along these gradients, surrounding land cover can also profoundly affect fly population levels by providing food and/or habitat for parasites/predators [30]. A relationship between population density change and elevation over time has also been found [31], suggesting that adult flies move actively through the landscape. In northern Greece [13], hot spots of OFF populations were observed in the summer months at mid-to-high altitudes (up to 700 m) where conditions were cooler, while the corresponding lowland populations were very small and the opposite phenomenon in the autumn months, when temperatures were closer to the OFF optimum were measured in the lowlands. In a similar study in Israel, apart from environmental factors (mainly temperature and availability of fruit), endogenous factors, such as reproductive quiescence [1,5,27], were found to be important in annual changes of population dynamics. Even in nearby plots (less than one km apart, with same cultivars, tree ages, planting distances, etc.), the different ground morphology yielded different levels of population and damage [27].

In this paper, we investigate spatial and temporal OFF population dynamics on Samos Island, Greece. A spatial Krigging model approach [32] was used on the island to determine spatial and temporal similarities and differences in OFF populations for three seasons. Here, we build on that model and consider applications of crop protection management (bait spraying) rarely studied in the literature. We separate after each spraying event the area into sprayed and unsprayed and compare the effects of spraying for three 10 day periods. This is an exploratory approach, with the use of data for three seasons (2017–2019) for the whole island, which integrates local meteorological (from our own network of temperature and relative humidity data loggers), insect, and spatial data to provide insights on the landscape ecology of OFFs and recommendations for area-wide plant protection management programs.

Our overall goal is to examine how climate, altitude, and pest management influence OFF population dynamics in the landscape. More specifically, the research objectives are:

(1) To monitor the changes in the OFF populations over space in the period June to October for three seasons, both within each season and among the seasons;

(2) To analyze the changes in OFF populations in relation to previous populations for all years and other climatic variables using a linear mixed-effects model and considering spraying events.

2. Materials and Methods

2.1. The Case Study Area: Samos Island

The study area is Samos Island, which is located in the North Aegean Region between circa 37.6 and 37.8 north latitudes and 26.6 and 27.1 east longitudes and covers an area of

477.395 km². The island has a varying topography with two mountain ranges, one in the center north ("Ampelos" or "Karvounis" at 1153 m) and the dry, rugged Kerkis mountain (1434 m) on the west coast (Figure 1). The climate is predominantly Mediterranean, with mild, wet winters and warm, dry summers and prolonged sunshine. Mean annual temperatures range from 10.0 °C in February to 28.5 °C in July. The coldest month is usually February (6.5 °C to 13.2 °C) and the warmest is July (22.2 °C to 32.5 °C). Depending on the local topography, rainfall is within the range of 700 to 900 mm per annum, with somewhat higher averages in the northern part. Summers are usually dry and hot. Almost 60% of the total rainfall occurs in winter, whereas rain is almost absent during the driest months of July and August. The prevailing wind direction in Samos is northerly (almost 60% of the total winds throughout the year). The rugged and diverse terrain and the steep slopes generate microclimates.

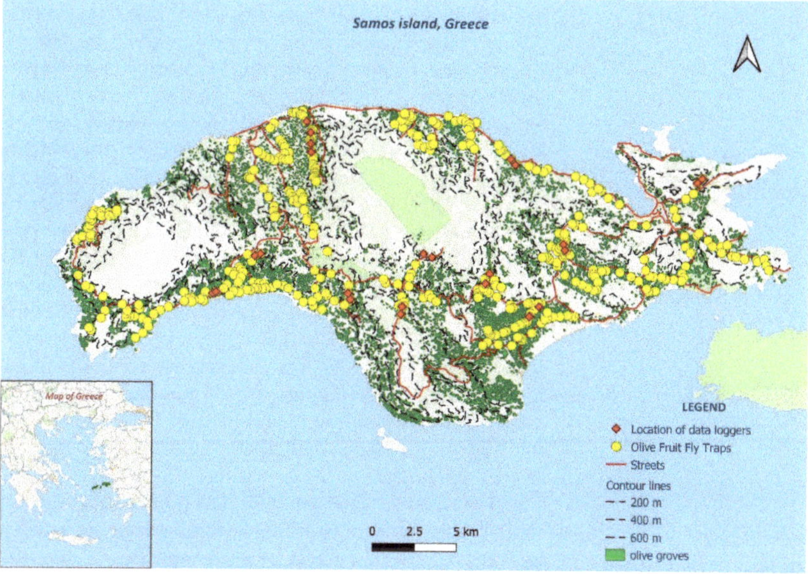

Figure 1. Map of locations for the OFF traps in Samos Island, Greece.

Land use is roughly 25% agricultural land (53% of which are olives and 10% vineyards); 54% shrublands, maquis, and phrygana; and 21% forests (pine and oak). Olive groves cover almost 10,000 ha with ca. 1.3 million olive trees, 55.6% of which are in lowlands (0–200 m), 27.4% in semi-mountainous areas (200–400 m), and 17% in mountainous areas (>400 m), and olive crops are not irrigated. An unknown proportion of fields are neglected (few cultivation practices, except for periodically collecting olives; [33] reports the same for neglected plantations on Lesvos Island) or abandoned, especially in remote and less accessible areas as a direct result of the depopulation of the island in the last 10 years (−36% from 1951 to 1981, +4% since), which was more pronounced in rural and mountainous settlements, and the declining olive oil prices. The most extended cultivar is non-irrigated "Throumpolia", followed by "Koroneiki", "Manaki", "Kalamon", etc. Throumpolia is classified as an intermediate-sized fruit cultivar (weight 2.7–4.2 gr) for mixed use (olive oil/table olives).

2.2. The Olive Fly Management Program on Samos

Twelve plots were selected to represent the island's olive groves according to altitude, percentage of olive cover in one km² zone around the trap, and ensuring that both northern and southern exposures were selected. The selected plots were two above 400 m, four in the 200–400 m zone, and six in the lowland area (<200 m). Two temperature and humidity loggers were set inside the tree canopy in each location [34,35].

Hourly minimum, mean, and maximum temperature and relative humidity measurements were taken from these loggers for the three seasons of the study (June to October). Daily maximum and minimum temperature and and relative humidity data were obtained from the Hellenic National Meteorological Service (HNMS, hereafter) reference station, which is stationed at the airport (latitude: 37.69, longitude: 26.91, altitude: 6.0 m a.s.l.). The long-term climatic series were utilized to define monthly means, identify extreme values, and present the average intra-monthly daily variations of maximum and minimum temperature and relative humidity. Long-term climatic averages for every day were calculated for the normal thirty-year period 1961–1990 for each month (June to October).

In Greece, as well as in our case study area, OFFs are controlled with administrative area pest management programs, recognizing the necessity of a area-wide approach. Insect populations are monitored and insecticide bait sprays are applied by tractors and pedestrian personnel, the frequency and intensity of which are determined by population levels, olive growth stage, and weather conditions. Within the framework of this program, a network of McPhail traps was installed. These traps are commonly used to monitor the OFF population, typically filled with 2% ammonium sulfate solution or protein food lures. On Samos, there is a dense network of around 400 (1 trap/2000 trees) McPhail traps (Figure 1), re-installed at the start of each season (end of May). Traps were checked, cleaned, and refilled every five days. Pest numbers in traps were regularly recorded in a database [36]. The first spray every year is applied at the pit hardening stage (usually starting in the second half of June) because it is when olives start to be susceptible to OFF attack, and then sprays are re-applied only in hot spots suggested by monitoring. The overall efficiency of these spraying programs has been questioned: (a) many fields, especially in remote and inaccessible areas, are abandoned and not sprayed, acting as sources for OFF populations [37]; (b) organic fields scattered throughout the plantations' mosaic are not sprayed; and (c) as the available funds of the pest management program decrease, spraying has to be more selective in areas with more intense pest pressure. Nevertheless, and despite these shortcomings, the analysis can shed some light on the effectiveness and the impact of these events on OFF populations.

Transformation of the OFF data according to spaying events was performed on a 10-day basis, which was deemed more appropriate to estimate spraying event impacts for two reasons: the first is that the actual spraying event typically lasts more than three days. The five-day measurement may be wholly during the event itself. The second is that it agrees with the spraying directions for all substances used. Sprayed areas were calculated based on the routes of the tractors. Traps that fell within a 100 m radius from the line were considered "sprayed", and the rest of the traps were "unsprayed". Therefore, the 399 traps were divided into two subsets every ten days: spayed and unsprayed areas.

The data from the McPhail trap network and spraying data (date, area, substance) in the OFF-management program for three seasons (2017–2018–2019) were used to monitor OFF populations' temporal and spatial dynamics. The trap network used to collect our data consisted of 399 McPhail traps (Figure 1) set on olive trees on June 1 and checked every five days until October 31. We used a mobile phone application [36] to automatically record and control the accuracy and reliability of measurements, which involved digitizing the trap location and the development of a geodatabase accessible via mobile internet where results were stored. The data for the insect populations were automatically stored in a geodatabase by the "trap-setters". Agronomists performed occasional blind controls. Spraying data were provided by the rural development authority using an Android application to record the routes of the tractors used for spraying [36]. All experimental plot data (trap results and meteorological data) are derived from trees over one hundred years old and of the Throumpolia variety to exclude a potential selection bias of the OFFs [38,39].

2.3. Statistical Analysis

The measured variables are listed in Table 1.

Table 1. List of variables used in the analysis.

Variable	Name	Definition	Data Availability
variable 1	Spatial location and altitude of the traps	Geographical coordinates of trap locations	once
variable 2	Hours with maximum temperature exceeding 32 °C	Hours per 10 days and month when maximum temperature exceeds 32 °C calculated by hourly measurements of temperature	10 days/monthly
variable 3	Hours with maximum temperature exceeding 35 °C	Hours per 10 days and month when maximum temperature exceeds 32 °C calculated by hourly measurements of temperature	10 days/monthly
variable 4	Hours with Relative Humidity below 60%	Hours per 10 days and month when RH is below 60% calculated by hourly measurements of temperature	10 days/monthly
variable 5	OFF population	Olive female fly populations as measured in each trap per 5 days, recalculated for 10 days	10 days/monthly, seasonal
variable 6	OFF population	Olive fly populations as measured in each trap per 5 days, recalculated for 10 days	10 days/monthly, seasonal
variable 7	% fertile infestations on the olive fruit/month	Percentage of fertile infestations of total infestations, measured in sampled olives per month for August, September, and October	monthly
variable 8	Spraying incident	If a spray is performed, the area is considered as sprayed for the particular 10 day period	random

Climatic analysis, where maximum and minimum temperature and relative humidity records from the data loggers were transformed into (1) the number of hours within the period used where the temperature exceeds 35 °C represents the absolute maximum temperature for the insect's movement; (2) the number of hours within the period used where the temperature exceeds 32 °C that hinders its movement and activity; and (3) the number of hours within the period where the relative humidity is below 60% as unfavorable conditions.

We performed linear regressions every ten days with (i) the population in the next 10-day period (T1) and (ii) populations for the next three 10-day periods (T1, T2, T3) for the 399-trap network. These regressions are performed for each calculation's spayed and unsprayed areas in T0. We also calculated % population differences for these periods (T1–T0%, T2–T1%, T3–T2%).

More details can be found in the Supplementary Materials. The efficacy of the treatments was verified by monthly olive fruit sampling and infestation control by the local Office for Rural Economy and Veterinary of Samos: Olives are collected randomly from a set number of trees for 22 local communities of the island, the number of trees is determined by the type of terrain (flat, intermediate, mountainous), and the total number of trees in each community is also determined (twenty trees are sampled and marked per 10,000 trees in total). Ten olives per tree are picked in the last 10 days of August, September, and October from different heights in each tree and are dissected to determine the ratio of infested to uninfested olives. The infested olives are further classified by the age of larvae and/or exit points on the fruit.

2.4. Analysis

We performed linear regressions for every 10 days with (i) the population in the next 10-day period (T1) and (ii) populations for the next three 10-day periods (T1, T2, T3) for the 399-trap network. These regressions are performed for the spayed and unsprayed areas in T0 of each calculation. We also calculated % population differences for these periods (T1–T0%, T2–T1%, T3–T2%).

OFF population fluctuations over longer time periods in the season per altitude zone were also performed with descriptive statistics and ANOVA tests per altitude zone for each month.

OFF numbers per trap numbers were log-transformed ($\log(x + 1)$) to achieve a normal distribution and the data were checked for homoscedasticity. We then built a linear mixed-effects model for non-negative count data in R [40] using the function lme in library nlme to test the effects of various explanatory variables on fly numbers in the traps on the basis of maximum likelihood. All fixed effect explanatory variables (altitude (continuous), sprayed (yes/no), olive infestation levels (continuous), fly numbers after 10 days (continuous), fly numbers after 20 days (continuous), and fly numbers after 30 days (continuous) were added to the full model as main factors. In addition, the variables fly numbers after 10 days, fly numbers after 20 days, and fly numbers after 30 days were also added as two- and three-way interactions. We used the following random effect structure in the model: Region/Trap ID/Year/Sampling date to account for the temporal and spatial non-independence of the fly counts and hence to avoid pseudoreplication. All correlations (Pearson's r) between the main factors were well below 0.7. Consequently, we used the function step (AIC) in the library MASS to simplify the model. During this process, the explanatory variables olive infestation, the three-way interaction fly numbers after 10 days–fly numbers after 20 days–fly numbers after 30 days, and the interaction fly numbers after 20 days–fly numbers after 30 days were eliminated from the full model. We then performed an ANOVA test of the simplified model using the function anova.

To estimate the importance and effect of spraying, we numbered all counts of each year in 10 day periods from one to thirteen to cover the 150 days of the season (130 days + 10 before the first count and 10 days after the last) and then divided the total number of traps into those that were in areas that were sprayed (and labeled "sprayed") and the rest (labeled "unsprayed").

3. Results

3.1. Measured Weather Parameters

The weather conditions on Samos Island according to the measurements of our sensors appear rather similar concerning T_{mean} (Table 2): 2019 was slightly hotter in most months of the season than the other two years except June, but there is no overall pattern detected, and most early differences were ironed out later in the season. Relative humidity% values indicate again that 2018 was cooler on average and with higher humidity in the start of the season than both 2017 and 2019, and this trend continued until late in the season. Lower average humidity was recorded in 2017 for almost all months of the season (Table 2). This picture of relative uniformity is not confirmed though when the average hours with temperatures over 32 °C are examined: 2018 is characterized by significantly fewer hours of higher temperatures than the 2017 and 2019 in June, a fact that could potentially lead to more favorable conditions for early development of OFF populations (RH% is also high, which also is expected to favor early growth), while in hotter 2019, lower OFF numbers are expected early in the season. July was significantly hotter in 2017 (20 more hours of high temperatures on average than 2019, Table 2), but then the trend reversed, and August 2019 was hotter than 2017 and 2018. September was again cooler in 2019, which seems to be a year of temperature extremes. These lower temperatures late in the season are expected to favor OFF populations significantly. What becomes evident is that average temperatures and relative humidity figures appear to be unsuitable to describe some trends that could be of importance for OFFs.

3.2. OFF Populations: Temporal and Spatial Differences

OFF populations measured in our trap network were not statistically significantly different in the three reference years. (Table S1). The season in 2019 starts with significantly lower OFF adult counts in the traps, compared to 2017 and 2018 (Figure 2), even though maximum values do not differ so much due to the presence of so-called "hot spots". This pattern continues into July, with the counts of 2019 being as low as 25% of those in other years but with a very high maximum value. In August, OFF counts in the traps in 2019 catch up with those of two previous years, while the highest maximum values of arrests are also recorded in 2019 (one count is more than 1350 OFFs). The lowest counts for August are recorded in 2017. This high growth rate continues in September 2019 with very high average trap arrests, as well as the absolute maximum value of arrests. Counts in October 2019 drop again below the average values of the two previous years. These developments seem to follow the weather conditions of the month in question and those of the previous month.

Table 2. Mean temperature (T_{mean}), mean relative humidity% (RH%), and average hours with temperatures over 32 °C per month and year on Samos.

Month	Average Hours with Temp > 32 °C			Tmean °C			RHmean%		
	2017	2018	2019	2017	2018	2019	2017	2018	2019
6	24.9	7.7	30.1	25.8	24.9	27.2	52.5	61.4	55.8
7	52.3	34.5	30.4	28.2	27.7	27.0	46.5	56.2	50
8	36.2	33.2	47.3	27.7	27.7	28.3	52.4	58.3	49.9
9	13.2	11.9	4.7	24.3	24.7	23.5	51.8	58.1	60.2
10	0.7	0	0.2	18.6	19.1	20.7	62.1	72.7	70.9

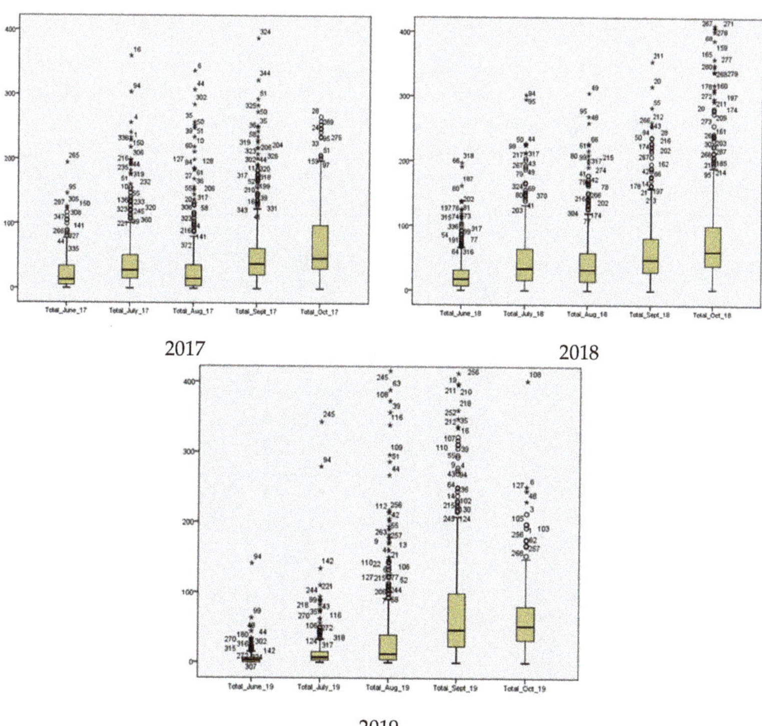

Figure 2. Boxplots of annual and monthly OFF populations on Samos (2017–2019).

A closer look at the monthly differences with elevation classes reveals similar patterns in population changes over the season in different altitudes despite the significant differences in each year and especially 2019 in our case study. In the first month of each season, OFF counts in the lowland traps and 200–400 m altitudes are typically larger, but in hotter

July, counts at higher altitudes increase (Table 3). In August, most counts are reported in middle altitudes, and in the final two months, lower temperatures and higher relative humidity favor lowland populations.

Table 3. Average counts of OFFs per month and year for different altitude zones (asterisks indicate statistically significant differences—between values with the same indication e.g., *1, *2 etc.—of the average values with ANOVA tests and $p < 0.05$ for each month per year and altitude zone).

Altitude Zone	Year/Month	June	July	August	September	October
0–200 m	2017	23.7	38.8 *2	25.3 *3	53.7 *4	82.2 *7
	2018	25.5	49.0	49.6	69.4 *5	100.6 *6
	2019	6.0 *1	13.5	53.4	83.6	59.7
200–400 m	2017	18.6	51.5	66.7 *3	70.6 *4	39.9 *7
	2018	26.0	50.6	52.8	50.1 *5	51.7 *6
	2019	4.0	13.6	68.2	102.0	68.6 *8
>400 m	2017	27.8	74.4 *2	33.8 *3	36.8 *4	28.4 *7
	2018	21.6	42.4	49.0	51.8	41.2 *6
	2019	1.1 *1	27.9	35.9	54.3	45.2 *8
Total	2017	22.9	43.2	33.8	56.0	71.0
	2018	25.4	49.0	50.2	64.7	87.8
	2019	5.4	14.3	55.2	85.5	60.6

For the importance and effect of spraying, the findings vary significantly (Table 4), but overall and especially in the period of the first 10 days after the event, counts in areas that were sprayed do not seem to correspond to decreases in the OFF populations. In some cases, the effect of the spraying seems to be present in the second 10-day period. But, in many cases, the third 10-day period marks a rapid recovery of the OFF population. The timing within the season also seems to be important, as spraying events in the hotter months and until roughly the 6th period (end of July) appear more effective than in the last part of the season in the autumn (10th–13th period). Finally, infestation results from olive fruit sampling indicate that infestation rates rise as the fruits mature and grow in September and October each season, and climatic conditions also become more favorable for OFF populations.

Table 4. The 30-day differences per number of trap counts for traps in sprayed and unsprayed areas for three 10-day periods: 10 days after spraying (T1), immediately before spraying (T0)%, 20 days after spraying T2–T1 (%), and 30 days after spraying T3–T2 (%).

10-Day Periods for 3 Years	T1−T0 (%)		T2−T1 (%)		T3−T2 (%)	
	Unsprayed	Sprayed	Unsprayed	Sprayed	Unsprayed	Sprayed
1st	69.8		−26.0		4.1	
2nd	−28.7	20.5	−1.4	65.1	−7.4	11.4
3rd	5.7	2.6	−27.4	19.7	−14.7	29.3
4th	0.5	−34.8	17.1	−34.5	36.6	90.1
5th	−8.4	98.7	40.1	39.1	14.1	40.6
6th	42.5	9.5	22.4	32.1	14.0	−24.4
7th	52.8	−17.3	12.5	8.5	−10.5	18.5
8th	21.9	2.3	11.4	−16.7	−0.4	−4.2
9th	−3.7	18.0	−3.4	17.1	−15.2	−28.1
10th	2.1	−9.1	−28.1	6.6	46.6	−0.5
11th	−3.4	−41.3	22.6	37.8	−16.4	3.2
12th	15.7	51.0	−9.0	−15.6		
13th	−11.8	−10.2				

The findings of the linear mixed-effects models (Table 5) that combined many of the aforementioned explanatory variables for OFF numbers in the traps of our network did not show some significant results when all explanatory variables were included (altitude, sprayed, olive infestation levels, fly numbers after 10 days, fly numbers after 20 days, fly numbers after 30 days, and their interactions) (Table S2). The removal of the 20- and 30-day fly numbers and infestation levels improved the reliability of the results.

Table 5. R-core Data Analysis Results.

	Value	Std. Error	DF	t-Value	*p*-Value
Intercept	0.5037640	0.11655140	718	4.322248	<0.001
Altitude	0.0032617	0.00037427	346	8.714936	<0.001
Spray	0.3988375	0.06502091	718	6.133989	<0.001
1st 10-days	0.0282857	0.00176229	718	16.050507	<0.001
2nd 10-days	0.0131370	0.00153348	718	8.566800	<0.001
3rd 10-days	0.0176605	0.00211540	718	8.348552	<0.001
1st: 2nd 10-days	−0.0000568	0.00000792	718	−7.174855	<0.001
1st: 3rd 10-days	−0.0002365	0.00003094	718	−7.644837	<0.001

4. Discussion

4.1. Relevance of the Approach

In this paper, we investigate OFF population dynamics in relation to applications of crop protection management (bait spraying), with an approach that separates areas after each spraying event into sprayed and unsprayed and compares the effects of spraying for three ten-day periods. This approach is descriptive and exploratory, considering only macroscopic or landscape level effects of management practices. It does not take into account tree-level and field-level factors (see e.g., [13,30,31]). Nevertheless, its landscape scale is the actual scale at which OFF management is practiced, and we therefore argue that it should be practiced at that scale. The use of data for three seasons (2017–2019) for the whole island ensures that meteorological differences are considered, along with spatial data. We discuss some of the most important findings and lessons from and for better management.

4.2. Temporal Patterns of OFF Populations

The findings of the counts of OFF populations show that while weather conditions have a strong impact, conventional indicators such as the average daily temperature, or even the maximum daily temperature, are not suitable to describe or explain differences in OFFs over the different seasons or even different altitudes. Cumulative indicators though seem to correspond better to these differences in OFF populations. The average hours with temperatures over 32 °C are the indicator that best describes both higher OFF counts in cooler early 2018 and lower early 2019 counts during the hot start of the season. The values of RH% have to be also taken into account as they may favor or hinder insect movement and growth [41,42]. Later in the seasons, the hours over 32 °C again seem to describe the rapid growth of OFF counts in September of 2019. But, overall for the three seasons, despite the seasonal and the spatial differences of OFF counts, the overall populations counted in our trap network are not significantly different. Even in seasons such as 2019, where OFF numbers initially lagged, the populations quickly reached average levels. This seems a fundamental result and has to be taken into account in any efforts for collective management. At the same time, counts are not necessarily infestations. The data collected by the local office on fruit infestations do not indicate significant differences within each year. Infestation rates rise as the fruits mature and grow in September and October and climatic conditions favor OFF population growth. It has to be stressed though that many other factors affect infestation rates, such as unsuitable fruit (shrunken) due to low precipitation, higher predation rates, and others [43]. Differences are observed though, during and between the seasons, from north to south, and at high to low altitudes. This could be attributed to humid and cool northerly winds that prevail in the Aegean in the summer months that lower temperatures. OFF populations in the south increase later in the season due to hotter conditions in the southern lowlands and flatter terrain.

These findings also suggest that adults actively move in the landscape, as has been documented elsewhere [13,27,31,35]. Changes in favorable conditions that the climatic analysis has revealed may also affect reproductive success and population growth rates [41,44,45]. In fact, both factors could explain the differences observed here, and other authors have suggested that these factors are related: movement towards areas with more favorable conditions affects population dynamics [6,13,27,46]. The inclusion of more seasons would help in making sense of the climatic changes in the landscape.

4.3. Spraying

The analysis for spraying seems to reveal that the results are at best mixed in terms of the effectiveness of the management approaches used. This can be explained by a combination of different factors. The first, and most important one, is that they represent local events, typically one or two communities and their fields. To be ordered, OFF populations in the area have to be high and/or rising. Therefore, they are focused on "hot-spots" of OFF populations and a proper comparison has to take into account not just comparing these areas with other areas where OFF populations were lower or decreasing but with the same area populations if the spraying did not take place [27].

Another factor is the type of substances used: a typical example is the second period of the season, where counts in columns T2–T1 (%) and T3–T2 (%) are increasing and then decreasing due to the use of pyrethroid insecticides (deltamethrin) in the spray. These have a rapid effect but reduced effectiveness over time, and therefore OFF populations recover fast. Kampouraki et al. corroborate this [24], but Varikou et al. (mostly in 2018, but also in their 2017 and 2015 publications) question such claims and state that all attractants used are not very efficient, especially late in the season [47–49].

One more issue of concern for spraying seems to be its effectiveness in areas where populations are consistently high. The findings seem to suggest that spraying is more effective in areas of relatively low populations, even if it was found that the OFF population is vulnerable to insecticides on Samos island [24].

Finally, the changes must be weighed against the eventual movements of OFF populations due to weather conditions and the need to find food. These movements are a well-documented fact [13,27,31,50–53]. On Samos, the presence of many organic fields (almost 20% of the total area that is not sprayed) and abandoned fields (around 15% of the total area) in the margins or within the landscape matrix of the olives allows the movement of flies from unsprayed fields to the ones that were sprayed.

4.4. OFF Population Management

One of the objectives of our approach is to provide insights for OFF management. Four issues stand out:

(a) Spraying seems to be more effective in areas of average or relatively low overall populations but less so in areas where populations are consistently high;

(b) Spraying late in the season seems to be less effective in the lowlands, where weather conditions support a rapid increase in the population;

(c) Even in seasons like 2019, where OFF counts were low at first and infestation rates were also low, in late September and especially October in all periods, OFF populations increase significantly. This highlights the need of integrated pest management and use of alternative pest control methods, like repellents or mechanisms to confuse host recognition [54], which is more necessary considering the late stage of the fruit maturation, the greater damage inflicted, and the requirement of no residues in olive oil.

(d) The management regime of the olive plantation and most importantly the presence of neglected or abandoned plantations needs to be recorded. Mapping of these sites will support the effectiveness of the pest management program. Such fields are mostly located in uplands and intermediate altitude zones and can serve as "reservoirs" for flies as they are unsprayed, and fruits are not harvested.

The findings also indicate some shortcomings of the current management system. Possible improvements include:

(i) Mapping of abandonment hotspots;

(ii) Zone sprays between treated and untreated olive plantations that reduce pest movements, which is an old practice (Greek Ministry of Agriculture, 1949), discontinued in recent decades due to: (a) the decrease in available funds for spraying and (b) the false feeling that collective management can be successful in local areas due to extremely active substances that were used in the past (all out of circulation now) and air applications;

(iii) Grazing in abandoned fields can reduce pest population by consumption of the infested fruits;

(iv) Ground cover of (organic) olive groves with mixtures of selected flowering plants can enhance the presence of natural enemies of pests (Braconidae, Miridae, Chrysopidae) by providing them habitat and food and could be part of a sustainable management system [55];

(v) Mass trapping of fruit flies, especially in organic olive orchards, could decrease the OFF population [56] and improve the efficacy of management programs;

(vi) Improvement of bait sprays and frequent treatment in hot spots could be applied depending on the active substance and lure duration [49,57], as reduced efficacy of the usually applied food attractants mixed with registered plant protection products has been reported [58], along with reduced attraction of hydrolyzed proteins [47], indicating the need for refreshing bait spots of the treated olive trees only with attractants reducing the insecticide doses per hectare by two- or three-fold [48].

5. Conclusions

Insect dynamics in the field are a complex topic due to a plethora of interrelated factors. In this paper, we investigated the landscape-level dynamics of OFFs to provide insights for more effective management at the area level. We advocate for landscape-level management to account for the landscape-level factors that need to be considered. Weather conditions and pest management regimes were proven important, but more factors will also have to be considered, such as host (fruit) availability; inhibition factors, like natural enemies, symbiotic agents, or other food supplies; and cultivation practices that were not measured in our approach.

The results nevertheless indicate that the landscape level can indeed be used to describe changes in detail that add depth to explanations at the individual-plot level. These descriptions seem to confirm movements in the population. The shortcomings of management were also identified in our approach, the most important of which refer to the presence of neglected or abandoned plantations interspersed in the olive mosaic that underline the need for more comprehensive area-level management and the incorporation of the landscape matrix in the estimations, which can reduce the need for spraying and overall costs.

Supplementary Materials: The following supporting information can be downloaded at: https://www.mdpi.com/article/10.3390/su151914466/s1, Table S1: Statistical analysis OFF in traps, for 3 years period; Table S2: Statistical analysis of R-core Data Analysis results.

Author Contributions: Conceptualization, T.T. and T.K.; Formal analysis, G.K., D.K., T.T. and T.K.; Investigation, G.K.; Methodology, D.K., T.T. and T.K.; Supervision, T.K.; Visualization, G.K.; Writing—original draft, G.K.; Writing—review & editing, D.K., T.T. and T.K. All authors have read and agreed to the published version of the manuscript.

Funding: This research received no external funding.

Institutional Review Board Statement: Not applicable.

Informed Consent Statement: Not applicable.

Data Availability Statement: All data are available to anyone interested on the server of the Geography Department of the University of the Aegean.

Conflicts of Interest: The authors declare no conflict of interest.

References

1. Daane, K.M.; Johnson, M.W. Olive Fruit Fly: Managing an Ancient Pest in Modern Times. *Annu. Rev. Entomol.* **2010**, *55*, 151–169. [CrossRef]
2. Ordano, M.; Engelhard, I.; Rempoulakis, P.; Nemny-Lavy, E.; Blum, M.; Yasin, S.; Lensky, I.M.; Papadopoulos, N.T.; Nestel, D. Olive Fruit Fly (*Bactrocera oleae*) Population Dynamics in the Eastern Mediterranean: Influence of Exogenous Uncertainty on a Monophagous Frugivorous Insect. *PLoS ONE* **2015**, *10*, e0127798. [CrossRef]
3. Neuenschwander, P.; Michelakis, S. Olive Fruit Drop Caused by *Dacus oleae* (Gmel.) (Dipt. Tephritidae). *Z. Angew. Entomol.* **1981**, *91*, 193–205. [CrossRef]
4. Kapatos, E.T.; Fletcher, B.S. The Phenology of the Olive Fly, *Dacus oleae* (Gmel.) (Diptera, Tephritidae), in Corfu. *Z. Angew. Entomol.* **1984**, *97*, 360–370. [CrossRef]
5. Gutierrez, A.P.; Ponti, L.; Cossu, Q.A. Effects of Climate Warming on Olive and Olive Fly (*Bactrocera oleae* (Gmelin)) in California and Italy. *Clim. Chang.* **2009**, *95*, 195–217. [CrossRef]
6. Petacchi, R.; Marchi, S.; Federici, S.; Ragaglini, G. Large-Scale Simulation of Temperature-Dependent Phenology in Wintering Populations of *Bactrocera oleae* (Rossi). *J. Appl. Entomol.* **2015**, *139*, 496–509. [CrossRef]
7. Tsiropoulos, G.J. Reproduction and Survival of the Adult *Dacus oleae* Feeding on Pollens and Honeydews1. *Environ. Entomol.* **1977**, *6*, 390–392. [CrossRef]
8. Malheiro, R.; Casal, S.; Baptista, P.; Pereira, J.A. A Review of *Bactrocera oleae* (Rossi) Impact in Olive Products: From the Tree to the Table. *Trends Food Sci. Technol.* **2015**, *44*, 226–242. [CrossRef]
9. Tzanakakis, M.E. *Insects and Mites Feeding on Olive: Distribution, Importance, Habits, Seasonal Development, and Dormancy*; BRILL: Leiden, The Netherlands, 2006; ISBN 978-90-474-1846-7.
10. Malheiro, R.; Casal, S.; Baptista, P.; Pereira, J.A. Physico-Chemical Characteristics of Olive Leaves and Fruits and Their Relation with *Bactrocera oleae* (Rossi) Cultivar Oviposition Preference. *Sci. Hortic.* **2015**, *194*, 208–214. [CrossRef]
11. Tzanakakis, M. Seasonal Development And Dormancy of Insects And Mites Feeding on Olive: A Review. *Neth. J. Zool.* **2003**, *52*, 87–224. [CrossRef]
12. Burrack, H.J.; Connell, J.H.; Zalom, F.G. Comparison of Olive Fruit Fly (*Bactrocera oleae* (Gmelin)) (Diptera: Tephritidae) Captures in Several Commercial Traps in California. *Int. J. Pest Manag.* **2008**, *54*, 227–234. [CrossRef]
13. Kounatidis, I.; Papadopoulos, N.T.; Mavragani-Tsipidou, P.; Cohen, Y.; Tertivanidis, K.; Nomikou, M.; Nestel, D. Effect of Elevation on Spatio-Temporal Patterns of Olive Fly (*Bactrocera oleae*) Populations in Northern Greece. *J. Appl. Entomol.* **2008**, *132*, 722–733. [CrossRef]
14. Baratella, V.; Pucci, C.; Paparatti, B.; Speranza, S. Response of *Bactrocera oleae* to Different Photoperiods and Temperatures Using a Novel Method for Continuous Laboratory Rearing. *Biol. Control.* **2017**, *110*, 79–88. [CrossRef]
15. Genç, H.; Nation, J.L. Survival and Development of *Bactrocera oleae* Gmelin (Diptera:Tephritidae) Immature Stages at Four Temperatures in the Laboratory. *Afr. J. Biotechnol.* **2008**, *7*, 2495–2500.
16. Tsitsipis, J.A. Effect of Constant Temperatures on Larval and Pupal Development of Olive Fruit Flies Reared on Artificial Diet. *Env. Entomol.* **1980**, *9*, 764–768. [CrossRef]
17. Fletcher, B.S. The Biology of Dacine Fruit Flies. *Annu. Rev. Entomol.* **1987**, *32*, 115–144. [CrossRef]
18. Tsitsipis, J.A. Effect of Constant Temperatures on The Eggs of The Olive Fruit Fly, Dacus oleae (Diptera, Tephritidae). 1977.
19. Kapatos, E.T.; Fletcher, B.S. Mortality Factors and Life-Budgets for Immature Stages of the Olive Fly, *Dacus oleae* (Gmel.) (Diptera, Tephritidae), in Corfu. *J. Appl. Entomol.* **1986**, *102*, 326–342. [CrossRef]
20. Margaritopoulos, J.T.; Skavdis, G.; Kalogiannis, N.; Nikou, D.; Morou, E.; Skouras, P.J.; Tsitsipis, J.A.; Vontas, J. Efficacy of the Pyrethroid Alpha-Cypermethrin against *Bactrocera oleae* Populations from Greece, and Improved Diagnostic for an IAChE Mutation. *Pest Manag. Sci.* **2008**, *64*, 900–908. [CrossRef]
21. Kakani, E.G.; Zygouridis, N.E.; Tsoumani, K.T.; Seraphides, N.; Zalom, F.G.; Mathiopoulos, K.D. Spinosad Resistance Development in Wild Olive Fruit Fly *Bactrocera oleae* (Diptera: Tephritidae) Populations in California. *Pest Manag. Sci.* **2010**, *66*, 447–453. [CrossRef]
22. Skouras, P.J.; Margaritopoulos, J.T.; Seraphides, N.A.; Ioannides, I.M.; Kakani, E.G.; Mathiopoulos, K.D.; Tsitsipis, J.A. Organophosphate Resistance in Olive Fruit Fly, *Bactrocera oleae*, Populations in Greece and Cyprus. *Pest Manag. Sci.* **2007**, *63*, 42–48. [CrossRef]
23. Vontas, J.G.; Cosmidis, N.; Loukas, M.; Tsakas, S.; Hejazi, M.J.; Ayoutanti, A.; Hemingway, J. Altered Acetylcholinesterase Confers Organophosphate Resistance in the Olive Fruit Fly *Bactrocera oleae*. *Pestic. Biochem. Physiol.* **2001**, *71*, 124–132. [CrossRef]
24. Kampouraki, A.; Stavrakaki, M.; Karataraki, A.; Katsikogiannis, G.; Pitika, E.; Varikou, K.; Vlachaki, A.; Chrysargyris, A.; Malandraki, E.; Sidiropoulos, N.; et al. Recent Evolution and Operational Impact of Insecticide Resistance in Olive Fruit Fly *Bactrocera oleae* Populations from Greece. *J. Pest Sci.* **2018**, *91*, 1429–1439. [CrossRef]

25. Tscharntke, T.; Bommarco, R.; Clough, Y.; Crist, T.O.; Kleijn, D.; Rand, T.A.; Tylianakis, J.M.; van Nouhuys, S.; Vidal, S. Reprint of "Conservation Biological Control and Enemy Diversity on a Landscape Scale" [Biol. Control 43 (2007) 294–309]. *Biol. Control.* **2008**, *45*, 238–253. [CrossRef]

26. Tscharntke, T.; Brandl, R. Plant-Insect Interactions in Fragmented Landscapes. *Annu. Rev. Entomol.* **2004**, *49*, 405–430. [CrossRef] [PubMed]

27. Nestel, D.; Carvalho, J.; Nemny-Lavy, E. The Spatial Dimension in the Ecology of Insect Pests and Its Relevance to Pest Management. In *Insect Pest Management: Field and Protected Crops*; Horowitz, A.R., Ishaaya, I., Eds.; Springer: Berlin/Heidelberg, Germany, 2004; pp. 45–63. ISBN 978-3-662-07913-3.

28. Pontikakos, C.M.; Tsiligiridis, T.A.; Drougka, M.E. Location-Aware System for Olive Fruit Fly Spray Control. *Comput. Electron. Agric.* **2010**, *70*, 355–368. [CrossRef]

29. Pontikakos, C.M.; Tsiligiridis, T.A.; Yialouris, C.P.; Kontodimas, D.C. Pest Management Control of Olive Fruit Fly (*Bactrocera oleae*) Based on a Location-Aware Agro-Environmental System. *Comput. Electron. Agric.* **2012**, *87*, 39–50. [CrossRef]

30. Ortega, M.; Pascual, S. Spatio-Temporal Analysis of the Relationship between Landscape Structure and the Olive Fruit Fly *Bactrocera oleae* (Diptera: Tephritidae): Effect of Landscape on *B. oleae*. *Agric. For. Entomol.* **2014**, *16*, 14–23. [CrossRef]

31. Castrignanò, A.; Boccaccio, L.; Cohen, Y.; Nestel, D.; Kounatidis, I.; Papadopoulos, N.T.; De Benedetto, D.; Mavragani-Tsipidou, P. Spatio-Temporal Population Dynamics and Area-Wide Delineation of *Bactrocera oleae* Monitoring Zones Using Multi-Variate Geostatistics. *Precis. Agric.* **2012**, *13*, 421–441. [CrossRef]

32. Kavroudakis, D.; Kizos, T.; Tscheulin, T.; Katsikogiannis, G.; Stavrianakis, G.; Tsalta, L. Spatial Analysis of Olive Fly on Samos Island. *Int. J. Pest Manag. in press.*

33. Kizos, T.; Koulouri, M. Same Land Cover, Same Land Use at the Large Scale, Different Landscapes at the Small Scale: Landscape Change in Olive Plantations on Lesvos Island, Greece. *Landsc. Res.* **2010**, *35*, 449–467. [CrossRef]

34. Blum, M.; Lensky, I.M.; Nestel, D. Estimation of Olive Grove Canopy Temperature from MODIS Thermal Imagery Is More Accurate than Interpolation from Meteorological Stations. *Agric. For. Meteorol.* **2013**, *176*, 90–93. [CrossRef]

35. Blum, M.; Lensky, I.M.; Rempoulakis, P.; Nestel, D. Modeling Insect Population Fluctuations with Satellite Land Surface Temperature. *Ecol. Model.* **2015**, *311*, 39–47. [CrossRef]

36. Michalakis, V.I.; Kopsachilis, V.; Katsikogiannis, G.; Vaitis, M.; Kizos, T. Integrated Geo-Spatial Information System for Olive Fruit Fly Management. In Proceedings of the 18th Panhellenic Entomological Congress, Athens, Greece, 15–18 October 2019.

37. Bournakas, V. *Management of Olive Fruit Fly with Bait Sprays Method in Integrated Pest Management*; GAIA ΕΠΙΧΕΙΡΕΙΝ S.A.: Athens, Greece, 2017; ISBN 978-618-81642-6-0.

38. Navrozidis, E.; Zartaloudis, Z.; Thomidis, T.; Karagiannidis, N.; Roubos, K.; Michailides, Z. Effect of Soil Plowing and Fertilization on the Susceptibility of Four Olive Cultivars to the Insect *Bactrocera oleae* and the Fungi *Sphaeropsis dalmatica* AndSpilocaea Oleagina. *Phytoparasitica* **2007**, *35*, 429–432. [CrossRef]

39. Malheiro, R.; Casal, S.; Pinheiro, L.; Baptista, P.; Pereira, J.A. Olive Cultivar and Maturation Process on the Oviposition Preference of *Bactrocera oleae* (Rossi) (Diptera: Tephritidae). *Bull. Entomol. Res.* **2019**, *109*, 43–53. [CrossRef] [PubMed]

40. R Core Team R: A Language and Environment for Statistical Computing. 2021. Available online: https://www.r-project.org/ (accessed on 13 December 2021).

41. Gonçalves, M.F.; Torres, L.M. The Use of the Cumulative Degree-Days to Predict Olive Fly, *Bactrocera oleae* (Rossi), Activity in Traditional Olive Groves from the Northeast of Portugal. *J. Pest Sci.* **2011**, *84*, 187–197. [CrossRef]

42. Yokoyama, V.Y. Olive Fruit Fly (Diptera: Tephritidae) in California: Longevity, Oviposition, and Development in Canning Olives in the Laboratory and Greenhouse. *J. Econ. Entomol.* **2012**, *105*, 186–195. [CrossRef]

43. Kombargi, W.S.; Michelakis, S.E.; Petrakis, C.A. Effect of Olive Surface Waxes on Oviposition by *Bactrocera oleae* (Diptera: Tephritidae). *J. Econ. Entomol.* **1998**, *91*, 993–998. [CrossRef]

44. Benhadi-Marín, J.; Santos, S.A.P.; Baptista, P.; Pereira, J.A. Distribution of *Bactrocera oleae* (Rossi, 1790) throughout the Iberian Peninsula Based on a Maximum Entropy Modelling Approach. *Ann. Appl. Biol.* **2020**, *177*, 112–120. [CrossRef]

45. Gonçalves, F.; Torres, L. The Use of Trap Captures to Forecast Infestation by the Olive Fly, *Bactrocera oleae* (Rossi) (Diptera: Tephritidae), in Traditional Olive Groves in North-Eastern Portugal. *Int. J. Pest Manag.* **2013**, *59*, 279–286. [CrossRef]

46. Burrack, H.J.; Bingham, R.; Price, R.; Connell, J.H.; Phillips, P.A.; Wunderlich, L.; Vossen, P.M.; O'Connell, N.V.; Ferguson, L.; Zalom, F.G. Understanding the Seasonal and Reproductive Biology of Olive Fruit Fly Is Critical to Its Management. *Calif. Agric.* **2011**, *65*, 14–20. [CrossRef]

47. Varikou, K.; Garantonakis, N.; Birouraki, A. Residual Attractiveness of Various Bait Spray Solutions to *Bactrocera oleae*. *Crop Prot.* **2015**, *68*, 60–66. [CrossRef]

48. Varikou, K.; Garantonakis, N.; Birouraki, A.; Gkilpathi, D.; Kapogia, E. Refreshing Bait Spots in an Olive Orchard for the Control of *Bactrocera oleae* (Diptera: Tephritidae). *Crop Prot.* **2017**, *92*, 153–159. [CrossRef]

49. Varikou, K.; Garantonakis, N.; Marketaki, M.; Charalampous, A.; Anagnostopoulos, C.; Bempelou, E. Residual Degradation and Toxicity of Insecticides against *Bactrocera oleae*. *Env. Sci. Pollut. Res.* **2018**, *25*, 479–489. [CrossRef] [PubMed]

50. Eber, S.; Brandl, R. Ecological and Genetic Spatial Patterns of Urophora Cardui (Diptera: Tephritidae) as Evidence for Population Structure and Biogeographical Processes. *J. Anim. Ecol.* **1994**, *63*, 187–199. [CrossRef]

51. Nestel, D.; Nemny-Lavy, E. Nutrient Balance in Medfly, Ceratitis Capitata, Larval Diets Affects the Ability of the Developing Insect to Incorporate Lipid and Protein Reserves. *Entomol. Exp. Appl.* **2007**, *126*, 53–60. [CrossRef]

52. Shelly, T.; Epsky, N.; Jang, E.B.; Reyes-Flores, J.; Vargas, R. (Eds.) *Trapping and the Detection, Control, and Regulation of Tephritid Fruit Flies*; Springer: Dordrecht, The Netherlands, 2014; ISBN 978-94-017-9192-2.
53. Rey, P.J.; Manzaneda, A.J.; Valera, F.; Alcántara, J.M.; Tarifa, R.; Isla, J.; Molina-Pardo, J.L.; Calvo, G.; Salido, T.; Gutiérrez, J.E.; et al. Landscape-Moderated Biodiversity Effects of Ground Herb Cover in Olive Groves: Implications for Regional Biodiversity Conservation. *Agric. Ecosyst. Environ.* **2019**, *277*, 61–73. [CrossRef]
54. Delrio, G.; Deliperi, S.; Lentini, A. Experiments for the Control of Olive Fly Using a "Push-Pull" Method. *IOBC/WPRS Bull.* **2010**, *59*, 89–92.
55. Karamaouna, F.; Kati, V.; Volakakis, N.; Varikou, K.; Garantonakis, N.; Economou, L.; Birouraki, A.; Markellou, E.; Liberopoulou, S.; Edwards, M. Ground Cover Management with Mixtures of Flowering Plants to Enhance Insect Pollinators and Natural Enemies of Pests in Olive Groves. *Agric. Ecosyst. Environ.* **2019**, *274*, 76–89. [CrossRef]
56. Broumas, T.; Haniotakis, G.; Liaropoulos, C.; Tomazou, T.; Ragoussis, N. The Efficacy of an Improved Form of the Mass-Trapping Method, Forthe Control of the Olive Fruit Fly, *Bactrocera oleae* (Gmelin) (Dipt., Tephritidae): Pilot-Scale Feasibility Studies. *J. Appl. Entomol.* **2002**, *126*, 217–223. [CrossRef]
57. Varikou, K.; Garantonakis, N.; Birouraki, A.; Ioannou, A.; Kapogia, E. Improvement of Bait Sprays for the Control of *Bactrocera oleae* (Diptera: Tephritidae). *Crop Prot.* **2016**, *81*, 1–8. [CrossRef]
58. Varikou, K.; Garantonakis, N.; Birouraki, A. Comparative Field Studies of *Bactrocera oleae* Baits in Olive Orchards in Crete. *Crop Prot.* **2014**, *65*, 238–243. [CrossRef]

 *sustainability*

Article

Decentralized Wetland-Aquaponics Addressing Environmental Degradation and Food Security Challenges in Disadvantaged Rural Areas: A Nature-Based Solution Driven by Mediterranean Living Labs

Fatima Yahya [1], Antoine El Samrani [2], Mohamad Khalil [1], Alaa El-Din Abdin [3], Rasha El-Kholy [3], Mohamed Embaby [3], Mohab Negm [3], Dirk De Ketelaere [4], Anna Spiteri [4], Eleanna Pana [5] and Vasileios Takavakoglou [5,*]

[1] Azm Research Center, EDST, Lebanese University, Tripoli P.O. Box 100, Lebanon; fatimayahya91@hotmail.com (F.Y.); mohamad.khalil@ul.edu.lb (M.K.)
[2] Laboratory of Geoscience, Georesources and Environment-L2GE, Faculty of Science, Lebanese University, Campus Fanar EDST, Beirut P.O. Box 90656, Lebanon; antoineelsamrani@ul.edu.lb
[3] National Water Research Center, Egyptian Chinese University, Cairo 19346, Egypt; alaa_ea_abdin@yahoo.com (A.E.-D.A.); rasha.elkholy@nwrc.gov.eg (R.E.-K.); mohamed_embaby@nwrc.gov.eg (M.E.); mohabnegm@hotmail.com (M.N.)
[4] Integrated Resources Management Company, Ltd., 24 Pope Benedict XV Square, 1083 Senglea, Malta; dirk@environmentalmalta.com (D.D.K.); anna@environmentalmalta.com (A.S.)
[5] Hellenic Agricultural Organization "DIMITRA", Soil and Water Resources Institute, 57001 Thessaloniki, Greece; eleannapana@gmail.com
* Correspondence: v.takavakoglou@swri.gr; Tel.: +30-2310-473-429

Citation: Yahya, F.; El Samrani, A.; Khalil, M.; Abdin, A.E.-D.; El-Kholy, R.; Embaby, M.; Negm, M.; De Ketelaere, D.; Spiteri, A.; Pana, E.; et al. Decentralized Wetland-Aquaponics Addressing Environmental Degradation and Food Security Challenges in Disadvantaged Rural Areas: A Nature-Based Solution Driven by Mediterranean Living Labs. *Sustainability* **2023**, *15*, 15024. https://doi.org/10.3390/su152015024

Academic Editor: Dimitris Skalkos

Received: 12 August 2023
Revised: 14 October 2023
Accepted: 16 October 2023
Published: 18 October 2023

Abstract: The Mediterranean region is highly vulnerable to climate change, soil and water resource degradation, and biodiversity loss. These challenges disproportionately affect disadvantaged rural areas, impacting both natural resources and the livelihoods of local agricultural societies. Urgent transformative measures are essential to address land and water management as well as food security challenges in these disadvantaged areas. Living labs are being called upon to play a key role in addressing these challenges through the development of Nature-based Solutions (NbSs) that are able to provide environmental and socioeconomic benefits towards the achievement of Sustainable Development Goals. The aim of this work is to provide insights on an open innovation ecosystem of Mediterranean Living Labs for the synergetic development and participatory assessment of decentralized wetland-aquaponics, as NbSs are able to address environmental and food security challenges in disadvantaged rural areas. The study addresses the knowledge gap of Living Labs contribution to the development of decentralized wetland-aquaponics and the limited research on small-scale aquaponics systems in rural Mediterranean settings, while revealing the role of public participation in ascertaining the solution and evaluating its feasibility and impacts in light of the local social values and interests in the mountainous area of Akkar al-Atika in Lebanon.

Keywords: living-labs; cross-border cooperation; participatory; aquaponics; constructed floating wetlands; food security; environmental management; impact assessment; Nature-based Solutions; pollution control; water resources; climate change; Mediterranean

1. Introduction

The Mediterranean is one of the most vulnerable regions in the world to the impacts of climate change, water scarcity, land degradation and biodiversity loss [1]. These threats are especially severe for Mediterranean Areas that face Natural or other specific Constraints (ANCs), such as mountainous areas, islands and low-income areas with clear biophysical limitations. In these areas, climate change affects not only natural resources but also local societies and economies that depend on agricultural production. As indicated by recent remote sensing studies, the combined impacts of both climate and human interventions in Land Use and/or Land Cover changes may not only harm water-dependent natural ecosystems such as lakes and wetlands [2], but also artificial ecosystems related to agricultural production (e.g., paddy fields) [3], and thus restrict the access of rural communities to fundamental ecosystem services for their survivability [4]. In such disadvantaged areas, land degradation, along with unsustainable exploitation of water resources and water scarcity, lead to increased use of agrochemicals and high production costs, food insecurity, economic decline and eventually depopulation. Therefore, there is an evident need for transformative measures of land and water management to sustaining the livelihoods of local societies [1,5].

The implementation of Nature-based Solutions (NbSs) in the food production process can contribute to sustainable environmental management while also providing multiple socioeconomic benefits [6]. In particular, solutions that rely on ecosystem services could be integrated with modern food production processes through an ecological approach with a dual objective of pollution control and safe food production. In this perspective, decentralized, low-cost aquaponic systems that take advantage of wetland ecosystem functions and services could be considered an effective solution for food production based on circular economy concepts that minimize inputs and waste. In the last decade, food production systems such as closed-loop aquaponics systems based on floating wetlands principles for primary production and/or using constructed wetlands as biofilters [7] have attracted considerable attention from academics and practitioners [8,9]. These systems are part of the circular bioeconomy, which is an emerging field of research [10], but their global adoption is described as "small and limited" [11,12], while the assessment of their positive environmental and socio-economic impact is still limited [13].

The adaptation of such NbSs to the needs of local societies, especially in disadvantaged areas, is of prominent importance for the Mediterranean. In this process, the Living Labs approach can be part of an institutional transformative change that combines top-down and bottom-up strategies for achieving sustainability [14]. Living Labs provide a collaborative environment that enables the integration of research and innovation processes in real-life communities and environments [15]. According to the United Nations General Assembly held in 2015, Living labs are seen as assets to achieve the Sustainable Development Goals (SDGs) and mainstream NbSs. A key factor in Living Labs' success is the engagement and active participation of stakeholders in both the development and assessment of NbSs, as these are two critical steps for their endorsement and adoption by local societies. However, recent studies indicate that the way in which Living Labs contribute to the development of operational solutions and sustainability transitions has yet to be explored in more detail [16,17]. Furthermore, there is a research gap in the participatory assessment of the feasibility and impact of NbSs [18] and especially of aquaponics as a Nature-based Solution in the Mediterranean context. Thus, there is an evident need for evaluations of decentralized small-scale systems tailored for disadvantaged rural areas and the opening of research outcomes and potential up to participatory approaches in real-world contexts [19].

The aim of this work is to provide insights on an open innovation ecosystem of Mediterranean Living Labs for the synergetic and participatory development and assessment of decentralized aquaponic systems that integrate floating wetlands services as NbSs for disadvantaged rural areas. The specific objectives of this work are: (a) to illustrate the critical role of Living Labs and the added value of cross-border cooperation and synergies in the development of decentralized wetland-aquaponics as operational agri-environmental NbSs for disadvantaged areas; and (b) to assess the feasibility of adoption and the potential environmental and socioeconomic impacts of decentralized systems using a bottom-up approach in the mountainous and low-income area of Akkar al-Atika in Lebanon.

2. Methodological Framework

2.1. Living Labs Deployment and Operation

This work has been realized within the framework of the Mara-Mediterra project, a research and innovation action funded by the PRIMA foundation under the EU Horizon 2020 Programme. The project demonstrates an open innovation ecosystem of six Living Labs in hotspots of land and water degradation around the Mediterranean, which include the Djelfa area in Algeria, the coastal area of the Nile Delta in Egypt, the North Aegean islands in Greece, the Akkar al-Atika mountainous area in Lebanon, the Lake Marmara area in Turkey, and a peri-urban area in Malta, which provides the pathways for the Living Labs operation.

In a local context, the Living Labs of Mara-Mediterra operate as user-centered innovation platforms based on a systematic approach to co-creating users and integrating R&D and innovation processes in real-life communities [20,21]. A multidisciplinary group of farmers, local community representatives and key stakeholders form a roundtable that guides the operations of the local living lab. These operations include experimental trials and demonstration actions for the co-deployment and co-assessment of eco-engineering and agro-ecological NbSs in real-life settings. At the international cooperation level, the Mara-Mediterra project provides a cross-border facilitation network that allows for the cross-fertilization of local knowledge. This is achieved through scheduled, dedicated events of knowledge and experience exchange between the coordinators of the Living Labs, the deployment of open-access audiovisual materials, and knowledge exchange missions in mirror hotspot areas. The above activities, at the local and cross-border level, create a favorable environment for participatory knowledge development and transfer that allow not only the deployment and validation of NbSs but also the customization and scaling out of best practices, as in the case of wetland-aquaponics, in order to integrate societal challenges and nature conservation across different scales and landscapes.

2.2. Bottom-Up Feasibility and Impact Assessment

The participatory feasibility assessment of wetland-aquaponics took place in the municipality of Akkar al-Atika in Lebanon. The population of this mountainous area is approximately 14,000, with a considerably low density (168 hab/km^2). The local economy is mainly based on forestry and agriculture. However, between 2001 and 2019, Akkar al-Atika showed a significant decrease in forest productivity due to forest fires and climate change impacts. Furthermore, the agricultural areas are impacted by the limited availability of water, while the groundwater is extensively overexploited due to uncontrolled drilling activities, resulting in the degradation of water resources and a decreasing level of the water table. To address these challenges, there is a need for transformative changes in the primary sector. Akkar al-Atika, as the case study area of the present work, is a representative disadvantaged rural area in Lebanon in which the introduction of decentralized wetland-aquaponics as NbSs through the local Living Lab could be a promising alternative to primary production.

To assess the feasibility and the potential environmental and socioeconomic benefits of wetland-aquaponics in Akkar al-Atika, a participatory bottom-up approach was applied through the Living Lab as a novel way to assess decentralized aquaponics systems, in contrast to common expert-driven assessments. In line with similar studies [22], structured interviews of the local stakeholders with the use of questionnaires have been used to assess the acceptance and potential benefits of decentralized wetland-aquaponics in the targeted area. The questionnaires consisted of 4 parts addressing social, technical, economic, and environmental factors related to the wetland-aquaponics system, and basic statistical tools were used for the exploratory analysis of the data.

In terms of data collection and characterization, the survey took place in the Municipality of Akkar Al-Atika, during the period June–September 2023 and consisted of 100 participants. Amongst the interviewed persons, 95% live in Akkar Al-Atika village all year and 5% live in the area only during the summer. The gender balance was 48% male and 52% female, while most of the participants had a secondary and higher education degree. The majority of the respondents had an average annual income of less than 600 USD (48%), and only 5% had an average annual income of over 2000 USD.

3. Synergies of Mediterranean Living Labs for Agri-Environmental Nature-Based Solutions: The Case of Wetland-Aquaponics

NbSs have the potential to offer long-term transformative pathways towards the sustainability of rural Mediterranean landscapes. Using the Living Labs approach, the overall ambition of Mara-Mediterra is to open up the NbSs innovation process to all active players so that new ideas can circulate more freely and eventually be transformed into tools, services and practices that address key environmental challenges in rural Mediterranean areas and foster a stronger culture of environmental stewardship and green entrepreneurship, thereby safeguarding sustainable rural landscapes and boosting the rural economy. In this effort, the synergistic effect and the added value of Living Labs cooperation across borders, within the open innovation ecosystem of Mara-Mediterra are illustrated in the development and evolution of the decentralized wetland-aquaponics concept as a NbS for areas that are facing natural or other specific constraints (Figure 1).

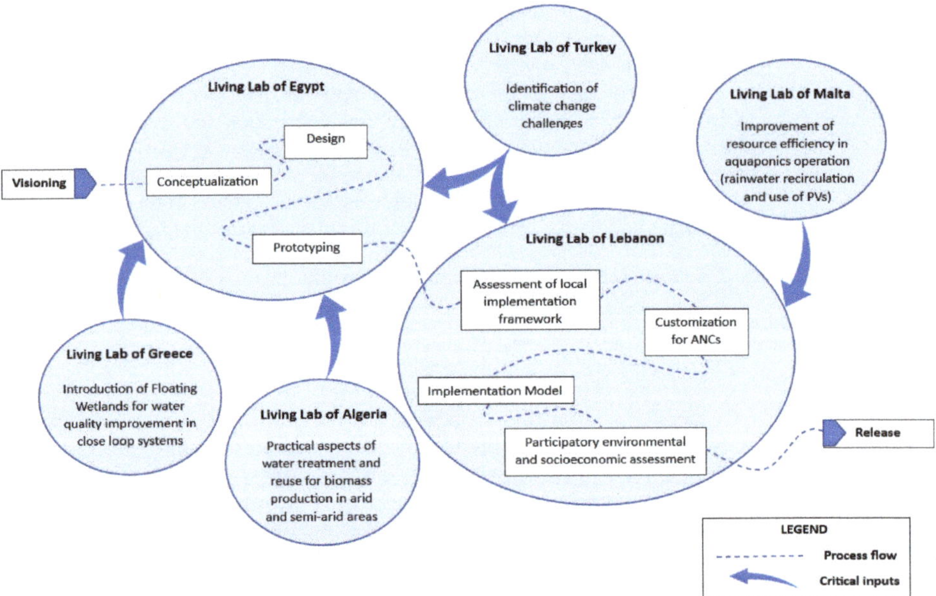

Figure 1. Cross border synergies of Living Labs within the open innovation ecosystem of the PRIMA funded Mara-Mediterra project. Visual representation of Living Labs contribution in the development process of decentralized wetland-aquaponics for Areas facing Natural or other specific Constraints (ANCs).

Aquaponics, which combines fish and plant production in a controlled environment, has been visioned by the participants in the Egyptian Living Lab as a possible solution for food production in the seawater intrusion-affected area of the upper Nile Delta, where biophysical limitations due to the degradation of soil and water quality have deteriorated agricultural production. The conceptualization and novelty of the design were achieved with the collaboration and critical contributions of three other Living Labs. Their key inputs included: (a) the introduction of floating wetlands simulating the provisional and regulatory wetland ecosystems' services for growing plants on floating rafts and for algae-based water quality improvement in a closed loop system; (b) the identification of climate change challenges for the operation of the system; and (c) the practical aspects of water treatment and reuse for biomass production in arid and semi-arid areas, from the Living Labs of Greece, Turkey and Algeria, respectively. Following a genuine participatory approach, the design of the wetland-aquaponics system was finalized in cooperation with the actors involved in the Egyptian Living Lab and the prototype unit was established by the Egyptian Chinese University, which coordinates the Living Lab in Egypt.

The prototype system triggered the interest of the Lebanese Living Lab for sustainable food production in the disadvantaged (mountainous and low-income) area of Akkar al-Atika, while the open ecosystem of the Living Labs network facilitated the cross-fertilization of knowledge and results across borders. Through the assessment of the local implementation particularities and priorities in the Lebanese Living Lab, an alternative system configuration and implementation model was formulated. Small, decentralized units were defined as more appropriate for the area in order to be used at the household/neighborhood level, while the system configuration was optimized to reduce construction and operation costs. The added value from other Living Labs was delivered in the form of scientific expertise for the assessment of water resource availability (Living Lab of Turkey) and in the form of ideas for the improvement of resource efficiency in the operation of the system, through the recycling of rainwater and the use of Photovoltaics (PVs) for energy autonomy (Living Lab of Malta).

The resulting design of the small-scale decentralized aquaponic system for Lebanon is aimed at increasing the farm productivity and profitability without significant water consumption, supporting poor families in producing their own fresh foods, and keeping farmers active during winter seasons, especially during periods of limited access to their lands. The proposed decentralized solution was finalized with the active involvement of stakeholders in the Living Lab of Lebanon, including both the co-design and the co-assessment of impacts on the local society (presented in Chapter 4).

The final system design (Figure 2) includes a 63 m^2 greenhouse (L × W × H/ 9 × 7 × 4.5 m) made from polyethylene films, which has the lowest cost in the market, a small-scale hydroponic system (16 pipes × 24 growing crop positions/pipe) and a small-scale aquaponic system (16 pipes × 24 growing crop positions/pipe). In each set-up, 4 inches of growing pipes, 2 inches of recuperation pipes, and net cups for growing crops of upper and lower diameter, respectively, 7.5 and 4.5 cm, are used. The distance between two horizontal positions is 22.5 cm and the distance between two vertical positions is 37 cm.

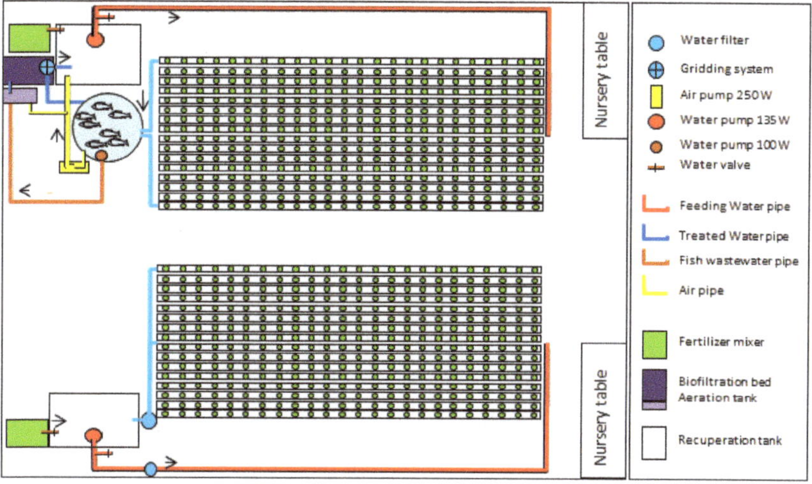

Figure 2. Conceptual design of aquaponic system layout.

The hydroponic growing pipes have a triangular prism shape and are connected to a main feeder tank of 1000 L by a 135 W circulating pump. The water is recirculated to the tank by gravity. A 50-L fertilizer mixing tank is connected to the feeder tank. The aquaponic, having the same prismatic shape, is directly connected to a 1000-L tank receiving treated water coming from the fish-rearing tank reservoir of 1000-L; the two tanks are aerated continuously. For the treatment of aquaculture effluents, a wetland biofilter is introduced in the design for pollution control since several studies have demonstrated that constructed wetlands can efficiently remove the major pollutants from catfish, shrimp and milkfish pond effluents under low hydraulic loading rates (between 0.018 and 0.135 m) and long hydraulic retention times (1–12.8 days) [23–25]. As physical and biochemical processes take place, the wetland functions as a natural filter for water quality improvement [26]. Effluents coming from the fish tank are treated inside an aeration tank connected to the biofiltration bed; the treated water outlet is connected to a gridding setup before being conducted to the feeder tank and to the fish rearing tank. A fertilizer mixing tank is connected to the feeder tank manually, and if needed. The whole system is operated by using three submersible pumps, two of which are 135 W and one of which is 100 W. Normally, the temperature in Akkar al-Atika fluctuates between 8 and 30 °C; thus, the temperature must be controlled. For this reason, the feeding tank is laid underground to minimize temperature fluctuation, while the power setup can support the operation of the system on or off-grid through photovoltaic panels if necessary to feed the heating water system when needed during extreme cold days. In addition to the economic aspects, stand-alone photovoltaics present additional climate-beneficial effects on reducing greenhouse gas emissions that should also be positively evaluated [27].

In addition, a nursery has been included in the design for a two-fold purpose: (a) to feed the hydro and aquaponic systems, and (b) to distribute the small plants to the farmers for free. This last objective is designed to attract farmers to attend the Living Lab, interact with the system, observe its functioning, and compare the growth in these two systems with their own traditional production. This strategy allows for corrective actions to be taken in the Living Lab, free access for continuous training, integration of farmers in decision-making and corrective procedures, and finally encouraging the farmers to reach the appropriate conclusions within their own particular context. Throughout this process, the farmers are comforted by the knowledge that they will be served by the know-how to find the right balance in the application of these practices.

Overall, it is highlighted that although the Living Labs in each country operate as intermediaries between local stakeholders, research organizations, key actors and decision makers [15], the establishment of a transnational network provides a cooperation platform for cross-border transfer of knowledge and experience for joint value co-creation, prototyping, or validation to expand innovation [28] and thus reveal and promote solutions across borders in order to address common problems. This approach, as adopted in the Mara-Mediterra project and supported by the PRIMA Foundation, creates the opportunity to re-establish meaningful connections between societies and ecosystems across borders. Thus, with the support of EU it demonstrates the strengthening of cooperation and relationships between Southern-Southern Mediterranean Partner States through direct interaction, while promoting their collaboration with Northern Mediterranean countries and actors towards sustainability. In this perspective, it allows for socio-ecological interventions that can provide, across borders, viable solutions [21] for areas that are facing natural or other specific constraints and common challenges, and thus promote the safeguarding of the environment and the livelihood of Mediterranean rural communities.

4. Participatory Assessment of Feasibility and Potential Impacts

4.1. Public Perception and Feasibility

The participatory feasibility assessment focused on appraising public perception, attitude, and social acceptance level of the decentralized wetland-aquaponics as NbSs in Akkar al-Atika. Based on the outcomes of the survey in Akkar al-Atika, a considerable part of local society (83% of the local stakeholders) considered decentralized wetland-aquaponics as an interesting economic activity for their area (Figure 3). This result indicates that the social acceptance of such systems is very high and paves the way for their potential future adoption in the area. This perception is in line with the FAO statement, according to which aquaponics finds its niche within the realm of sustainable yet intensive agriculture, particularly in applications tailored for family-scale operations [29].

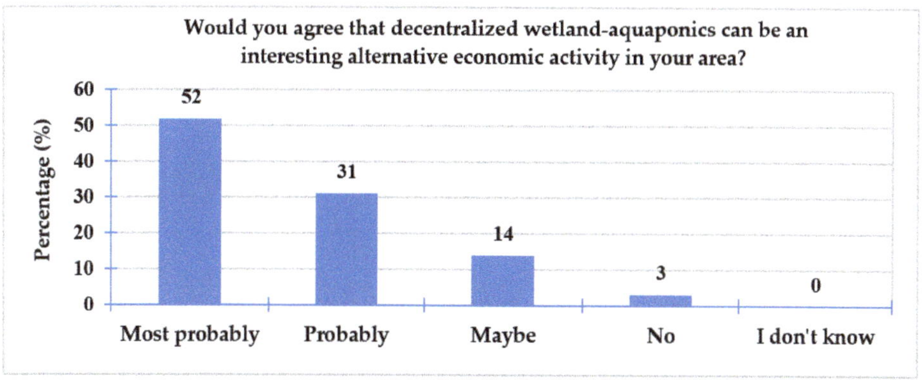

Figure 3. Public interest for decentralized wetland-aquaponics as an alternative economic activity in Akkar al-Atika.

The main reasons behind the acceptance and interest in decentralized wetland-aquaponics are the vision of a system that offers supportive and collaborative methods of vegetable and fish production and thus the opportunity to grow food in locations and situations where soil-based agriculture has been proven to be very challenging. The main reason for interest in the Akkar al-Atika area is related to food security for the local families, that is, the increase in food availability and quality for self-consumption (Figure 4).

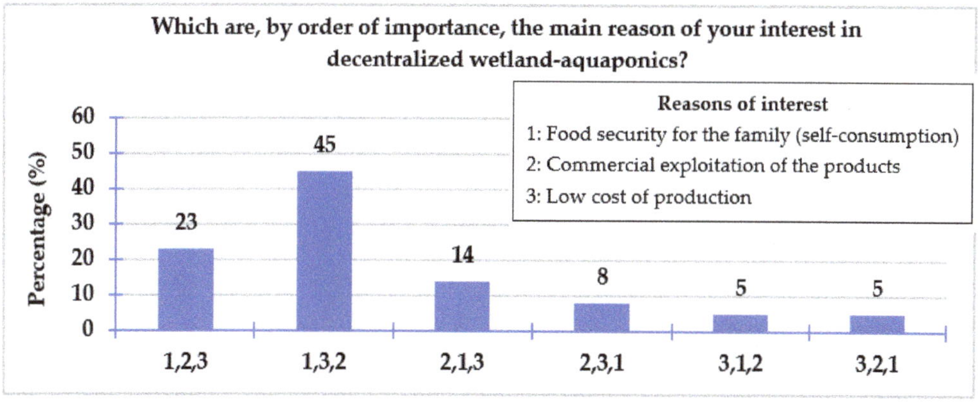

Figure 4. The main reasons of interest in decentralized wetland-aquaponics, according to their order of importance for the stakeholders in Akkar al-Atika. In the y axis, the 3 digits per class represent the order of importance for the stakeholders (first the most important reason and third the least important reason).

This reason has been identified as the first priority (highest importance) for 68% of the stakeholders. For most stakeholders (45%), the first, in order of importance, reason was food security; the second was the lower cost of production that can be achieved using lower inputs of water and fertilizers; and the less important reason was the potential commercial exploitation of products.

In terms of the decentralized implementation model, two equally balanced trends have been identified in the area. On the one hand, 52% of stakeholders prefer cooperative systems operating at the community/neighborhood level that will allow the sharing of workload, risks and profits. On the other hand, 47% prefer slightly smaller systems operating at the household/farm level as an independent family business (Figure 5).

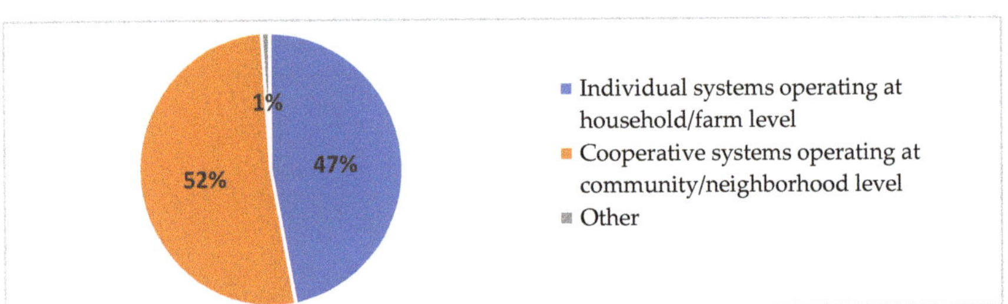

Figure 5. Public perception regarding the most beneficial level of decentralization for the wetland-aquaponics systems Akkar al-Atika.

However, there remain several challenges that need to be addressed at the operational level in order for this technology to be adopted by local stakeholders. In the case of Akkar al-Atika, the vast majority of stakeholders (68%) are mainly concerned about the capital costs of construction, as this is a low-income area. Similar concerns were also reported in a similar study with small-scale farmers in São Carlos, Brazil [20]. Furthermore, the lack of knowledge regarding construction and operation is considered the main difficulty for 35% of stakeholders, while 21% are concerned by the lack of scientific/technical guidance and support (Figure 6). From this perspective, the mobilization and cooperation with administrative authorities, the scientific community, NGOs, market networks and potential investors, the exploitation of financing mechanisms and tools, as well as the promotion of social innovation initiatives, could be explored to address these multidimensional challenges.

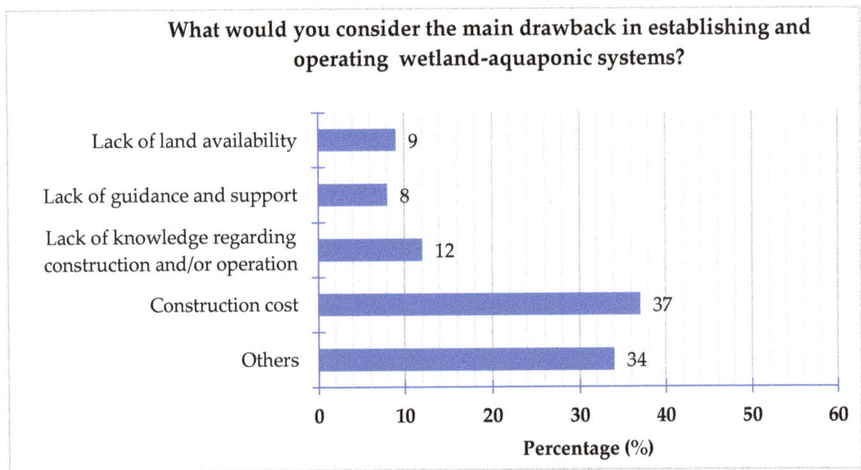

Figure 6. Stakeholders' assessment of challenges/drawbacks for the establishment and operation of decentralized wetland-aquaponic systems in Akkar al-Atika.

4.2. Environmental and Socioeconomic Impact Assessment

The participatory impact assessment is exploring the sustainability of aquaponics considering the environmental, economic and social dynamics [29] through the prism of local society in Akkar-al Atika. From an economic point of view, the initial investment cost is probably the main disadvantage of small aquaponic systems. Beyond this, the systems are usually characterized by low operational costs as well as viable returns from the production and commercial exploitation of fish and vegetables, which is of particular importance for small-income households. In the case of Akkar al-Atika, this is clearly reflected in the public perception of economic benefits since 77% of the stakeholders consider that the establishment of decentralized wetland-aquaponics systems could improve the economy of the area (Figure 7).

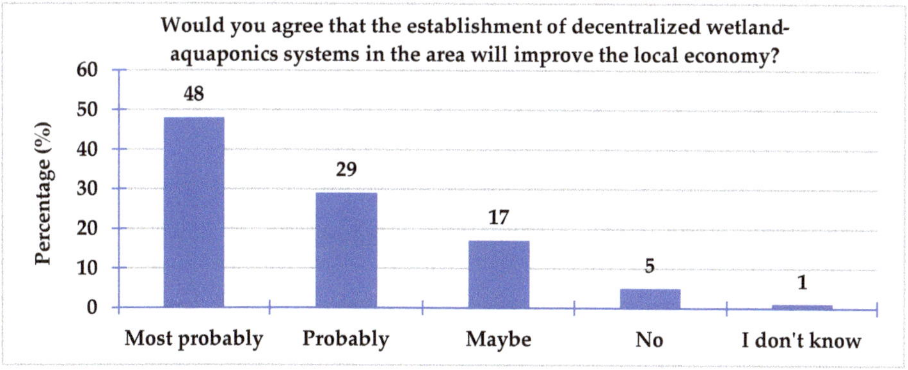

Figure 7. Public expectations regarding the impact of decentralized systems on the local economy of Akkar al-Atika.

The above expectations have been explored through an empirical economic analysis using stakeholders and experts' opinions, for the preliminary assessment of potential economic benefits from the operation of decentralized systems in Akkar al-Atika (Table 1). Running costs are divided between variable and fixed costs. Since the operational cost associated with fish production is quite higher than the cost of vegetable production [30], a higher ratio of plant to the fish revenue model was considered, providing a cost-effective solution for the smallholder operators of Akkar al-Atika. It is assumed that each water pump is working between 10 and 14 h per day and consuming 16 to 18 kW per month, while the air pump working 24/7 consumes about 45 kW per month. Total monthly power consumption is about 100 kW, and 1 kW is rated at 0.1 USD. Tilapia production is estimated to be 10 kg per month, with a mean price rate of 8 USD per kg. The net profit for commercial purposes can reach 255 USD per month, or a yearly profit of 3060 USD, assuming a monthly production efficiency of 95%. However, this high efficiency rate is usually reachable in hydroponic practices. Returning to the objectives that led to this study, small-scale aquaponics is proposed to be decentralized and is designed to help farmers and the rural population. Thus, in the economic model, it becomes more profitable by removing the worker's salary cost (100 USD per month), which leads to an enormous improvement in net profit of about 28%, which becomes 355 USD per month. Thus, for the vast majority of the local stakeholders with low annual income (<600 USD/year for 48% of the population, as mentioned in Section 2), the small-scale aquaponic system may provide a significant increase in their yearly income of at least five to seven times.

Table 1. Empirical analysis of operational Cost and Profit for a mothy production of 770 crops of lettuce and 10 Kg of Tilapia fish.

	Cost Per Month in USD	
Variable costs	• Seeds • Germination compost • Fertilizers and nutrients, packaging, * Fish feed for Tilapia • * Tilapia Fish	80 USD/Month
Fixed costs	1 Worker Electricity (15–20 kW)	100 USD/Month 10 USD/Month
	Total cost:	**190 USD/Month**
Production of lettuce (french lettuce) as lolloroso, lolloverde, oak leaf lettuce (*Lactuca sativa* var. *crispa*) red and green, endive, kale, etc…)	**Gross profit per month in USD** Carton trays number to be sold on a on monthly basis: 128 • 6 crops in each carton tray • Sales ratio 95% • 1 head of lettuce 0.5 USD • 3 USD/tray	365 USD/Month
* Production of Tilapia Fish	10 kg of Tilapia fish −1 kg of Tilapia 8 USD	80 USD/Month
	Gross profit:	**445 USD/month**
	Net Monthly Profit:	**255 USD/month**

* The fish feed amount and tilapia production is calculated based on the following data: (1) at the beginning of fish rearing: 200 fish/m^3 are put in the tank (80–100 g/fish, 16–20 kg/m^3, 0.5 USD/fish (2) Feeding: (a) Starting cultivation period <150 g/fish (0.75 kg/day); (b) Mid cultivation period >200 g/fish (1 to 1.2 kg/day); (c) Fast growing period >300 g/fish (1.7 to 2 kg/day), at this stage it is assumed that the rearing tank contains 150 fish having weight greater than 600 g/fish and the 1 m^3 tank has 130 kg of fish. (3) Protein content in fish feed fluctuate between 27 and 35% as a function of the cultivation period. (4) Total rearing cycle is around 6 months.

What remains to be resolved is the problem of the implementation cost of such systems for families having very low incomes. In this context, the intervention of financing programs developed by world organizations, or communal installations, would appear to be realistic solutions.

Environmentally, aquaponics may play a crucial role in the sustainable management of natural resources in Mediterranean watersheds. According to the stakeholders of Akkar al-Atika, the establishment of decentralized systems may benefit both the soil and water resources of the area and thus address multiple environmental challenges that are related to climate change as well as human activities. Specifically, the main potential environmental benefit (Figure 8) identified by 37% of stakeholders is the prevention of soil and water pollution due to closed-loop water treatment and the reduced use of agro-chemicals in food production. As near-zero discharge systems, aquaponics drastically reduces the problem of nutrient and emerging contaminants pollution and runoff from soil-based agricultural systems while preventing the eutrophication and degradation of water resources and thus addressing environmental challenges, which are particularly evident in Mediterranean watersheds and coastal zones [31]. In addition, a significant percentage of stakeholders (30%) anticipate that aquaponics will effectively contribute to the conservation of water due to the recirculation of water and the lower water demand for primary production. The water scarcity problem in arid and semi-arid regions is of primary importance, especially in food production. It is highlighted that in such areas, the water recirculation in aquaponics can reach a water reuse efficiency of 95–99% [32].

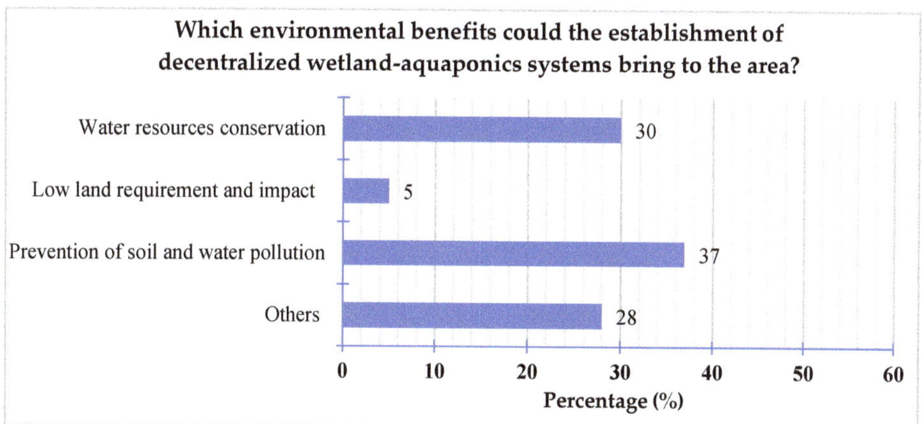

Figure 8. Public perception regarding the main potential environmental benefits of decentralized systems in Akkar al-Atika.

Socially, aquaponics can contribute to the livelihood of societies in disadvantaged areas in multiple ways. In the case of Akkar al-Atika, the potential positive impact on local society has been identified by a vast percentage (95%) of stakeholders (Figure 9), in terms of increased skills and future employment opportunities, green job creation in the area during construction and operation, as well as the willingness to stay in the area and thus prevent the depopulation of Akkar al-Atika.

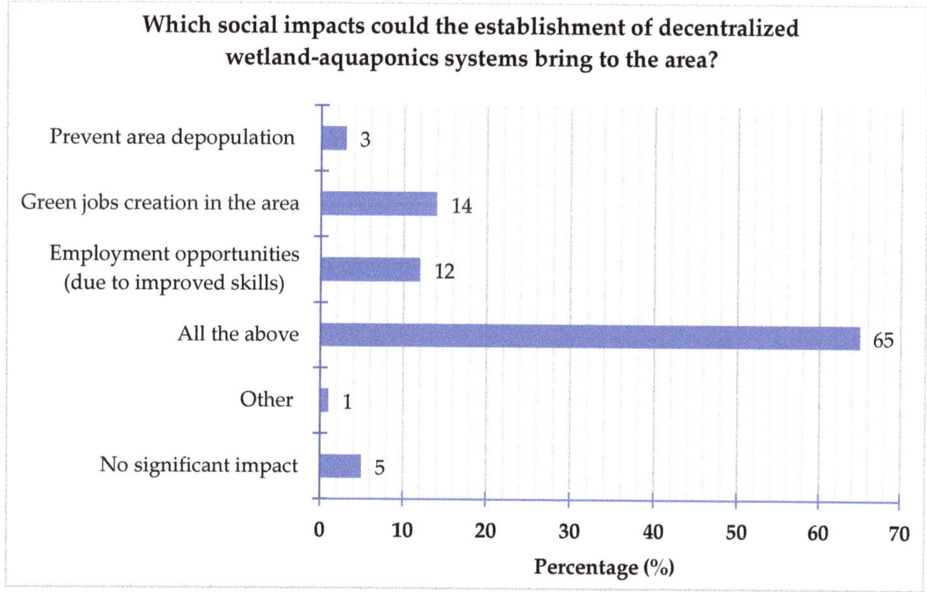

Figure 9. Public perception regarding the main potential benefits of decentralized systems for the local society in Akkar al-Atika.

The food that is grown locally, with culturally appropriate and socially acceptable crops, taking into account the local preferences and characteristics, strengthens the links of ownership and pride between the land, the people and their products. Most important, local food production, along with access to markets and the development of skills, are invaluable tools for promoting the emancipation of women and the empowerment of vulnerable groups of society. In this perspective, the creation of green jobs and the strengthening of employment opportunities for people with increased skills provide a solid basis for fair, equitable, and sustainable socio-economic growth [33].

Furthermore, local society will also benefit in terms of food security and health. In terms of food safety, and given the fact that aquaponics are controlled systems with biosecurity measures, they require far fewer agro-chemicals (e.g., chemical fertilizers and pesticides) for plant production [31], which makes the food safer to consume while safeguarding against potential contamination.

In terms of food security, aquaponics can be considered a constant source of crops (e.g., vegetables) and fish protein throughout the year. Especially the fish protein is a valuable addition to the dietary needs of several people, particularly in remote mountainous areas. More than 47% of the stakeholders in Akkar al-Atika are willing to reduce their meat consumption and replace it with fish protein produced locally (Figure 10), as long as the price of fish is not higher than the price of meat. Such dietary improvement may result in multiple health benefits since fish is considered a primary source of omega-3 fatty acids, while several studies document the lower risk of heart attacks and strokes for people who regularly consume fish protein [34].

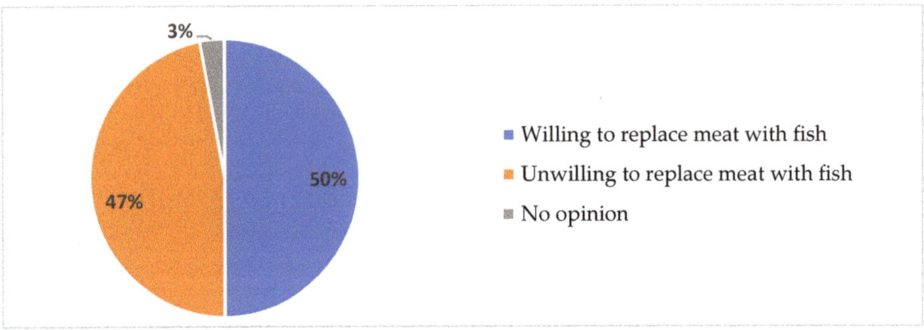

Figure 10. Public willingness of dietary modification, from meat to fish protein in Akkar al-Atika.

5. Discussion and Recommendations

The main challenge for the Akkar al-Atika area was to find an effective way for the customization and introduction of a system that would be able to address key societal challenges and needs in the area. In this perspective, the Living Labs in the open innovation ecosystem of Mara-Mediterra, acted as a participatory mechanism, facilitating stakeholders to: (a) co-create and interact with an operational system; (b) assess the potential and usefulness of such a solution; and (c) transfer their experience to other areas.

The participatory development and evaluation of NbSs, such as decentralized wetland-aquaponics, creates a favorable environment for their adoption and mainstreaming, as the stakeholders are actively involved in the deployment and assessment of solutions that address their needs in a real-life context. The findings of this work address the lack of field evaluations of small-scale aquaponics with constructed wetlands in rural Mediterranean areas. Furthermore, reveal the stakeholder perspectives that provide new insights compared to techno-economic feasibility studies performed by scientists and accredited professionals, which are likely cost prohibitive to marginalized regions and local grassroots initiatives and can have a transformative impact on sustainability [35]. The participatory assessment engages diverse actors in local society in scrutinizing and appraising knowledge on the specific issue. It does not aspire to directly influence political decisions but rather to enlighten them with social values and interests and to accelerate the transition process towards sustainable development with respect to local societies and the environment.

However, it is underlined that decentralized systems may have different economic viability and sustainability implications compared to large-scale commercial aquaponics. Thus, there is a need for more assessments of small-scale decentralized aquaponics installations in real-life settings to evaluate sustainability performance, as well as field testing for the quantification of their environmental benefits (e.g., water conservation).

The case of wetland-aquaponics within the framework of the Mara-Mediterra project illustrates the cross-border cooperation potential of Living Labs in the development of novel as well as practical solutions within a sustainability context. However, implementing a practice such as decentralized wetland-aquaponics, will not be easily affordable for small farmers in disadvantaged rural and mountainous areas. Quantified performance assessment, technical assistance, training programs and financial support are needed to safeguard the successful implementation of such a practice. Investments by local authorities and NGOs are essential companions for starting up low-cost and small-scale wetland-aquaponics systems. It is clear that decentralized small-scale wetland-aquaponics are not miracle solutions for eradicating famine and food security problems, but they may certainly provide access to food both in terms of quantity and quality, increase incomes for farmer families, and promote the empowerment of vulnerable groups of society while contributing to sustainable environmental management by preventing the degradation of soil and water resources, maintaining soil fertility, and reducing the overexploitation of fresh water supplies for crop production in arid and semi-arid climates [36]. The major challenges remain the cost of investment and the energy consumption, which can be reduced by using renewable energy. In this context, the local stakeholders in Akkar al-Atika were skeptical about the investment to be made and the return that they could have. However, they showed enthusiasm regarding the decentralized, small-scale aquaponics implementation if funded and initiated through SDG programs.

Overall, based on the outcomes of the Living Labs and participatory assessment survey and taking into account recent scientific studies [30] as well as the relevant FAO recommendations [29], the potential key factors of success for decentralized systems in disadvantaged areas include:

- Size between 50 and 150 m^2, depending on the preferred level of operation (farm, neighborhood or small community).
- Construction that uses locally sourced materials and workforce to minimize capital costs and maximize green job opportunities.
- Exploitation of harvested rainwater in the production process if and where possible.
- Promotion of energy autonomy through the use of renewable energy sources (e.g., photovoltaics, biogas from waste).
- Use of natural systems and processes for the quality improvement of the recirculated water (e.g., constructed wetlands)
- Efficient management systems that will lower the operational costs while ensuring the delivery of quality products able to satisfy both the consumption needs of the owner as well as the market needs for commercial exploitation of production.
- Mobilization and engagement of local society in demonstration actions through participatory processes of co-design, co-development and co-evaluation of pilot systems (e.g., Living Labs) that will allow the community to embrace and support relevant initiatives for the benefit of the local economy.
- Deployment of a multi-actor cooperation network in and beyond the area, supporting the establishment and operation of the decentralized systems through communication and publicity actions, knowledge transfer, market exploitation, capital leverage, entrepreneurship support, lifelong learning, and skills development.

6. Conclusions

Living labs are participatory platforms that enable the co-creation and co-evaluation of sustainable Nature-based Solutions to address rural challenges in the Mediterranean region. At the local level, Living Labs bring together different actors from the quadruple helix model, such as public authorities, academic institutions, the private sector and civil society, to collaborate on common goals and accelerate the development and adoption of solutions at the operational level. Thus, the Living Labs approach could be expanded to optimize the design of both technological and nature-based solutions, assess their feasibility across different contexts, and promote their mainstreaming. At the cross-border level, Living Labs may play a critical role not only in the cross-fertilization of applied innovations that are aligned with the United Nations Sustainable Development Goals (SDGs), but also in bringing corresponding communities together and contributing to the establishment of bridges between countries and societies that face common challenges. In this direction, the strengthening of cross-border cooperation pathways, especially in Mediterranean, needs to be further enhanced through enabling governance schemes and mechanisms of collaboration.

The increasing environmental consciousness within society, coupled with the progress in Nature-based Solutions (NbSs) and the demand for dependable yet cost-effective solutions, foster a conducive environment for the emergence of decentralized wetland-aquaponic systems. These systems present themselves as an appealing eco-technology capable of effectively tackling environmental challenges and ensuring sustainable food security. However, more real-life assessments are needed to evaluate their sustainability performance. In light of this, it becomes imperative to assess and quantify through field testing both the environmental (e.g., water conservation) and economic performance of these systems in the rural Mediterranean context and enhance their design and construction. This optimization process will unlock their scale-out potential and mainstreaming.

Author Contributions: Conceptualization, V.T. and A.E.S.; methodology, A.E.S.; validation, A.E.-D.A. and D.D.K.; investigation, F.Y.; data curation, F.Y.; writing—original draft preparation, V.T., A.E.S. and F.Y.; writing—review and editing, F.Y., A.E.S., E.P., M.K., R.E.-K., M.E., M.N., A.S. and D.D.K.; visualization, E.P. All authors have read and agreed to the published version of the manuscript.

Funding: This research was conducted within the framework of the Mara-Mediterra project. The project is part of the PRIMA programme supported by the European Union, under Horizon 2020 Research and Innovation Framework Programme, with Grant Agreement Number [2121]. This work reflects only the author's view, and the PRIMA Foundation is not responsible for any use that may be made of the information it contains.

Institutional Review Board Statement: Not applicable.

Informed Consent Statement: Not applicable.

Data Availability Statement: The data presented in this study are available on request from the corresponding author.

Acknowledgments: Authors greatly appreciate Israa Yahya and the Municipal authority of Akkar al-Atika for their assistance and facilitation of this research.

Conflicts of Interest: The authors declare no conflict of interest.

References

1. Antonelli, M.; Basile, L.; Gagliardi, F.; Isernia, P. The future of the Mediterranean agri-food systems: Trends and perspectives from a Delphi survey. *Land Use Policy* **2022**, *120*, 106263. [CrossRef]
2. Aslam, R.W.; Shu, H.; Yaseen, A.; Sajjad, A.; Ul Abidin, S.Z. Identification of time-varying wetlands neglected in Pakistan through remote sensing techniques. *Environ. Sci. Pollut. Res.* **2023**, *30*, 74031–74044. [CrossRef]
3. Yu, L.; Liu, T. The Impact of Artificial Wetland Expansion on Local Temperature in the Growing Season—The Case Study of the Sanjiang Plain, China. *Remote Sens.* **2019**, *11*, 2915. [CrossRef]
4. Gunacti, M.C.; Gul, G.O.; Cetinkaya, C.P.; Gul, A.; Barbaros, F. Evaluating Impact of Land Use and Land Cover Change Under Climate Change on the Lake Marmara System. *Water Resour. Manag.* **2023**, *37*, 2643–2656. [CrossRef]

5. Harmanny, K.S.; Malek, Ž. Adaptations in irrigated agriculture in the Mediterranean region: An overview and spatial analysis of implemented strategies. *Reg. Environ. Chang.* **2019**, *19*, 1401–1416. [CrossRef]
6. Takavakoglou, V.; Pana, E.; Skalkos, D. Constructed Wetlands as Nature-Based Solutions in the Post-COVID Agri-Food Supply Chain: Challenges and Opportunities. *Sustainability* **2022**, *14*, 3145. [CrossRef]
7. Chen, R.Z.; Wong, M.H. Integrated wetlands for food production. *Environ. Res.* **2016**, *148*, 429–442. [CrossRef]
8. Mchunu, N.; Odindo, A.; Muchaonyerwa, P. The effects of urine and urine-separated plant nutrient sources on growth and dry matter production of perennial rye-grass (*Lolium perenne.* L.). *Agric. Water Manag.* **2018**, *207*, 37–43. [CrossRef]
9. Yep, B.; Zheng, Y. Aquaponic trends and challenges—A review. *J. Clean. Prod.* **2019**, *228*, 1586–1599. [CrossRef]
10. Kotzen, B.; Emerenciano, M.G.C.; Moheimani, N.; Burnell, G.M. Aquaponics: Alternative types and approaches. In *Aquaponics Food Production Systems*; Springer: Cham, Switzerland, 2019; pp. 301–330.
11. Love, D.C.; Fry, J.P.; Li, X.; Hill, E.S.; Genello, L.; Semmens, K.; Thompson, R.E. Commercial aquaponics production and profitability: Findings from an international survey. *Aquaculture* **2015**, *435*, 67–74. [CrossRef]
12. Love, D.C.; Genello, L.; Li, X.; Thompson, R.E.; Fry, J.P. Production and consumption of homegrown produce and fish by noncommercial aquaponics gardeners. *J. Agric. Food Syst. Community Dev.* **2015**, *6*, 161–173. [CrossRef]
13. Adeleke, B.; Cassim, S.; Taylor, S. Pathways to low-cost aquaponic systems for sustainable livelihoods and economic development in poor communities: Defining critical success factors. *Aquac. Int.* **2022**, *30*, 1575–1591. [CrossRef]
14. Purcell, W.M.; Henriksen, H.; Spengler, J.D. Universities as the engine of transformational sustainability toward delivering the SDGs: "Living labs" for sustainability. *Int. J. Sustain. High. Educ.* **2019**, *20*, 1343–1357. [CrossRef]
15. Leal Filho, W.; Ozuyar, P.G.; Dinis, M.A.P.; Azul, A.M.; Alvarez, M.G.; da Silva Neiva, S.; Salvia, A.L.; Borsari, B.; Danila, A.; Vasconcelos, C.R. Living labs in the context of the UN sustainable development goals: State of the art. *Sustain. Sci.* **2023**, *18*, 1163–1179. [CrossRef]
16. Von Wirth, T.; Fuenfschilling, L.; Frantzeskaki, N.; Coenen, L. Impacts of urban living labs on sustainability transitions: Mechanisms and strategies for systemic change through experimentation. *Eur. Plann. Stud.* **2019**, *27*, 229–257. [CrossRef]
17. Kok, P.W.K.; Van der Meij, M.G.; Wagner, P.; Cesuroglu, P.; Broerse, E.W.J.; Regeer, J.B. Exploring the practice of Labs for sustainable transformation: The challenge of 'creating impact'. *J. Clean. Prod.* **2023**, *388*, 135994. [CrossRef]
18. Van der Jagt, A.P.N.; Buijs, A.; Dobbs, C.; van Lierop, M.; Pauleit, S.; Randrup, T.B.; Wild, T. An action framework for the participatory assessment of nature-based solutions in cities. *Ambio* **2023**, *52*, 54–67. [CrossRef]
19. Gott, J.; Morgenstern, R.; Turnšek, M. Aquaponics for the Anthropocene: Towards a 'Sustainability First' Agenda. In *Aquaponics Food Production Systems*; Goddek, S., Joyce, A., Kotzen, B., Burnell, G.M., Eds.; Springer: Cham, Switzerland, 2019. [CrossRef]
20. Compagnucci, L.; Spigarelli, F.; Coelho, J.; Duarte, C. Living Labs and user engagement for innovation and sustainability. *J. Clean. Prod.* **2021**, *289*, 125721. [CrossRef]
21. Fischer, J.; Riechers, M.; Loos, J.; Martin-Lopez, B.; Temperton, V.M. Making the UN decade on ecosystem restoration a social-ecological endeavour. *Trends Ecol. Evol.* **2021**, *36*, 20–28. [CrossRef]
22. Brewer, A.; Alfaro, J.F.; Malheiros, T.F. Evaluating the capacity of small farmers to adopt aquaponics systems: Empirical evidence from Brazil. *Renew. Agric. Food Syst.* **2021**, *36*, 1–9. [CrossRef]
23. Schwartz, M.F.; Boyd, C.E. Constructed wetlands for treatment of channel catfish pond effluents. *Prog. Fish Cult.* **1995**, *57*, 255–267. [CrossRef]
24. Lin, Y.F.; Jing, S.R.; Lee, D.Y.; Wang, T.W. Removal of solids and oxygen demand from aquaculture wastewater with a constructed wetland system in the start-up phase. *Water Environ. Res.* **2002**, *74*, 136–141. [CrossRef] [PubMed]
25. Turcios, A.E.; Papenbrock, J. Sustainable Treatment of Aquaculture Effluents—What Can We Learn from the Past for the Future? *Sustainability* **2014**, *6*, 836–856. [CrossRef]
26. Takavakoglou, V.; Georgiadis, A.; Pana, E.; Georgiou, P.E.; Karpouzos, D.K.; Plakas, K.V. Screening Life Cycle Environmental Impacts and Assessing Economic Performance of Floating Wetlands for Marine Water Pollution Control. *J. Mar. Sci. Eng.* **2021**, *9*, 1345. [CrossRef]
27. Baiyin, B.; Tagawa, K.; Gutierrez, J. Techno-Economic Feasibility Analysis of a Stand-Alone Photovoltaic System for Combined Aquaponics on Drylands. *Sustainability* **2020**, *12*, 9556. [CrossRef]
28. European Network of Living Labs (ENoLL). What Are Living Labs. 2023. Available online: https://enoll.org/about-us/ (accessed on 12 May 2023).
29. Somerville, C.; Cohen, M.; Pantanella, E.; Stankus, A.; Lovatelli, A. *Small-Scale Aquaponic Food Production. Integrated Fish and Plant Farming*; FAO Fisheries and Aquaculture Technical Paper No. 589; FAO: Rome, Italy, 2014; p. 262.
30. Adeleke, B.; Robertson-Andersson, D.; Moodley, G.; Taylor, S. Economic viability of a small scale low-cost aquaponic system in South Africa. *J. Appl. Aquac.* **2023**, *35*, 285–304.
31. Joyce, A.; Goddek, S.; Kotzen, B.; Wuertz, S. Aquaponics: Closing the Cycle on Limited Water, Land and Nutrient Resources. In *Aquaponics Food Production Systems*; Goddek, S., Joyce, A., Kotzen, B., Burnell, G.M., Eds.; Springer: Cham, Switzerland, 2019. [CrossRef]
32. Dalsgaard, J.; Lund, I.; Thorarinsdottir, R.; Drengstig, A.; Arvonen, K.; Pedersen, P.B. Farming different species in RAS in Nordic countries: Current status and future perspectives. *Aquac. Eng.* **2013**, *53*, 2–13. [CrossRef]
33. Milliken, S.; Stander, H. Aquaponics and Social Enterprise. In *Aquaponics Food Production Systems*; Goddek, S., Joyce, A., Kotzen, B., Burnell, G.M., Eds.; Springer: Cham, Switzerland, 2019; pp. 607–619. [CrossRef]

34. Chowdhury, R.; Stevens, S.; Gorman, D.; Pan, A.; Warnakula, S.; Chowdhury, S.; Ward, H.; Johnson, L.; Crowe, F.; Hu, F.B.; et al. Association between fish consumption, long chain omega 3 fatty acids, and risk of cerebrovascular disease: Systematic review and meta-analysis. *BMJ* **2012**, *345*, e6698. [CrossRef]

35. Fritz, L.; Binder, C.R. Participation as Relational Space: A Critical Approach to Analysing Participation in Sustainability Research. *Sustainability* **2018**, *10*, 2853. [CrossRef]

36. Verner, D.; Vellani, S.; Goodman, E.; Love, D.C. Frontier Agriculture: Climate-Smart and Water-Saving Agriculture Technologies for Livelihoods and Food Security. In *New Forms of Urban Agriculture: An Urban Ecology Perspective*; Diehl, J.A., Kaur, H., Eds.; Springer: Singapore, 2021. [CrossRef]

 *sustainability*

Article

From Farm to Fork: Irrigation Management and Cold Storage Strategies for the Shelf Life of Seedless Sugrathirtyfive Table Grape Variety

Vittorio Alba [1],*, Alessandra Russi [1], Giovanna Forte [1], Rosa Anna Milella [1], Sabino Roccotelli [1], Pasquale Campi [2], Anna Francesca Modugno [2], Vito Pipoli [3], Giovanni Gentilesco [1], Luigi Tarricone [1] and Angelo Raffaele Caputo [1]

[1] Research Centre for Viticulture and Enology, CREA-Council for Agricultural Research and Economics, via Casamassima 148, 70010 Turi, BA, Italy; alessandra.russi@crea.gov.it (A.R.); giovanna.forte@crea.gov.it (G.F.); rosaanna.milella@crea.gov.it (R.A.M.); sabino.roccotelli@crea.gov.it (S.R.); giovanni.gentilesco@crea.gov.it (G.G.); luigi.tarricone@crea.gov.it (L.T.); angeloraffaele.caputo@crea.gov.it (A.R.C.)

[2] Research Centre for Agriculture and Environment, CREA-Council for Agricultural Research and Economics, Via Celso Ulpiani, 5, 70125 Bari, BA, Italy; pasquale.campi@crea.gov.it (P.C.); francesca.modugno@crea.gov.it (A.F.M.)

[3] Ionic Consortium of Fruit and Vegetable Growers Coop, SP240, 70018 Rutigliano, BA, Italy; vito.pipoli@cjo.it

* Correspondence: vittorio.alba@crea.gov.it

Citation: Alba, V.; Russi, A.; Forte, G.; Milella, R.A.; Roccotelli, S.; Campi, P.; Modugno, A.F.; Pipoli, V.; Gentilesco, G.; Tarricone, L.; et al. From Farm to Fork: Irrigation Management and Cold Storage Strategies for the Shelf Life of Seedless Sugrathirtyfive Table Grape Variety. *Sustainability* **2024**, *16*, 3543. https://doi.org/10.3390/su16093543

Academic Editor: Dimitris Skalkos

Received: 26 March 2024
Revised: 19 April 2024
Accepted: 22 April 2024
Published: 24 April 2024

Abstract: Background: Sustainable water management for table grape has the primary goal of optimizing irrigation through Smart Irrigation (SI) approaches, particularly in Mediterranean regions. In addition, extending the shelf life of table grapes through effective cold storage practices is crucial to meet consumer demands year-round. This research examined the journey "from farm to fork" of Sugrathirtyfive variety (Autumn Crisp® brand), exploring the combined effects of Irrigation Volumes (IV), SO_2-Generating Pads (SGPs) and Cold Storage Duration (CSD) on the quality of grapes. Methods: Normal Irrigation (NI—based on the farmer's experience) and SI (100% vine evapotranspiration restored) were supplied in 2023 to Sugrathirtyfive variety white table grape, trained to an overhead tendone system. Yield and quality parameters, berry texture, CIELAB colour coordinates, phenolic content, flavonoids, antioxidant activity and sensory attributes were evaluated on grapes subjected to different times and methods of cold storage. Results: SI grapes showed higher Total Soluble Solids (TSSs) and nutraceutical content, as well as improved CIELAB coordinates with interesting improved berry texture parameters. No differences emerged between single- or dual-release SGPs after 15 days (T1) and 40 days (T2) of CSD. Conclusions: Under our cold storage conditions (3 °C, 85% U.R.), 40 days represent the maximum temporal limit for the cold storage of Sugrathirtyfive variety, regardless of IV, provided they are refrigerated with the aid of SGPs.

Keywords: sustainable water management; table grapes; cold storage; SO_2-Generating Pads; shelf life

1. Introduction

Limited natural water resources are the primary constraint for table grape cultivation, particularly in the Mediterranean region, where the ambient evapotranspirative demand exceeds the modest precipitation levels, resulting in a water deficit extending from spring to early autumn [1,2]. In a Mediterranean climate with hot and dry summers, irrigation is absolutely necessary for grapevines to secure production [3], especially in Southern Italy. Precipitation often does not exceed the threshold of 500 mm/year [4]. Moreover, the rainfall is mostly concentrated in the autumn–winter period and is not usable during phenological phases with higher water requirements, such as the flowering–beginning of berry ripening period [5].

Sustainability in water use in agriculture thus becomes a priority, achievable through the optimization of the irrigation variables involved in the water balance. In addition, the scarcity of water resources in these environments must avoid using empirical irrigation scheduling. This methodology could overestimate irrigation volumes, resulting in unnecessary water losses due to runoff and drainage [2].

Adopting solutions capable of correctly determining the crop water requirement through the losses due to evapotranspiration is necessary. The use of soil water balance integrated with dedicated sensors (Smart Irrigation) is a sustainable solution. Theoretically, many Decision-Support Systems (DSSs) can be generated to satisfy the above exigencies, and these DSSs have become available in the scientific literature over the past 30 years [6]. The most widely used DSSs are based on evapotranspirative methods.

The literature reveals a growing interest in the impact of irrigation practices on grape quality attributes, underscoring the intricate relationship between irrigation levels and qualitative and quantitative traits of table grapes [7–11].

The production of table grapes in Puglia is increasingly diverse, covering the period from June to December, due to the cultivation of various early and late varieties and the adoption of plastic film covering and agronomic techniques to force early ripening or to delay harvest [12]. In addition to this, the table grape viticulture in Puglia is undergoing a phase of profound renewal in response to increased global competition from new competitors in both the northern hemisphere (Spain, Egypt, India) and the southern hemisphere (Chile, South Africa, Peru) [13], offering seedless varieties that cater to European consumer preferences. In particular, the varietal landscape of table grapes has undergone a significant evolution in recent years. In newly established vineyards, greater emphasis is placed on incorporating novel seedless grape varieties, reflecting a progressive shift towards aligning production with market demands. Among these new table grape varieties, Sugrathirtyfive is a patented (commercial name Autumn Crisp®—United States Plant Patent USOOPP20491 P2) late-season white seedless table grape variety, with extra-large, oval, milky-green berry with excellent flavor, firmness and berry attachment (Figure 1). As consumer demand for fresh products transcends seasonal boundaries, the need to extend the shelf life of table grapes through effective cold storage practices becomes paramount. Moreover, offering the consumer grapes with high nutraceutical properties even many days after harvesting is essential, considering that consuming fresh grapes significantly benefits human health [14–16]. The intricate balance between maintaining optimal conditions for grape preservation and the inherent perishability of this fruit poses a fascinating challenge [17].

To fulfil market demands and ensure a year-round supply of high-quality grapes to consumers, it is essential to employ techniques that enhance grape shelf life. Thanks to table grape's low sensitivity to chilling, minimal respiration rates and low ethylene production, cold storage is a widely employed post-harvest method, proven to be effective in extending fruit shelf life, significantly mitigating mass loss, and managing the occurrence of pathogens like grey mold induced by Botrytis cinerea [18,19]. Grapes exhibit diverse responses to cold storage, depending on the cultivar and storage duration, which is constrained by specific factors, necessitating effective methods for handling, packaging and specialized cooling to ensure the optimal condition of grapes upon delivery, ranging from a few days immediately after harvest up to even a month away [20]. Several papers report the impact of different storage times and conditions on table grape cultivars like Thompson Seedless [21], Italia and Red Globe [19,22], Kyoho [23], Regal Seedless [24] and Benitaka [18]. Cold storage combined with the utilization of sulfur dioxide (SO_2)-Generating Pads has exhibited promising outcomes in controlling post-harvest diseases, presenting a convenient and effective alternative. This combination facilitates gas circulation within the storage container, preventing mass loss while ensuring the desired preservation outcomes [25,26]. In this sense, the storage of table grapes represents a critical juncture in ensuring the provision of high-quality, flavorful grapes to consumers year-round. The delicate nature of table grapes demands a nuanced understanding of the interplay between storage duration, cold storage conditions and the resulting impact on grape quality.

Figure 1. Sugrathirtyfive seedless table grape ready to harvest grown in a private commercial farm trained using an overhead covered tendone trellis system (**left**), clusters, berries and their section (**right**).

Until now, research on table grapes has considered irrigation factors, methods and storage duration individually, or, at most, by separating the phases related to vineyard irrigation management from the subsequent post-harvest phase. Therefore, integrating irrigation effects with post-harvest storage conditions, especially concerning a newly introduced seedless grape variety on the market, represents a research frontier that merits deeper investigation. This research focuses on an exploration of the interplay between two different Irrigation Volumes (IVs), different post-harvest types of SO_2-Generating Pads (SGPs) and the Cold Storage Duration (CSD). In particular, their collective influence was investigated from field to table by evaluating grape quality and productive traits, texture, color and nutraceutical content (polyphenols, flavonoids and antioxidant activity) of berries over time, with a final sensory evaluation of the grapes—emerging high-quality seedless Sugrathirtyfive table grapes.

2. Material and Methods

2.1. Field Trial and Irrigation Volumes

The experimental trial was conducted in 2023 on a private commercial vineyard that was 9 years old, situated in Adelfia (BA), Southern Italy (latitude: 40°59′14″ N, longitude: 16°51′34″ E, elevation: 172). *Vitis vinifera* cv. Sugrathirtyfive (Autumn Crisp® brand), grafted onto *Vitis berlandieri* × *Vitis rupestris* 34 E.M. rootstock, was spaced at 2.50 × 2.50 m (1600 vines ha^{-1}). The vines were pruned to 30 buds per vine, trained using an overhead tendone system (Apulia type) and subjected to drip irrigation. Additionally, the vineyard was covered with netting and a polyethylene plastic film with a 200 μm sheet thickness from budbreak to harvest to protect the canopy and clusters from adverse effects of wind, rain, and hail.

According to the United States Department of Agriculture (USDA) classification, soil texture was clay. At 0.5 m of depth, there was a parent rock that reduced the capacity of the root systems to expand beyond this layer. Soil water content in volume at field capacity (fc, −0.03 MPa) and wilting point (wp, −1.5 MPa) were 0.34 and 0.26 m^3 m^{-3}, respectively (measured in the Richards chambers).

Irrigation was supplied by a drip irrigation system having 3 drippers per vine and a flow rate of 16 L h^{-1} per dripper. Two Irrigation Volumes (IV) were considered:

- Normal Irrigation (NI): empirical irrigation management based on the knowledge and experience of the farmer, tendentially at fixed intervals approximately every 7 days, depending on the occurrence of rain, starting from 24 June (175th Julian day) until the last irrigation intervention on 10 October (283rd Julian day), for a total of 14 watering rounds;
- Smart Irrigation (SI), which restored 100% of crop evapotranspiration. Irrigation occurred when ready water availability was exhausted, according to the methodology of Allen et al. [27]. In particular, the tabulated crop coefficients (Kcinit = 0.15; Kcmed = 0.80; Kcend = 0.40) and depletion fraction value of 0.45 were adopted. Correction of Kcini (for precipitation events), Kcmed and Kcend (for climatic conditions and crop height) was performed according to the methodology of Allen et al. [27].

Soil water content in volume (SWC) was measured by capacitive probes 10HS (Meter Group Inc., Pullman, WA, USA). For each treatment, three vines were monitored. At each point, two capacitive probes were installed horizontally into the soil profile and transversely to the row, at −0.125 and −0.375 m from the soil surface, to intercept the dynamics of SWC below the dripping lines. All sensors were connected to data-loggers (TECNO.EL srl, Roma, Italy) and data were transferred to a web server via GPRS mode. Daily soil water content for the soil profile (0.5 m) was determined as an average of the values measured for each depth.

The farm did not have its own well, and water was supplied on a rotational basis from consortium irrigation systems. For this reason, the study focused on defining the irrigation volume rather than the irrigation timing.

2.2. Yield and Grapes Quality Parameters

Grapes were commercially harvested on 19 October 2023 when they reached ~18°Brix. Five clusters for each IV were considered and the following parameters were recorded: Cluster Weight, 20 Berry Weight, Equatorial Diameter, Total Soluble Solids (TSSs), pH, Titratable Acidity (TA).

A total of 100 berries per treatment were collected and pooled and a sample of 20 berries was employed to determine the color coordinates and texture attributes. Berry color was determined by a chromameter CM-5 (Konica Minolta, Chiyoda, Tokyo, Japan) using the CIELAB color system. The CIELAB, or CIE L* a* b*, system is a three-dimensional color-space consisting of three axes: L* axis (Lightness)—a grey scale with values from 0 (black) to 100 (white), a*axis—a red/green axis with positive (red) and negative (green) values and b* axis—a yellow/blue axis with positive (yellow) and negative (blue) values.

Compression and tensile tests were performed on the 20 berries/cluster/thesis using a Zwick Roell ver. Z 0.5 Materials Testing Machine (Woonsocket, RI, USA). A 2-cycle compression test was carried out on each whole berry in the equatorial position under a deformation of the berry of 20%, with waiting time between the two bites of 1 s, using a crosshead speed of 3.334 mms^{-1}, with a standard force of 0.1 N and a 0.02 m diameter cylindrical probe. Typical berry texture parameters scored were Hardness (N), Cohesiveness (adimensional), Gumminess (N − Hardness × Cohesiveness), Springiness (mm) and Chewiness (mJ, Gumminess × Springiness).

2.3. Preparation of Grape Skin Extracts (GSEs) and Total Phenolic Content (TPF), Total Flavonoids (FLV) and Antioxidant Activity (DPPH)

Skins from 10 frozen berries were manually separated from the pulp and extracted, according to Di Stefano and Cravero [28] with slight modifications. Briefly, skins were incubated overnight in the dark in 25 mL of 70% ethanol containing 1% chloridric acid. Then, the extracts were filtered through a 0.45 μm syringe cellulose filter and stored at −20 °C until further analysis.

TPF in GSEs was determined by the Folin–Ciocalteu colorimetric method described by Waterhouse [29]. Briefly, 1 mL of water, 0.02 mL of extract sample, 0.2 mL of the Folin-Ciocalteu reagent and 0.8 mL of 10% sodium carbonate solution were mixed and brought to 4 mL. The mixture was stored for 90 min at room temperature in the dark, and the absorbance was measured at 760 nm with a spectrophotometer Agilent 8453 (Agilent Technologies, Santa Clara, CA, USA). Results were expressed as milligrams of gallic acid equivalent/kg (mg GAE/Kg fw) of fresh grape based on a gallic acid calibration curve (50 to 500 mg/L with $R_2 = 0.998$).

FLV was determined by the aluminum chloride method [30] with some modifications. First, 1 mL of the GSE (diluted 1:10 with ethanol) was mixed with 1 mL of 2% aluminum chloride and incubated at 25 °C for 30 min. Then, the absorbance of the mixture was measured at 402 nm. Results were expressed as µg of rutin equivalent per kg (µg RE/Kg fw) of fresh grape using the calibration curve of quercetin (0–150 mg/L).

The antioxidant activity was evaluated by DPPH (2,2 O-diphenyl-1-picrylhydrazyl) assays, a radical scavenging assay based on single-electron transfer. The DPPH assay was conducted according to the technique of Brand-Williams et al. [31] with some modifications. A free-radical working solution was prepared by dissolving 2.5 mg of DPPH stock solution in 100 mL ethanol. The solution absorbance was adjusted at 0.7 ± 0.02 in 515 nm using a UV–Vis spectrophotometer Agilent 8453 (Agilent Technologies, Santa Clara, CA, USA). An aliquot of 200 µL of the sample, appropriately diluted, was mixed with 2 mL of DPPH solution (A_{sample}). A solution without grape extract was used as a blank (A_{blank}). The decrease in absorbance at 515 nm was measured after 30 min of incubation at 37 °C. Calibration curves were prepared using Trolox (Sigma-Aldrich, St. Louis, MO, USA). DPPH values were expressed as µM Trolox equivalents/kg of fresh grape (µg TE/Kg fw).

2.4. Times and Methods of Cold Storage

In order to test the storage suitability of the Sugrathirtyfive variety subjected to two different IVs, at harvest, grapes were refrigerated in fruit crates at 3 °C and 85% U.R. Three treatments for each of the two IVs were defined. Specifically, a Control (C) thesis was refrigerated without SO_2-Generating Pads (SGPs), while the other two theses were treated with the following:

SmartPac® bags (SPB) (Sodium Metabisulphite 12.5% *w/w*) (Serroplast, Rutigliano, Italy) are patented single-release SO_2-Generating Pads composed of a single multilayer film that allows the fruit's natural moisture to circulate through the inner layers of the coating, enabling linear preservation of the product for extended periods;

DECCO Grapage® (DECCO), (DECCO ITALIA S.R.L., Belpasso, Italy) a dual release SO_2-Generating Pad (5 g Sodium Metabisulphite 50%, Inert Technical Coadjuvants 50%);

The grapes were evaluated for quality parameters at different values of Cold Storage Duration (CSD): harvest (T0), after 15 days (T1) and after 40 days (T2) of cold storage.

2.5. Sensory Evaluation

To evaluate the sensory attributes and resilience to CSD of Sugrathirtyfive grapes cultivated under different IVs and subjected to two distinct SGP treatments, they underwent sensory assessment at 15 days (T1) and 40 days (T2) post-harvest. The sensory evaluation was conducted on blind samples within specially equipped individual workstations with neutral-colored walls and odor-neutral surfaces. The environmental temperature was maintained at a comfortable 22 °C, ensuring optimal conditions for evaluation. Brightness within the room was adjusted to an appropriate level, and extraneous noise or distractions were minimized, adhering to the guidelines outlined by [32]. ISO 2007. The taster panel was composed of 20 trained judges from the Research Centre for Viticulture and Enology, Council for Agricultural Research and Economics. The judges were requested not to smoke or eat for 1 h prior to the sensory sessions. The grapes were evaluated based on 23 OIV descriptors for table grape sensory analysis [33] for visual, olfactive, taste and tactile traits on cluster, stem, berries, skin and pulp.

Judges scored each attribute on a preference scale structured from 1 (low perception of the descriptor) to 10 (maximum perception of the descriptor).

2.6. Statistical Analysis

A three-way ANOVA with interactions between factors was performed on a total of 14 theses derived by the combination of the three factors (IV, SGP and CSD) as follows: NI-C-T0, SI-C-T0, NI-C-T1, NI-SPB-T1, NI-DECCO-T1, SI-C-T1, SI-SPB-T1, SI-DECCO-T1, NI-C-T2, NI-SPB-T2, NI-DECCO-T2, SI-C-T2, SI-SPB-T2, SI-DECCO-T2. Means were firstly by Tukey test, while a subsequent Dunnett's test was employed to compare the values of each individual trait for each thesis against the control sample, which, in our case, was NI-C-T0. Furthermore, a multivariate approach by means of a biplot PCA was performed at T0, T1 and T2. In addition, the differences in the perception of each descriptor during sensory evaluation of grapes were statistically analyzed by Non-Parametric Kruskall–Wallis test, and a Box Plot for the descriptors that resulted in statistically significant differences is provided. All the statistical analyses were performed using R Statistical Software v4.3.2.

3. Results

3.1. Soil Water Content (SWC) and Irrigation Volumes (IVs)

In the Smart Irrigation (SI) treatment, the irrigation scheduling allowed the optimization of the SWC (from -0.10 m to -0.50 m soil depth) within the RAW threshold (0.296 m^3 m^{-3}), avoiding any water stress. In particular, the SWC reached the field capacity, after irrigation or consistent precipitations. In August, irrigation was carried out before the SWC reached the RAW threshold, as irrigation was provided rotationally. In the Normal Irrigation (NI), SWC exceeded the field capacity almost throughout the entire vine cycle (Figure 2). This resulted in only water losses due to drainage because the flat ground and drip irrigation system avoided runoff losses. In this case, NI was excessive. Seasonal IVs were 335 and 264 mm for NI and SI treatments, respectively, with the number of irrigations during the 2023 season being 14. Thus, with SI treatment, 21% of the irrigation water was spared.

3.2. Univariate Analysis

This manuscript presents a comprehensive investigation into the impact of two different Irrigation Volumes (IVs)—Normal Irrigation (NI) and Smart Irrigation (SI), distinct SO_2-Generating Pads (SGPs)—Control (C), SmartPac® Bag (SPB), and DECCO Grapage® (DECCO) and three different Cold Storage Durations (CSDs)—harvest (T0), 15 days post-harvest (T1) and 40 days post-harvest (T2) on various parameters related to carpometry, must composition, berry skin colorimetric coordinates, berry texture, nutraceutical traits and cluster damages induced by cold storage on Sugrathirtyfive table grape. Table 1 presents the outcomes of a three-way ANOVA, illustrating interactions among the three factors and means separated by post hoc Tukey tests for each factor individually.

Regarding IVs, no differences in carpometric data were observed, indicating a substantial equality in the size and weight of berries between the two IV levels. Similar observations were noted for the other two factors, SGP and CSD. However, a statistically significant interaction was identified between SGP and CSD, specifically in relation to the 20 berries' weight. Additionally, a significant interaction was found regarding berry diameters, expressed as Equatorial Diameter, between IV and SGP. Cluster weight and its related weight loss over time (Figure 3) were analyzed independently of the other parameters. Clusters were weighed at harvest before packaging and cold storage. The direct monitoring of this parameter on the same clusters allowed for a paired sample *t*-test analysis, unlike other indices and parameters that, due to the destructive nature of the relief methods, did not permit time-dependent measurements on the same biological sample.

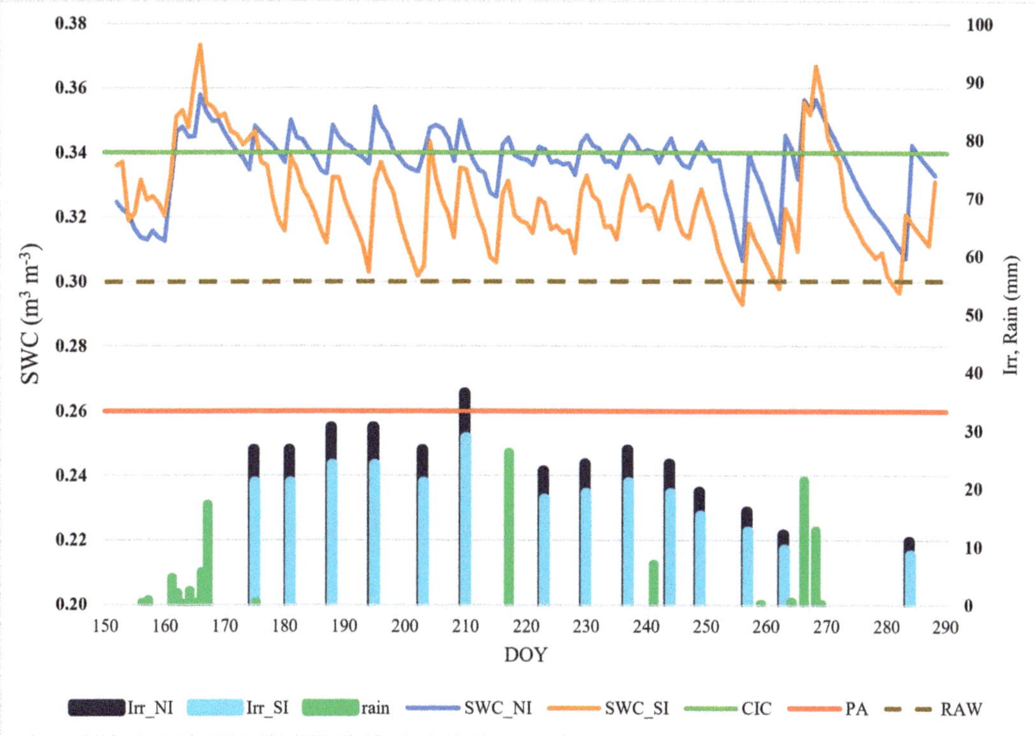

Figure 2. Soil water content (SWC) and Irrigation Volumes (IVs) of Normal Irrigation (NI) and Smart Irrigation (SI) provided to Sugrathirtyfive in 2023 on a private commercial farm trained using an overhead covered tendone trellis system.

By T2, grapes from the cold storage treatment without SGP were entirely covered in mold, rendering them unprocessable. Comparing cluster weight between T0 and T1, no statistical difference was found, except for the theses NI-C and SI-DECCO. At T2, cluster weight loss appeared generalized in all theses, except for those treated with SPB, regardless of IV. Concerning berry juice composition, Total Soluble Solids (TSSs) were significantly influenced by IV, with SI treatment showing higher values than NI and the employed SGP. Treatments with SPB or DECCO preserved sugar content integrity compared to the Control. CSD had no isolated effect on TSSs, except when combined with the other two factors. Juice pH behaved similarly, influenced by IV and SGP, with CSD also affecting it. Titratable Acidity (TA) remained relatively constant across all factor levels, with significant interactions.

IV statistically influenced the CIELAB coordinates, with an increase in Lightness (L*) in SI, where grapes also exhibited lower greenness (a*), indicating berries with a less intense green color compared to those under NI. Likewise, concerning SGP, SPB and DECCO demonstrated opposing effects. DECCO appeared to preserve L* better but lost a* compared to both SPB and the Control treatment. Additionally, it is noteworthy that L* tended to decrease with increasing CSD, while a* remained relatively unchanged over time. In contrast, yellowness (b*) was unaffected by IV and SGP but increased from T0 to T1 and T2, suggesting a shift towards yellow coloration over time. Importantly, no significant interactions were found among the combination of the three factors for all CIELAB coordinate parameters.

Table 1. Means of carpometry, juice berry composition, colorimetric coordinates, texture, nutraceutical traits and cold storage damages of Sugrathirtyfive table grapes grown under two different Irrigation Volumes (IVs) (NI = Normal Irrigation; SI = Smart Irrigation), different SO_2-Generating Pads (SGP) (C = Control; SPB = SmartPac® bags; DECCO = DECCO Grapage®), three different Cold Storage Durations (CSDs) (T0 = harvest; T1 = after 15 days; T2 = after 40 days) and relative interactions.

	Factors								Interactions			
	IV		SGP			CSD			IV × SGP	IV × CSD	SGP × CSD	IV × SGP × CSD
	NI	SI	C [†]	SPB	DECCO	T0	T1	T2 [†]				
20 berries' weight (g)	244.4	241.5	246.8	239.2	243.0	250.9	238.3	246.1	ns	ns	*	ns
Equatorial diameter (mm)	24.9	25.1	24.7	24.9	25.3	25.2	24.7	25.3	**	ns	ns	ns
Total Soluble Solids (°Brix)	15.7 [b]	17.3 [a]	15.7 [b]	17.2 [a]	16.6 [ab]	16.0	16.7	16.5	***	***	*	***
pH	3.75 [a]	3.63 [b]	3.63 [b]	3.70 [ab]	3.75 [a]	3.57 [b]	3.70 [b]	3.74[a]	*	ns	***	***
TA (g/L tartaric acid)	4.1	4.0	4.1	4.1	4.0	4.1	4.0	4.1	**	***	**	***
L*	39.90 [b]	40.91 [a]	40.26 [ab]	39.88 [b]	41.07 [a]	41.17 [a]	40.51 [ab]	39.86 [b]	ns	ns	ns	ns
a*	−3.48 [b]	−3.22 [a]	−3.35 [ab]	−3.24 [a]	−3.47 [b]	−3.36	−3.40	−3.27	**	ns	***	ns
b*	9.15	9.37	9.20	9.15	9.42	8.68 [b]	8.79 [b]	9.79 [a]	ns	ns	ns	***
Hardness (N)	14.38 [b]	16.25 [a]	15.74	14.81	15.40	17.61 [a]	15.13 [b]	14.44 [b]	ns	ns	ns	ns
Springiness (mm)	4.97	5.01	4.94	4.98	5.05	5.04	4.93	5.05	**	ns	ns	ns
Cohesiveness (adim)	0.65	0.65	0.62 [b]	0.67 [a]	0.66 [a]	0.63 [b]	0.65 [b]	0.68 [a]	ns	ns	ns	*
Chewiness (mj)	46.51 [b]	53.06 [a]	48.61	49.8	50.95	55.07	47.99	49.84	ns	**	ns	ns
Gumminess (N)	9.34 [b]	10.64 [a]	9.78	9.95	10.23	10.90	9.79	9.84	ns	*	ns	ns
TPF mg/kg fresh grape GAE	90.42 [b]	103.61 [a]	91.99 [b]	109.88 [a]	89.17 [b]	97.57	99.17	93.51	**	ns	ns	*
FLV g RE/kg	0.51 [b]	0.70 [a]	0.52 [ab]	0.78 [a]	0.50 [b]	0.44	0.73	0.49	ns	ns	**	ns
DPPH µmol TE /kg	452.86 [b]	502.40 [a]	437.20	502.29	463.41	468.13 [ab]	501.88 [a]	446.01 [b]	ns	ns	ns	ns
% berries damaged by SO_2	0.5%	0.5%	0.0% [b]	0.9% [a]	0.9% [a]	0.0% [b]	0.0% [b]	1.2% [a]	ns	ns	*	ns
% berries with gray rot/mold	16.3%	15.0%	36.3 [a]	0.6% [b]	1.9% [b]	0.0% [b]	7.9% [b]	27.6% [a]	ns	ns	*	ns
% stem browning	6.2%	4.7%	3.8%	7.0%	6.2%	0.0% [b]	2.1% [b]	10.2% [a]	ns	ns	ns	ns

TPF = Total Phenolic Content, FLV = Total Flavonoids, DPPH = 2,2-diphenyl-1-picrylhydrazyl. Means were separated by post hoc Tukey test in each factor singularly. Different letters correspond to statistically significative differences at $p < 0.05$. ns = not significant; * $p < 0.05$; ** $p < 0.01$; *** $p < 0.001$, [†] Means calculated excluding NI-C-T2 and SI-C-T2, if excepted for cold storage damages traits, for the unprocessability and non-marketability of grapes without SO_2-Generating Pads in T2, due to extensive development of mold on the berries.

Compression and texture tests on the berries provided insights into firmness and crunchiness. Hardness, representing the force required to achieve a given deformation, was significantly higher in the SI group compared to NI. SGP showed no significant effect on Hardness. Conversely, a notable decline in Hardness was observed with increasing CSD, with no significant interactions among the three factors. Springiness, representing the rate of material returning to its original state after deformation, remained constant and unaffected by individual factors. However, a significant interaction was observed between IV and SGP. Cohesiveness, reflecting a product's tendency to cohere, was unaffected by IV but was higher in grapes refrigerated with SGPs than Control. Similarly, CSD exerted an effect, progressively resulting in higher cohesiveness values. Only the interaction among the three factors was statistically significant in this case. Chewiness and Gumminess were influenced by IV, being higher in SI grapes, while remaining unaffected by the other two factors. Only the interaction between IV and SGP was statistically significant in both cases. In the analysis of the nutraceutical aspects of grapes, both Total Polyphenol Content (TPF) and Flavonoid (FLV) concentrations were influenced by IV. Within the scope of SGP, only SPB significantly preserved the concentration of both PFT and FLV, with no discernible effect from CSD. Significant interactions were observed for the combinations IV × SGP and IV × SGP × CSD for TPF, while FLV displayed a significant interaction for the combination SGP × CSD. Additionally, the radical-scavenging activity, assessed as the antioxidant power of extracts from grape skins using DPPH, corroborated the aforementioned trend. Grapes from SI exhibited a higher DPPH concentration, unaffected by SGP. Conversely, concerning CSD, there was an increase in DPPH until T1, followed by a decline at T2, returning to values comparable to those at harvest. No significant interaction was observed in this case, indicating an independent behavior of the three factors.

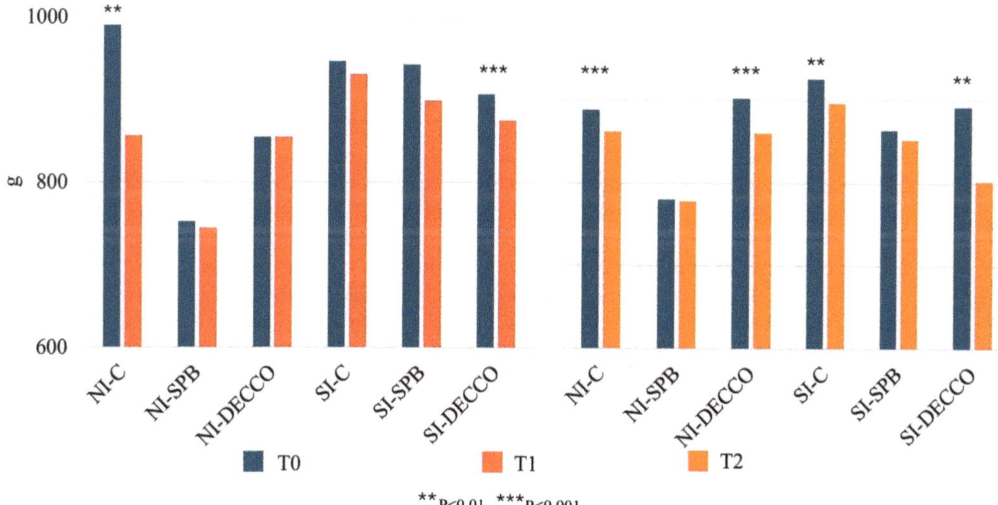

Figure 3. Comparison of cluster weight loss by means of paired sample t-test between clusters weighted at three different storage durations (T0 = at harvest; T1 = after 15 days; T2 = after 40 days) for each of the combined factors: Irrigation Volumes (IVs) (NI = Normal Irrigation; SI = Smart Irrigation) and SO_2-Generating Pads (C = Control; SPB = SmartPac® bags; DECCO = DECCO Grapage®).

Regarding parameters related to damages from storage, the percentage of berries damaged by SO_2 was minimal but present in modest quantities in the SGP treatments. Concerning CSD, damage from SO_2 onset was observed only at T2. Consequently, the only statistically significant interaction occurred in the combination of SGP × CSD. The percentage of berries with rot/mold exhibited a similar trend, with SGP use naturally reducing its incidence compared to the Control. As expected, a considerable increase was noted, particularly at T2, with values exceeding 27%. In this case, the combination of SGP × RSD also showed a statistically significant interaction. Regarding the percentage of stem browning, no effect of IV and SGP was recorded, while, though CSD showed no difference between T0 and T1, its effects were visible at T2, with stem browning values exceeding 10%. The data suggested that this issue arose only when a substantial CSD was reached, regardless of the other two factors.

In pursuit of a comprehensive assessment and identification of the most effective combination among the factor levels, these factors were consolidated into 14 overall theses. This amalgamation aimed to facilitate a Dunnett test (Table 2) for comparing each thesis with combined factors against the Control thesis. In our study, the Control thesis is represented by NI without SGP during cold storage and at CSD T0 (NI-C-T0). While no discernible differences were noted between the theses for berry weight and equatorial diameter compared to the Control, significant variations were observed for Total Soluble Solids (TSSs). The only theses that did not exhibit significant differences in TSSs were NI-C-T1, SI-C-T1 and NI-SPB-T2. In contrast, all other theses displayed higher TSSs values, particularly those derived from SI, regardless of the SGP employed and the considered CSD (SI-SPB-T1, SI-SPB-T2, SI-DECCO-T1, SI-DECCO-T2), with values close to or exceeding 18°Brix, compared to the 15.6 °Brix of NI-C-T0. At harvest (T0), juice pH was confirmed to be higher in the NI thesis, and a general increase was observed for all theses at T2, with SGP DECCO also showing an increase at T1. Furthermore, variations in Total Acidity (TA) were exclusive to the SI-DECCO, theses, lower at T1 and higher at T2 compared to the Control, affirming the stability of parameters among theses as reported in Table 1.

Few differences were identified in CIELAB coordinates. Specifically, Lightness (L*) was higher than the Control in the SI-C-T0 and SI-DECCO-T1 theses, and more stable in the other theses. Parameters a* and b* demonstrated relatively stable values, with SI-SPB-T1 exhibiting lower values of greenness, while, conversely, in the NI-DECCO-T1 theses, the berries displayed more pronounced green notes. Furthermore, higher yellowness (i.e., higher b* values) compared to the Control was observed for the NI-C-T1, SI-SPB-T1, and NI-DECCO-T1 theses. In terms of texture analysis parameters, including Hardness, Chewiness and Gumminess, SI-C-T0 was the only thesis showing significantly higher values compared to the Control. Conversely, regardless of IV or SGP, all other theses displayed similar values to the Control, even with the progression of CSD. On the contrary, Springiness remained generally constant, with no significant differences observed between the theses. Regarding nutraceutical parameters, in the SI-SPB-T1 thesis, both Flavonoid content (FLV) and 2,2-diphenyl-1-picrylhydrazyl (DPPH) radical-scavenging activity yielded significantly higher values compared to the reference thesis NI-C-T0. SI-C-T0 (at harvest), and SI-SPB at both T1 and T2 exhibited higher TPF contents, while all the theses from NI and those treated with SGP DECCO were identical to the Control thesis in all CSD. Finally, regarding cold storage damages, the percentage of berries damaged by SO_2 was negligible when no SGP was used. However, both SPB and DECCO systems showed mild signs of SO_2 scorching, never exceeding 1.8%. A distinct consideration must be made for the incidence of the percentage of berries with rot/mold, absent at harvest (T0) but progressively increasing from T0 to T1 and T2, exclusively in the theses without the SGP device, as expected. This phenomenon rendered the clusters at T2 entirely unusable for analysis, reaching stem browning percentages of 76.0% and 80.0%, respectively, in the NI-C and SI-C theses. Thus, it can be affirmed that both SPB and DECCO preserved the grapes from the onset of mold. The significant onset of percentage of stem browning occurred for all the theses at T2, irrespective of the IV. Additionally, the thesis treated with DECCO proved to be less effective than SPB in containing this phenomenon, already exhibiting issues of stem browning at T1.

Table 2. Means separation by Dunnet test of carpometry, berry juice composition, colorimetric coordinates, texture, nutraceutical traits and cold storage damages for merged factors on Sugrathirtyfive table grapes grown under two different Irrigation Volumes (IVs) (NI = Normal Irrigation; SI = Smart Irrigation), different SO_2-Generating Pads (SGPs) (C = Control; SPB = SmartPac® bags; DECCO = DECCO Grapage®), three different Cold Storage Durations (CSDs) (T0 = harvest; T1 = after 15 days; T2 = after 40 days).

	NI-C T0	NI-C T1	NI-C T2	SI-C T0	SI-C T1	SI-C T2	NI-SPB T1	NI-SPB T2	SI-SPB T1	SI-SPB T2	NI-DECCO T1	NI-DECCO T2	SI-DECCO T1	SI-DECCO T2
20 berry weight (g)	252.3	238.5	-	249.5	246.8	-	223.1	252.1	233.5	248.0	243.9	256.7	243.7	227.5
Equatorial diameter (mm)	25.4	24.2	-	25.1	24.3	-	24.1	24.6	25.1	25.9	25.6	25.5	24.8	25.3
TSSs (°Brix)	15.6	15.1	-	16.4 ***	15.9	-	16.2 ***	15.8	18.8 ***	18.0 ***	17.0 ***	14.4 ***	16.9 ***	17.9 ***
pH	3.61	3.77 ***	-	3.52 ***	3.62	-	3.64	3.84 ***	3.64	3.66 *	3.86 ***	3.78 ***	3.66 *	3.69 ***
TA (g/L tartaric acid)	4.1	4.2	-	4.2	4.0	-	4.2	4.2	4.0	3.9	4.1	3.7	3.5 **	4.7 ***
L*	39.94	38.73	-	42.40 *	39.98	-	39.79	39.57	40.64	39.51	41.68	39.66	42.24 *	40.69
a*	−3.33	−3.45	-	−3.40	−3.24	-	−3.52	−3.46	−2.93 *	−3.04	−3.78 **	−3.36	−3.51	−3.22
b*	8.20	10.12 **	-	9.16	9.34	-	8.82	9.23	10.25 **	8.31	10.29 **	8.24	9.80 *	9.37
Hardness (N)	13.67	13.54	-	21.55 ***	14.18	-	15.85	13.89	13.74	15.76	16.73	12.60	16.75	15.50
Springiness (mm)	5.08	4.83	-	5.01	4.85	-	4.81	4.90	5.02	5.17	5.12	5.08	4.96	5.05
Cohesiveness (adim)	0.63	0.62	-	0.62	0.62	-	0.66	0.67	0.65	0.70 ***	0.64	0.68	0.66	0.68
Chewiness (mI)	43.20	41.01	-	66.93 ***	43.30	-	50.96	45.42	45.34	57.48	55.29	43.20	52.07	53.25
Gumminess (N)	8.52	8.46	-	13.29 ***	8.87	-	10.53	9.26	8.92	11.10	10.77	8.52	11.17	10.49
DPPH umol TE /kg	451.23	436.58	-	485.03	495.98	-	513.43	444.31	538.90 **	472.50	473.12	398.49	513.26	468.74
FLV g RE/kg	0.55	0.62	-	0.34	0.58	-	0.77	0.40	0.88 ***	0.66	0.34	0.35	0.77	0.55
TPF mg/kg fresh grape GAE	82.10	96.47	-	113.03 ***	76.35	-	101.33	91.73	133.94 ***	112.53**	93.56	77.31	93.34	92.46
% berries damages SO_2	0.0%	0.0%	0.0%	0.0%	0.0%	0.0%	1.8% *	0.0%	1.8% *	0.0%	1.8% *	0.0%	1.8% *	0.0%
% berries with rot/mold	0.0%	32.0% ***	76.0% ***	0.0%	15.0%**	80.0% ***	0.0%	0.8%	0.0%	1.6%	0.2%	1.8%	0.2%	5.4%
% stem browning	0.0%	0.0%	13.0%**	0.0%	0.4%	8.0% **	2.0%	12.0% **	2.0%	12.0% **	7.0% *	8.0% **	1.2%	8.4% **

TPF = Total Phenolic Content, FLV = Total Flavonoids, DPPH = 2,2-diphenyl-1-picrylhydrazyl. * $p < 0.05$; ** $p < 0.01$; *** $p < 0.001$. The reference control is NI-C-T0, reported in bold and italics.

3.3. Multivariate Analysis

A Principal Component Analysis (PCA) biplot analysis was conducted at T0, T1 and T2 to gain a comprehensive understanding of the data. In Figure 4, the PCA at T0 focused on freshly harvested grapes from two levels of the IV factor. PC1 explained 52.47% of the variance, PC2 described 24.92%, totaling 77.39%, rendering further analysis on the third axis PC3 unnecessary. Notably, the NI and SI treatments were distinct, as illustrated by 95% confidence ellipses. The SI treatment was characterized by texture parameters, berry juice composition, colorimetric aspects, and significant contents of TPF and DPPH. In contrast, the NI treatment was predominantly characterized by FLV and pH, followed by Cohesiveness and, marginally, by Equatorial Diameter. Other variables contributed mainly to PC2, which explained variances within groups rather than distinguishing between NI and SI treatments.

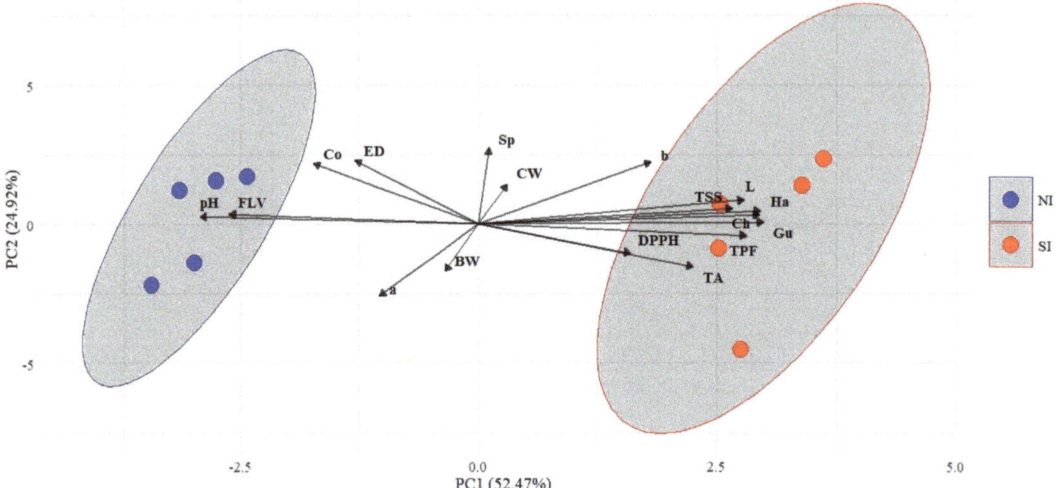

Figure 4. Biplot of Principal Component Analysis: eigenvalues, eigenvectors and percent of variation accounted for the first three principal components (PCs) of carpometric, must, colorimetric coordinates, texture and nutraceutical traits at harvest (T0) of Autumn Crisp table grapes grown with two different Irrigation Volumes (IVs) (Normal Irrigation = NI; Smart Irrigation = SI). Ellipse 95% is shown. CW = Cluster weight; BW = 20 berries' weight; ED = Equatorial Diameter; TSSs = Total Soluble Solids; TA = Titratable Acidity; Ha = Hardness; Sp = Springiness; Co = Cohesiveness; Ch = Chewiness; Gu = Gumminess; TPF = Total Polyphenolic Content; FLV = Flavonoids; DPPH = 2,2-diphenyl-1-picrylhydrazyl; DSO_2 = % berries damaged by SO_2; BRM = % berries with rot/mold; SB = % stem browning.

As expected, the PCA biplot at T1 significantly changed, necessitating consideration of IV levels, SGP and damages caused by CSD (Figure 5). PC1 explained 32.91% of the variance, PC2 contributed 20.32% and PC3 added 11.82%, totaling 76.32%. Variables contributing to PC1 included texturometric parameters, colorimetric aspects, berry dimensions and parameters derived from cold storage damages. Simultaneously, PC2 was strongly characterized positively by nutraceutical parameters, TSSs content and higher a* values, leading to a distinct separation of the SI-SPB treatment. Conversely, the NI-DECCO treatment exhibited opposite behavior, overlapping with treatments refrigerated without SGP, highlighting a significant incidence of the percentage of berries with post-harvest decay, as expected. Using PC3, correlated with carpometric variables, all treatments tended to overlap, except for NI-SPB.

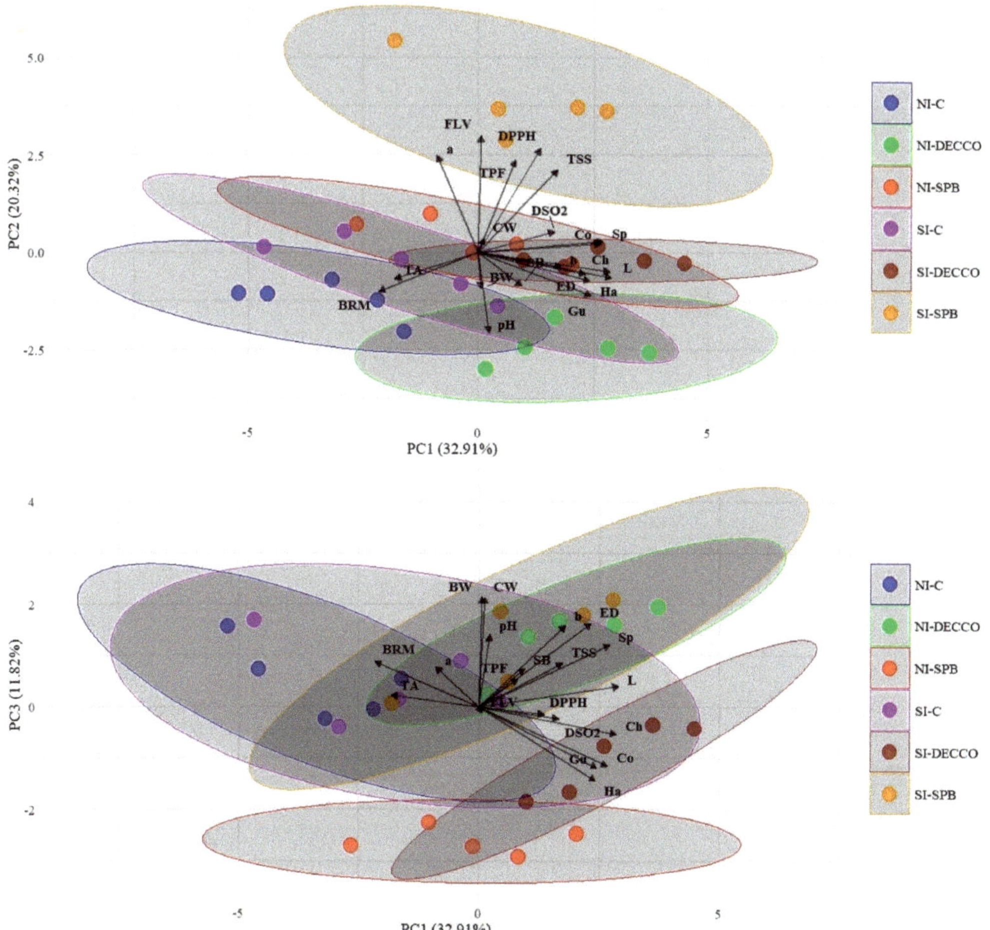

Figure 5. Biplot of Principal Component Analysis: eigenvalues, eigenvectors and percent of variation accounted for the first three principal components (PCs) of carpometric, must, colorimetric coordinates, texture and nutraceutical traits after 15 days' Cold Storage Duration (CSD) (T1) of Autumn Crisp table grapes grown with two different Irrigation Volumes (IVs) (Normal Irrigation = NI; Smart Irrigation = SI) and different SO_2-Generating Pads (SGPs) (C = Control; SPB = SmartPac® bags; DECCO = DECCO Grapage®). Ellipse 95% is shown. CW = Cluster weight; BW = 20 berries' weight; ED = Equatorial Diameter; TSSs = Total Soluble Solids; TA = Titratable Acidity; Ha = Hardness; Sp = Springiness; Co = Cohesiveness; Ch = Chewiness; Gu = Gumminess; TPF = Total Polyphenolic Content; FLV = Flavonoids; DPPH = 2,2-diphenyl-1-picrylhydrazyl; DSO_2 = % berries damaged by SO_2; BRM = % berries with rot/mold; SB = % stem browning.

The latest PCA biplot at T2 (Figure 6) excluded the NI-C and SI-C treatments due to their deterioration. Similar to T1, PC1 (46.24%) and PC2 (16.00%) explained the most variance, and PC3 (14.08%) was necessary, totaling 76.32%. As in T1, berry texture, colorimetric and nutraceutical variables contributed positively to PC1, while parameters related to cold storage damages correlated with PC2. SI treatments, whether SPB or DECCO, substantially overlapped, while NI-DECCO showed reduced cold storage damages, lower nutraceutical content, but good values of 20 berries' weight, while NI-SPB was positioned intermediately.

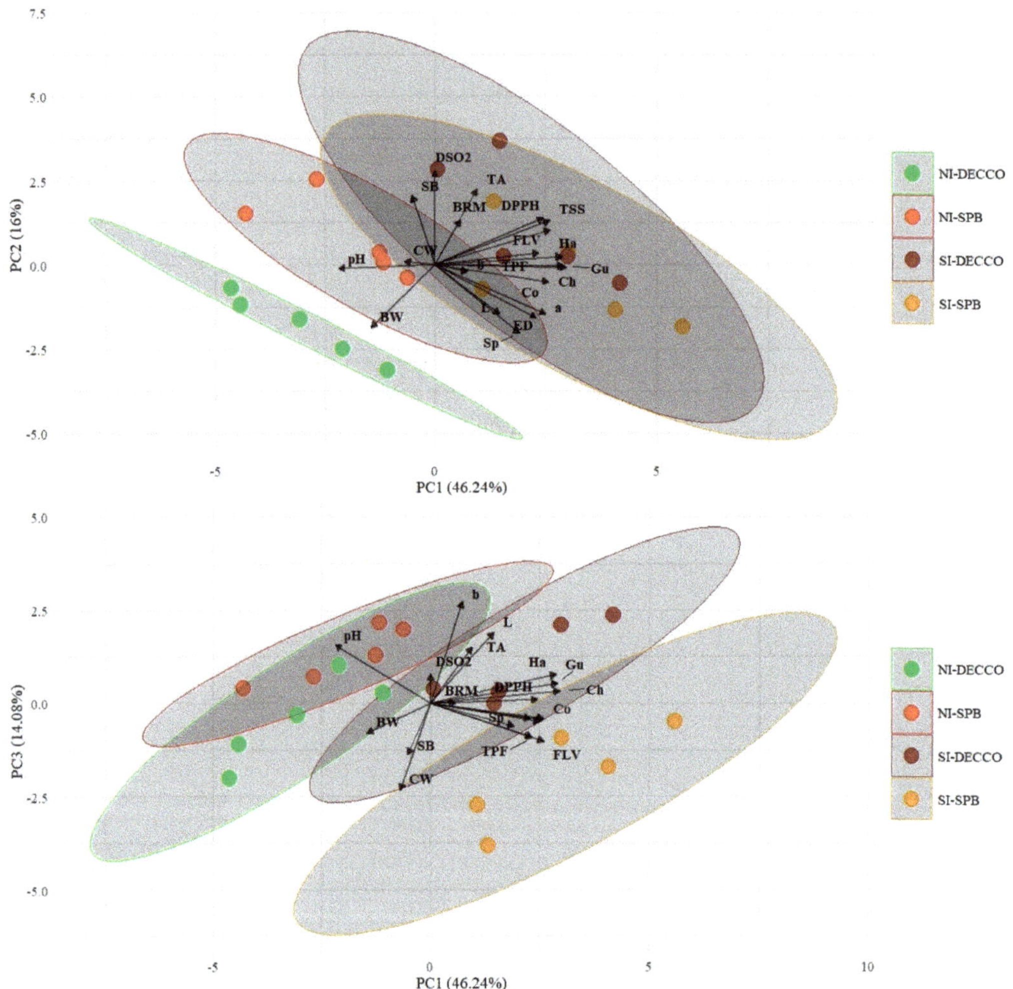

Figure 6. Biplot of Principal Component Analysis: eigenvalues, eigenvectors and percent of variation accounted for the first three principal components (PCs) of carpometric, must, colorimetric coordinates, texture and nutraceutical traits after 40 days' Cold Storage Duration (CSD) (T2) of Sugratable grapes grown with two different Irrigation Volumes (IVs) (Normal Irrigation = NI; Smart Irrigation = SI) and different SO_2-Generating Pads (SGPs) ((C = Control; SPB = SmartPac® bags; DECCO = DECCO Grapage®). Control (i.e., grapes with no SGP) was excluded in T2, due to extensive development of mold on the berries. Ellipse 95% is shown. CW = Cluster weight; BW = 20 berries' weight; ED = Equatorial Diameter; TSSs = Total Soluble Solids; TA = Titratable Acidity; Ha = Hardness; Sp = Springiness; Co = Cohesiveness; Ch = Chewiness; Gu = Gumminess; TPF = Total Polyphenolic Content; FLV = Flavonoids; DPPH = 2,2-diphenyl-1-picrylhydrazyl; DSO_2 = % berries damaged by SO_2; BRM = % berries with rot/mold; SB = % stem browning.

Analyzing PC1 and PC3, NI-SPB and NI-DECCO practically overlapped, characterized by higher pH values and greater yellowness (b*), positively correlated with PC3. SI-SPB and SI-DECCO stood out distinctly, showing higher nutraceutical contents, texture parameters and reduced effects of variables related to cold storage damages. This suggests good storage resilience for SI treatments, with better outcomes for SPB refrigerated treatments.

3.4. Sensory Analysis of Grapes

During the initial tasting session at T1 (Figure 7), all experimental treatments were included, encompassing grapes refrigerated without any SGP devices, which, as previously noted, remained in satisfactory condition both in terms of edibility and marketability at 15 days post-harvest. Beyond the inherent variations in descriptors due to the subjective nature of evaluation, the Kruskal–Wallis test unveiled statistically significant differences among treatments for the attributes "Berry crunchiness" ($p < 0.01$) and "Pulp consistency" ($p < 0.001$). Specifically, "Berry crunchiness" was notably higher in the NI-SPB and SI-SPB treatments compared to NI-C and SI-C, with NI-DECCO and SI-DECCO exhibiting intermediate values. Moreover, NI-C and SI-C displayed statistically lower values for "Pulp consistency", while all other treatments exhibited statistically similar results.

The sensory analysis was reiterated 40 days post-harvest (T2) (Figure 8). As previously mentioned, in this session, grapes from refrigerated treatments without SGP (NI-C and SI-C) were excluded due to their inedibility resulting from a high incidence of percentage of berries with rot/mold (Table 2). In this subsequent evaluation, descriptors that exhibited statistically significant differences among treatments were Stem coloration ($p < 0.05$); Stem turgidity ($p < 0.05$); Peduncle browning ($p < 0.01$); Berry color uniformity ($p < 0.05$) and Overall appearance ($p < 0.05$).

"Stem coloration" was notably lower in the NI-DECCO treatment compared to the NI-SPB and SI-DECCO treatments, with SI-SPB positioned in an intermediate position. Regarding "Stem turgidity," contradictory results were observed, with NI-SPB and SI-DECCO showing statistically higher values than NI-DECCO and SI-SPB, as rated identically by the panel. Concerning "Peduncle browning," the SI-SPB treatment notably better preserved the grapes for this descriptor, while the others were evaluated similarly. Additionally, "Berry colour uniformity" received positive evaluations for all treatments except SI-DECCO, which was statistically less favored. Moreover, "Overall Appearance" exhibited statistical differences among treatments, with the SI-DECCO treatment being judged to have the overall best appearance, followed by the NI-SPB and NI-DECCO treatments in an intermediate position and the SI-SPB treatment being the least appreciated.

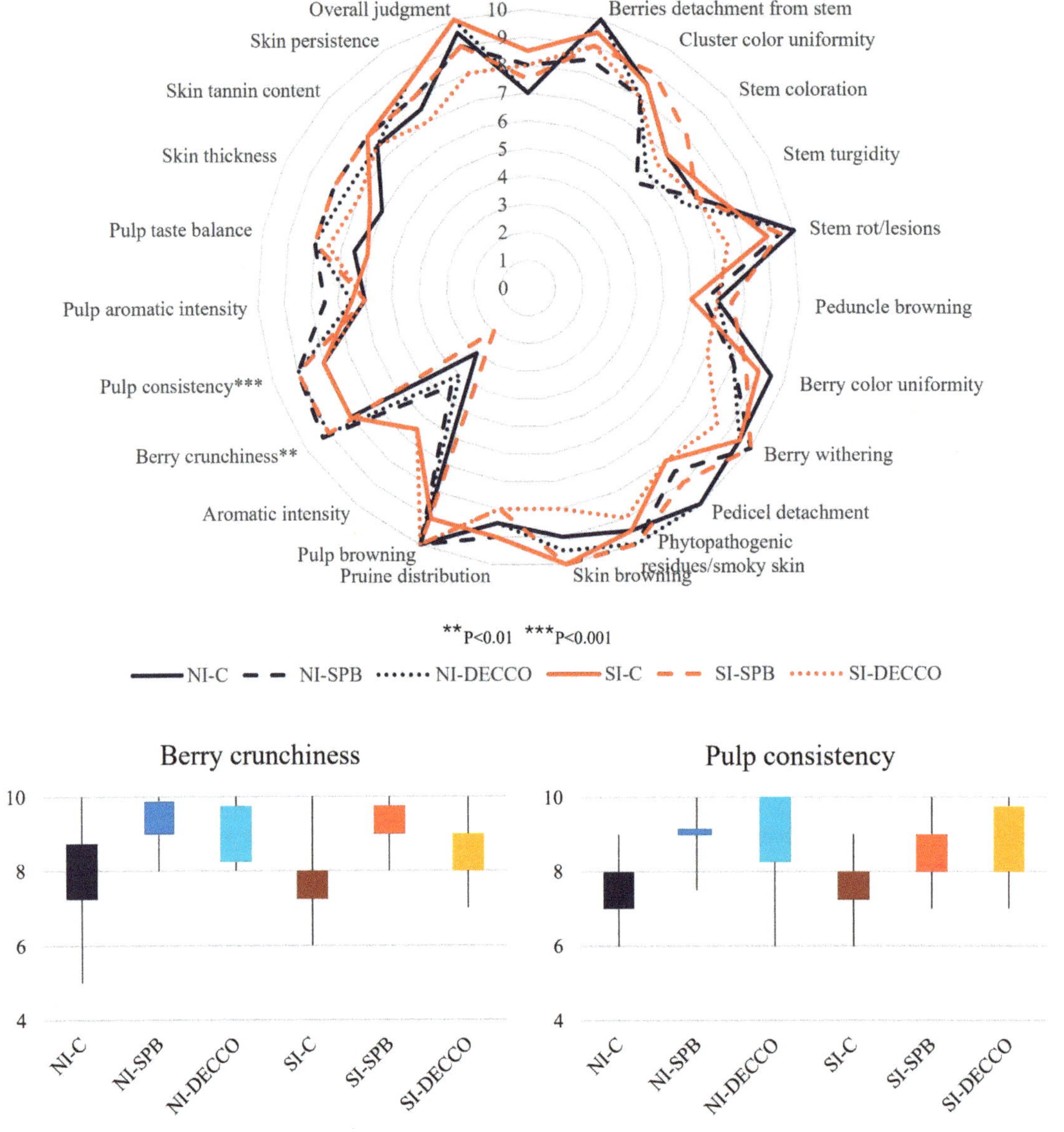

Figure 7. Spider chart of medians of sensory descriptors on Autumn Crisp table grapes grown with two different Irrigation Volumes (IVs) (Normal Irrigation = NI; Smart Irrigation = SI) and different SO_2-Generating Pads (SGPs) (C = Control; SPB = SmartPac® bags; DECCO = DECCO Grapage®) after 15 days' Cold Storage Duration (CSD) (T1). Box and Whisker plot of descriptors, showing statistically significant differences for the Kruskal–Wallis test, are reported on the right.

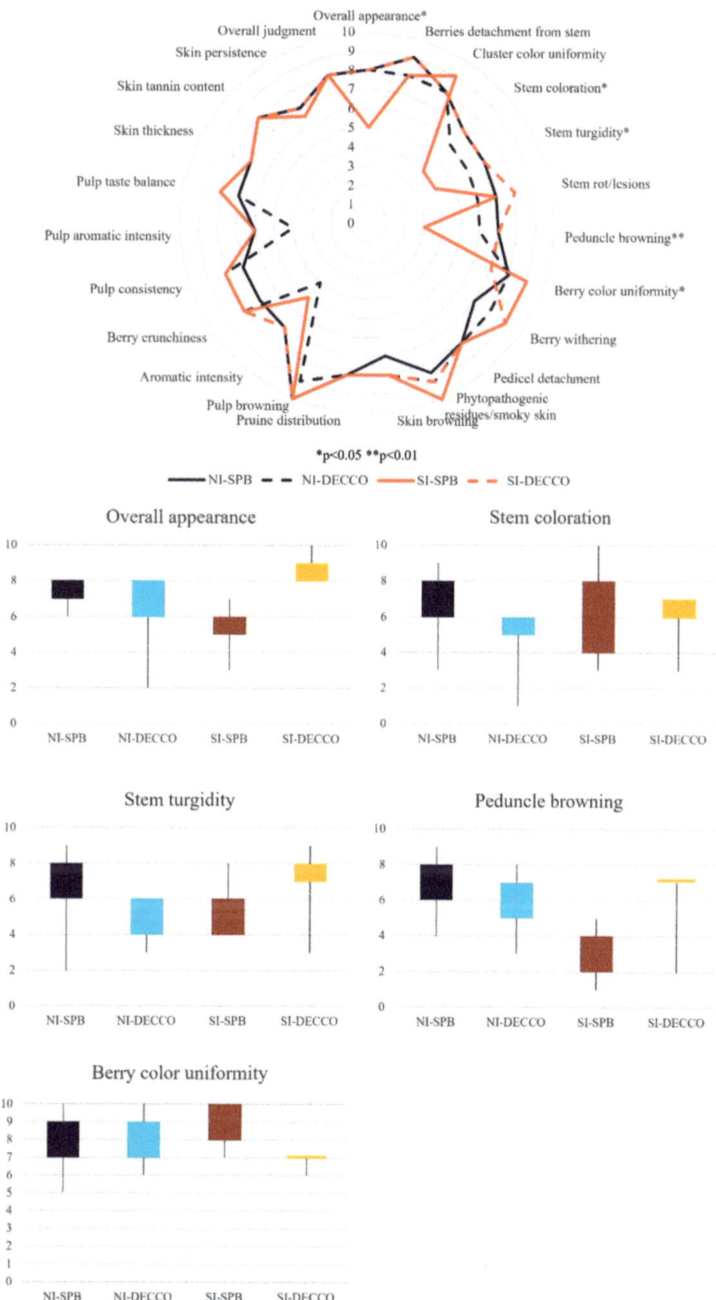

Figure 8. Spider chart of medians of sensory descriptors on table grapes grown with two different Irrigation Volumes (IVs) (Normal Irrigation = NI; Smart Irrigation = SI) and different SO_2-Generating Pads (SGPs) (SPB = SmartPac® bags; DECCO = DECCO Grapage®) after 40 days' Cold Storage Duration (CSD) (T2). Box and Whisker plot of descriptors, showing statistically significant differences for the Kruskal–Wallis test, are reported below the chart.

4. Discussion

This research aimed to monitor and evaluate a recently introduced seedless table grape cultivar, Sugrathirtyfive, throughout its journey from the field to cold storage and, ultimately, to the final consumer. The study investigated the potential for reducing irrigation water inputs to enhance agronomic and production sustainability, the ability to maintain premium quality characteristics of grapes through cold storage aided by SO_2-releasing devices and the sensory appreciation of the grape.

Reducing water inputs in table grape cultivation is a pressing objective, as evidenced by the publication of guidelines on the sustainable use of water in winegrape vineyards by the International Organization of Vine and Wine [34]. Among various strategies, Smart Irrigation (SI) in table grape cultivation represents a technological opportunity for growers, offering a simple and intuitive approach as part of Decision Support Systems (DSSs). SI is a well-established practice in both wine grapes [35–37] and table grapes [11,38], particularly in environments like Southern Italy, where growing without irrigation is impractical [3]. In a previous study conducted by Campi et al. [39] in the same area, IV calculated for Normal Irrigation (NI) by an empirical program was lower (296 mm) with respect to the value of 335 mm calculated in this trial. Meanwhile, Irrigation Volumes (IV) provided by SI were lower when compared to those provided by the deficit irrigation regime (300 mm) by Colak and Yazar [40] in Turkey. The IV saved with SI was about 80 mm higher than the water savings found by Vox et al. [41] for the cv. 'Crimson Seedless' that imposed a mild Deficit Irrigation (at 80% ETc).

Moderate water stress generally leads to improvements in grape quality, including increased Total Soluble Solids (TSSs), anthocyanins and phenolic concentrations, although berry weight and Titratable Acidity (TA) may decrease [42]. Instead, Conesa et al. [43] observed no significant differences in berry size and weight for another seedless variety, Crimson Seedless, under a 35% reduction in irrigation, indicating that production components were not compromised. These findings are in line with our data on grapes at harvest, for which TSSs and nutraceutical components were higher in SI, while berry weight and TA remained almost unchanged between NI and SI. In addition, Temnani et al. [44] reported that reducing irrigation by up to 40%, particularly post-veraison, enhanced water use efficiency and increased berry color and firmness. SI grapes exhibited higher berry firmness at harvest than NI grapes, particularly for parameters such as Hardness, Chewiness and glyming. Sugrathirtyfive generally revealed quite interesting firmness values when compared to other white berry varieties. As an example, Sugrathirtyfive showed similar values for Hardness and Gumminess compared to Regal seedless or Italia [45], while Springiness, Cohesiveness and Chewiness were even higher. Even the 10 white berry varieties analysed by Rolle et al. [46] provided results related to berry firmness that were absolutely in line with our values, except for Chewiness, which was significantly higher in Sugrathirtyfive. Chewiness is intended as the ability to measure the resistance to penetration of a given berry skin, and the very high values scored by Sugrathirtyfive suggest a skin thickness that makes it interesting for long refrigerated storage. In this sense, further investigation into the skin thickness of this variety should be undertaken.

At harvest, SI grapes showed significantly higher values of lightness (L*) and a greater tendency for berries to develop intense color (b*) compared to the greenness observed in NI grapes. In this sense, Pisciotta et al. [45] recorded slightly higher L* values around 40 in clusters of cv Regal seedless and around 37 for cv Italia, consistent with our values. The same authors also reported lower a* values for the same white berry varieties compared to Sugrathirtyfive, with a greater component of greenness and consistently higher b* (redness) values. These differences can be attributed to the training system (covered plastic film tendone or not), variety, vineyard management and environmental conditions. It is known that, in white grape varieties, color intensity and yellowness are primarily influenced by kaempferol, with minor contributions from quercetin and isorhamnetin [47,48]. These flavonols are part of the flavonoid group, and their biosynthesis is regulated by flavonol synthase (FLS), which studies have shown can be upregulated in response to water stress.

This upregulation is often a plant's defense mechanism to cope with stress by producing secondary metabolites that help mitigate its effects [49,50]. In our study, the higher flavonoid content in the SI treatment may be due to the upregulation of genes involved in their biosynthesis. Similarly, the total polyphenol content (TPF) was also stimulated to a greater extent in the SI treatment with reduced irrigation, which was expected, as polyphenol synthesis is generally triggered by plant defense mechanisms in response to abiotic stress [42]. Moreover, SI grapes showed significantly higher DPPH values than NI grapes. The antioxidant activity of grapes greatly depends on the quantitative and qualitative differences in phenolic compounds [51] and several classes of compounds (anthocyanins, phenolic acids and stilbenes) could contribute to the grape antioxidant activity, suggesting a synergic effect of these compounds. As well known, phenols are good antioxidants due to their susceptibility to oxidation resulting from the hydroxyl groups and unsaturated double bonds in their chemical structure [52].

Regarding the effects of SO_2-Generating Pads (SGPs) aimed at prolonged Cold Storage Duration (CSD), both SmartPac® bags (SPB) and DECCO Grapage® (DECCO) offered satisfactory results in preserving TSSs compared to the Control, and the berryic characteristics of the grapes such as Berry weight and Equatorial Diameter. For its part, the average cluster weight (Figure 3) mainly remained constant in the T0-T1 comparison, except for NI-C and SI-DECCO. At T2, SPB proved to be more effective than DECCO on both NI and SI grapes, contrasting with what was reported by Fernández-Trujillo et al. [53], who stated that the dual-phase SGP showed better performance for the long-term storage of grapes than the single-phase one. As for CIELAB coordinates, the dual release SO_2 system (DECCO) generally proved to be more effective in maintaining brightness (L*), which, however, decreased over the cold storage period. On the contrary, DECCO showed a decline in the a* index, which led the grapes to have more pronounced shades of green. However, L* significantly decreased over time, while the b* yellowness index increased. Ahmed et al. [25], in a study conducted on cv Italia, a white berry table grape variety that, although seed-containing, can serve as a benchmark with Sugrathirtyfive, reported average L* values around 30 at both 7 and 50 days of refrigerated storage. In our case, under all conditions and for all factors, L recorded values close to 40, indicating that grapes were still in commercially acceptable conditions even at 40 days post-harvest. Regarding the firmness of the berries, no significant difference was observed in the use of the different SGPs in all cold storage periods evaluated, in line with Roberto et al. [18], except for Cohesiveness, which increased over time, probably due to dehydration phenomena of the berry that, however, did not reflect, as mentioned, their variation in weight and size.

Regarding the nutraceutical aspect, the sensitivity to SO_2 generally differs among the various table grape varieties. Previous studies reported that phenolic compounds presented a different behavior post-harvest. After 54 days, phenolic content decreased for the Crimson Seedless or increased for new seedless table grape cultivars Timco™ and Krissy™ stored in perforated polyethene bags with an SO_2-generating mat [54]. In our study, the nutraceutical molecules TPF and FLV also did not suffer from the CSD effect, but rather from the SGP system considered, for which SPB proved to be more effective than the dual release SO_2 DECCO. However, the phytosanitary aspect of grapes is of fundamental importance for defining the commercial qualities of grapes over time after harvest. As known, SGPs have the function of preventing the incidence of grey mold, mainly caused by Botrytis cinerea [25]. This was also observed in our research, where both SGPs were effective in containing the percentage of berries with rot/mold compared to the Control, for which, as mentioned earlier, grapes at T2 were covered with mold to a rate exceeding 70% (Table 2) and thus deemed inedible. Moreover, the SGP, combined with cold storage, yielded appreciable results in terms of containing the phenomenon of stem browning, as reported in other studies [55]. It is also worth noting that the incidence of SO_2-induced damage caused by SGPs, while statistically significant compared to the Control, was marginal in terms of magnitude, with values averaging below 2%.

5. Conclusions

By examining the data in absolute terms through the combination of the levels of the three factors IV, SGP and CSD, the shelf life of Sugrathirtyfive grapes for periods exceeding 15 days (T1) requires the use of SGPs, under penalty of product loss. The use of SGPs allows grapes to still maintain commercially appreciable quality, with greater efficiency already at T1 in preserving the characteristics of firmness perceived by the panel in terms of Pulp Consistency and Berry Crunchiness with respect to Control. Subsequently, after 40 days of CSD (T2), regarding firmness aspects, no differences were observed among the treatments, describing a similar behavior despite the SGPs or IVs. On the contrary, some differences regarding stem and pedicel integrity began to emerge, with loss of berry color uniformity and the onset of phenomena such as peduncle browning and loss of stem turgidity. Some authors [54] reported that dual-phase release extends the shelf life of grapes by around 1 month. Under the storage conditions used in the study, the 40-day period may represent an appropriate time limit for the cold storage and consumption of Autumn Crisp grapes, even if grown more sustainably under SI, provided that they are refrigerated with the aid of SGPs.

Author Contributions: Conceptualization, V.A. and A.R.C.; Methodology, V.A. and A.R.C.; Software, A.R. and G.G; Validation, A.R.C. and L.T.; Formal Analysis, V.A., P.C., A.F.M., A.R., S.R., G.G., R.A.M. and G.F.; Investigation, V.A., S.R., P.C., L.T. and A.R.C.; Writing—Original Draft Preparation, V.A., R.A.M. and P.C.; Writing—Review and Editing, V.A. and L.T.; Supervision, V.P., A.R.C. and L.T.; Funding Acquisition, A.R.C. All authors have read and agreed to the published version of the manuscript.

Funding: This research was funded by PSR Regione Puglia 2014–2020, Measure 16—Cooperation, Sub-Measure 16.2 "Support for pilot projects and the development of new products, practices, processes, and technologies", Project Title "Sustainability and Innovation in Apulian Table Grape Farming", Project Acronym "INNOFRUIT".

Institutional Review Board Statement: Not applicable.

Informed Consent Statement: Not applicable.

Data Availability Statement: Data available on request. Information on the project can be found at www.innofruit.it accessed on 21 March 2024.

Conflicts of Interest: The authors declare no conflicts of interest.

References

1. Lionello, P.; Scarascia, L. The relation between climate change in the Mediterranean region and global warming. *Reg. Environ. Chang.* **2018**, *18*, 1481–1493. [CrossRef]
2. Pizarro, E.; Galleguillos, M.; Barría, P.; Callejas, R. Irrigation management or climate change? Which is more important to cope with water shortage in the production of table grape in a Mediterranean context. *Agric. Water Manag.* **2022**, *263*, 107467. [CrossRef]
3. Medrano, H.; Tomás, M.; Martorell, S.; Escalona, J.M.; Pou, A.; Fuentes, S.; Flexas, J.; Bota, J. Improving water use efficiency of vineyards in semi-arid regions. A review. *Agron. Sustain. Dev.* **2015**, *35*, 499–517. [CrossRef]
4. Cotecchia, V. Le Acque Sotterranee e l'Intrusione Marina in Puglia: Dalla Ricerca all'Emergenza Nella Salvaguardia Della Risorsa. In *Memorie Descrittive Della Carta Geologica d'Italia*; ISPRA Servizio Geologico d'Italia: Rome, Italy, 2014; Volume 92, pp. 338–369, ISBN 978-88-9311-003-7. Available online: https://www.isprambiente.gov.it/files2017/pubblicazioni/periodici-tecnici/memorie-descrittive-della-carta-geologica-ditalia/volume-92?b_start:int=0 (accessed on 18 April 2024).
5. Mabrouk, H. The use of water potentials in irrigation management of table grape grown under semiarid climate in Tunisia. *OENO One* **2014**, *48*, 123–133. [CrossRef]
6. Schilling, T.; Müller, R.; Ellwart, T.; Antoni, C.H. Context-dependent preferences for a decision support system's level of automation. *Comp. Hum. Behav. Rep.* **2024**, *13*, 100350. [CrossRef]
7. Centofanti, T.; Bañuelos, G.S.; Ayars, J.E. Fruit nutritional quality under deficit irrigation: The case of table grapes in California. *J. Sci. Food Agric.* **2019**, *99*, 2215–2225. [CrossRef] [PubMed]
8. Pinillos, V.; Ibáñez, S.; Cunha, J.M.; Hueso, J.J.; Cuevas, J. Postveraison Deficit Irrigation Effects on Fruit Quality and Yield of "Flame Seedless" Table Grape Cultivated under Greenhouse and Net. *Plants* **2020**, *9*, 1437. [CrossRef] [PubMed]
9. Jiang, X.; Liu, B.; Guan, X.; Wang, Z.; Wang, B.; Zhao, S.; Song, Y.; Zhao, Y.; Bi, J. Proper deficit irrigation applied at various stages of growth can maintain yield and improve the comprehensive fruit quality and economic return of table grapes grown in greenhouses. *Irrig. Drain.* **2021**, *70*, 1056–1072. [CrossRef]

10. Temnani, A.; Conesa, M.R.; Ruiz, M.; López, J.A.; Berríos, P.; Pérez-Pastor, A. Irrigation Protocols in Different Water Availability Scenarios for 'Crimson Seedless' Table Grapes under Mediterranean Semi-Arid Conditions. *Water* **2021**, *13*, 22. [CrossRef]

11. Conesa, M.R.; Berríos, P.; Temnani, A.; Pérez-Pastor, A. Assessment of the Type of Deficit Irrigation Applied during Berry Development in 'Crimson Seedless' Table Grapes. *Water* **2022**, *14*, 1311. [CrossRef]

12. de Palma, L.; Limosani, P.; Marasovic, I.; Pati, S.; Vox, G.; Schettini, E.; Novello, V. Vineyard protection with rain-shelter: Relationships between radiometric properties of plastic covers and table grape quality. *BIO Web Conf.* **2019**, *13*, 04007. [CrossRef]

13. OIV. *The Sustainable Use of Water in Winegrape Vineyards. OIV Collective Expertise Document*, 1st ed.; OIV Publications: Paris, France, 2021; ISBN 978-2-85038-023-5. Available online: https://www.oiv.int/the-sustainable-use-of-water-in-winegrape-vineyards (accessed on 18 April 2024).

14. Yu, J.M.; Ahmedna, M. Functional components of grape pomace: Their composition, biological properties and potential applications. *Int. J. Food Sci. Technol.* **2013**, *48*, 221–237. [CrossRef]

15. Park, E.; Edirisinghe, I.; Choy, Y.Y.; Waterhouse, A.; Burton-Freeman, B. Effects of grape seed extract beverage on blood pressure and metabolic indices in individuals with pre-hypertension: A randomised, double-blinded, two-arm, parallel, placebo-controlled trial. *Br. J. Nutr.* **2016**, *115*, 226–238. [CrossRef]

16. Lu, R.; Song, M.; Wang, Z.; Zhai, Y.; Hu, C.; Perl, A.; Ma, H. Independent flavonoid and anthocyanin biosynthesis in the flesh of a red-fleshed table grape revealed by metabolome and transcriptome co-analysis. *BMC Plant Biol.* **2023**, *23*, 361. [CrossRef]

17. Conesa, M.R.; de la Rosa, J.M.; Artés-Hernández, F.; Dodd, I.C.; Domingo, R.; Pérez-Pastor, A. Long-term impact of deficit irrigation on the physical quality of berries in 'Crimson Seedless' table grapes. *J. Sci. Food Agric.* **2015**, *95*, 2510–2520. [CrossRef]

18. Roberto, S.; Junior, O.; Muhlbeier, D.; Koyama, R.; Ahmed, S.; Dominguez, A. Post-harvest conservation of "Benitaka" table grapes with different SO$_2$−generating pads and plastic liners under cold storage. *BIO Web Conf.* **2019**, *15*, 01003. [CrossRef]

19. Piazzolla, F.; Amodio, M.L.; Pati, S.; Colelli, G. Evaluation of Quality and Storability of "Italia" Table Grapes Kept on the Vine in Comparison to Cold Storage Techniques. *Foods* **2021**, *10*, 943. [CrossRef]

20. Ginsburg, L.; Combrink, J.C.; Truter, A.B. Long and short term storage of table grapes. *Int. J. Refrig.* **1978**, *1*, 137–142. [CrossRef]

21. Burger, D.A.; Jacobs, G.; Huysamer, M.; Taylor, M.A. The Influence of Storage Duration and Elevation of Storage Temperature on the Development of Berry Split and Berry Abscission in *Vitis vinifera* L. cv. Thompson Seedless Table Grapes. *South Afr. J. Enol. Vitic.* **2005**, *26*, 68–70. [CrossRef]

22. Chironi, S.; Sortino, G.; Allegra, A.; Saletta, F.; Caviglia, V.; Ingrassia, M. Consumer assessment on sensory attributes of fresh table grapes cv 'italia' and 'red globe' after long cold storage treatment. *Chem. Eng. Trans.* **2017**, *58*, 421–426. [CrossRef]

23. Leng, F.; Wang, C.; Sun, L.; Li, P.; Cao, J.; Wang, Y.; Zhang, C.; Sun, C. Effects of Different Treatments on Physicochemical Characteristics of 'Kyoho' Grapes during Storage at Low Temperature. *Horticulturae* **2022**, *8*, 94. [CrossRef]

24. Ngcobo, M.E.K.; Delele, M.A.; Opara, U.L.; Meyer, C.J. Performance of multi-packaging for table grapes based on airflow, cooling rates and fruit quality. *J. Food Eng.* **2013**, *116*, 613–621. [CrossRef]

25. Ahmed, S.; Roberto, S.R.; Domingues, A.R.; Shahab, M.; Junior, O.J.C.; Sumida, C.H.; De Souza, R.T. Effects of Different Sulfur Dioxide Pads on Botrytis Mold in 'Italia' Table Grapes under Cold Storage. *Horticulturae* **2018**, *4*, 29. [CrossRef]

26. Yuan, Y.; Wei, J.; Xing, S.; Zhang, Z.; Wu, B.; Guan, J. Sulfur dioxide (SO$_2$) accumulation in postharvest grape: The role of pedicels of four different varieties. *Postharvest Biol. Technol.* **2022**, *190*, 111953. [CrossRef]

27. Allen, R.G.; Pereira, L.S.; Raes, D.; Smith, M. *Crop Evapotranspiration. Guidelines for Computing Crop Water Requirements*; FAO Irrigation and Drainage Paper No. 56; FAO—Food and Agriculture Organization of the United Nations: Rome, Italy, 1998. Available online: https://www.fao.org/3/x0490e/x0490e00.htm (accessed on 7 March 2024).

28. Di Stefano, R.; Cravero, M.C. Metodi per lo studio dei polifenoli dell'uva. *Riv. Vitic. Enol.* **1991**, *44*, 37–45.

29. Waterhouse, A.L. Determination of Total Phenolics. *Curr. Protoc. Food Anal. Chem.* **2002**, *6*, I1.1.1–I1.1.8.

30. Ayoola, G.A.; Ipav, S.S.; Solidiya, M.O.; Adepoju-Bello, A.A.; Coker, H.A.B.; Odugbemi, T.O. Phytochemical screening and free radical scavenging activities of the fruits and leaves of allanblackia floribunda olive (Guttiferae). *Int. J. Health Res.* **2008**, *1*, 81–93. [CrossRef]

31. Brand-Williams, W.; Cuvelier, M.E.; Berset, C. Use of a free radical method to evaluate antioxidant activity. *LWT Food Sci. Technol.* **1995**, *1*, 25–30. [CrossRef]

32. *ISO 8589:2007*; Sensory Analysis: General Guidance for Design of Test Rooms. International Organization for Standardization: Genèva, Switzerland, 2007.

33. OIV. OIV General form for the Sensorial Analysis of Table Grape. Resolution OIV/VITI 371/2010. 2010. Available online: https://www.oiv.int/public/medias/385/viti-2010-2-en.pdf (accessed on 30 January 2024).

34. OIV. Annual Assessment of the World Vine and Wine Sector in 2021. 2021. Available online: https://www.oiv.int/sites/default/files/documents/OIV_Annual_Assessment_of_the_World_Vine_and_Wine_Sector_in_2021.pdf (accessed on 18 April 2024).

35. Romero, P.; Navarro, J.M.; Botía Ordaz, P. Towards a sustainable viticulture: The combination of deficit irrigation strategies and agroecological practices in Mediterranean vineyards. A review and update. *Agric. Water Manag.* **2022**, *259*, 107216. [CrossRef]

36. Kang, C.; Diverres, G.; Karkee, M.; Zhang, Q.; Keller, M. Decision-support system for precision regulated deficit irrigation management for wine grapes. *Comput. Electron. Agric.* **2023**, *208*, 107777. [CrossRef]

37. Ribera-Fonseca, A.; Palacios-Peralta, C.; González-Villagra, J.; Diaz, M.R.; Serra, I. How Could Cover Crops and Deficit Irrigation Improve Water Use Efficiency and Oenological Properties of Southern Chile Vineyards? *J. Soil Sci. Plant Nutr.* **2023**, *23*, 6851–6865. [CrossRef]

38. Permanhani, M.; Costa, J.M.; Conceição, M.A.F.; de Souza, R.T.; Vasconcellos, M.A.S.; Chaves, M.M. Deficit irrigation in table grape: Eco-physiological basis and potential use to save water and improve quality. *Theor. Exp. Plant Phys.* **2016**, *28*, 85–108. [CrossRef]

39. Campi, P.; Modugno, F.; Palumbo, A.D.; Mastrorilli, M. Dimensioning the Irrigation Variables for Table Grape Vineyards in Litho-soils. *Ital. J. Agron.* **2010**, *4*, 315–321. [CrossRef]

40. Çolak, Y.B.; Yazar, A. Evaluation of crop water stress index on Royal table grape variety under partial root drying and conventional deficit irrigation regimes in the Mediterranean Region. *Sci. Hortic.* **2017**, *224*, 384–394. [CrossRef]

41. Vox, G.; Schettini, E.; Scarascia-Mugnozza, G.; Tarricone, L.; Gentilesco, G. Crimson seedless table grape grown under plastic film: Ecophysiological parameters and grape characteristics as affected by the irrigation volume. In Proceedings of the International Conference of Agricultural Engineering, Zurich, Switzerland, 6–10 July 2014; EurAgEng: Bedford, UK, 2014. Ref: C0354. pp. 1–8. Available online: https://www.geyseco.es/geystiona/adjs/comunicaciones/304/C03540001.pdf (accessed on 7 March 2024).

42. Mirás-Avalos, J.M.; Intrigliolo, D.S. Grape Composition under Abiotic Constrains: Water Stress and Salinity. *Front. Plant Sci.* **2017**, *8*, 851. [CrossRef] [PubMed]

43. Conesa, M.R.; Falagán, N.; de la Rosa, J.M.; Aguayo, E.; Domingo, R.; Pérez Pastor, A. Post-veraison deficit irrigation regimes enhance berry coloration and health-promoting bioactive compounds in 'Crimson Seedless' table grapes. *Agric. Water Manag.* **2016**, *163*, 9–18. [CrossRef]

44. Temnani, A.; Berríos, P.; Conesa, M.R.; Pérez-Pastor, A. Modelling the Impact of Water Stress during Post-Veraison on Berry Quality of Table Grapes. *Agronomy* **2022**, *12*, 1416. [CrossRef]

45. Pisciotta, A.; Planeta, D.; Giacosa, S.; Paissoni, M.A.; Di Lorenzo, R.; Rolle, L. Quality of Grapes Grown Inside Paper Bags in Mediterranean Area. *Agronomy* **2020**, *10*, 792. [CrossRef]

46. Rolle, L.; Giacosa, S.; Gerbi, V.; Novello, V. Comparative Study of Texture Properties, Color Characteristics, and Chemical Composition of Ten White Table-Grape Varieties. *Am. J. Enol. Vitic.* **2011**, *62*, 49–56. [CrossRef]

47. Castillo-Munoz, N.; Gomez-Alonso, S.; Garcia-Romero, E.; Hermosin-Gutierrez, I. Flavonol profiles of *Vitis vinifera* white grape cultivars. *J. Food Compos. Anal.* **2010**, *23*, 699–705. [CrossRef]

48. Šikuten, I.; Štambuk, P.; Andabaka, Ž.; Tomaz, I.; Marković, Z.; Stupić, D.; Maletić, E.; Kontić, J.K.; Preiner, D. Grapevine as a Rich Source of Polyphenolic Compounds. *Molecules* **2020**, *25*, 5604. [CrossRef] [PubMed]

49. Gambetta, G.A.; Herrera, J.C.; Dayer, S.; Feng, Q.; Hochberg, U.; Castellarin, S.D. The physiology of drought stress in grapevine: Towards an integrative definition of drought tolerance. *J. Exp. Bot.* **2020**, *71*, 4658–4676, Erratum in *J. Exp. Bot.* **2020**, *71*, 5717. [CrossRef]

50. Palai, G.; Caruso, G.; Gucci, R.; D'Onofrio, C. Berry flavonoids are differently modulated by timing and intensities of water deficit in *Vitis vinifera* L. cv. Sangiovese. *Front. Plant Sci.* **2022**, *13*, 1040899. [CrossRef] [PubMed]

51. Liu, Q.; Tang, G.Y.; Zhao, C.N.; Feng, X.L.; Xu, X.Y.; Cao, S.Y.; Meng, X.; Li, S.; Gan, R.Y.; Li, H.B. Comparison of Antioxidant Activities of Different Grape Varieties. *Molecules* **2018**, *23*, 2432. [CrossRef] [PubMed]

52. Giovinazzo, G.; Grieco, F. Functional Properties of Grape and Wine Polyphenols. *Plant Foods Hum. Nutr.* **2015**, *70*, 454–462. [CrossRef] [PubMed]

53. Fernández-Trujillo, J.P.; Obando-Ulloa, J.M.; Baró, R.; Martínez, J.A. Quality of two table grape guard cultivars treated with single or dual-phase release SO_2 generators. *J. Appl. Bot. Food Qual.* **2008**, *82*, 1–8.

54. Peña-Neira, A.; Cortiella, M.G.I.; Ubeda, C.; Pastenes, C.; Villalobos, L.; Contador, L.; Infante, R.; Gómez, C. Phenolic, polysaccharides composition, and texture properties during ripening and storage time of new table grape cultivars in Chile. *Plants* **2023**, *12*, 2488. [CrossRef]

55. de Aguiar, A.C.; Higuchi, M.T.; Yamashita, F.; Roberto, S.R. SO_2-Generating Pads and Packaging Materials for Postharvest Conservation of Table Grapes: A Review. *Horticulturae* **2023**, *9*, 724. [CrossRef]

sustainability

Article

How to Improve a Successful Product? The Case of "Asproudi" of the Monemvasia Winery Vineyard

Georgios Merkouropoulos [1,*,†], Dimitrios-Evangelos Miliordos [2,†], Georgios Tsimbidis [3], Polydefkis Hatzopoulos [4] and Yorgos Kotseridis [2]

[1] Department of Vitis, Institute of Olive Tree, Subtropical Crops and Viticulture, Hellenic Agricultural Organization DIMITRA, Leoforos Sofokli Venizelou 1, 14123 Lykovrysi, Greece
[2] Laboratory of Enology and Alcoholic Drinks, Department of Food Science and Human Nutrition, Agricultural University of Athens, 11855 Athens, Greece; dim.miliordos@gmail.com (D.-E.M.); ykotseridis@aua.gr (Y.K.)
[3] Monemvasia Winery, 23052 Monemvasia, Greece; info@monemvasiawinery.gr
[4] Molecular Biology Laboratory, Department of Biotechnology, Agricultural University of Athens, 11855 Athens, Greece; phat@aua.gr
* Correspondence: merkouropoulos@elgo.gr
† These authors contributed equally to this work.

Abstract: An interesting way to maintain genetic diversity in the vineyard could be based on selecting the desirable characters of each clone or variety in order to produce a high-quality poly-clonal or poly-varietal wine, according to the consumer's desire. The current study describes a holistic approach in viticulture towards wine production, applying a multidisciplinary methodology. Firstly, "Asproudi", a rare Greek variety, was analyzed molecularly. The initial hypothesis that "Asproudi" is a distinct variety was questioned; microsatellite analysis showed that "Asproudi" is a population of different genotypes, at least in the Monemvasia Winery vineyard. A targeted harvest of each genotype was performed during the same day and was followed by micro-vinifications. All standard analyses of must and wine were performed in the laboratory, while a sensory analysis by a professional team evaluated each of the produced wines, showing distinctive differences. The genetic relationship of some of the Monemvasia Winery "Asproudi" genotypes to the varieties maintained in the reference collection was revealed whereas some other genotypes remained unknown.

Keywords: *Vitis vinifera*; rare grapevine varieties; microsatellites; SSRs; micro-vinification; must analyses; wine analyses; sensory analysis

Citation: Merkouropoulos, G.; Miliordos, D.-E.; Tsimbidis, G.; Hatzopoulos, P.; Kotseridis, Y. How to Improve a Successful Product? The Case of "Asproudi" of the Monemvasia Winery Vineyard. *Sustainability* **2023**, *15*, 15597. https://doi.org/10.3390/su152115597

Academic Editor: Dimitris Skalkos

Received: 9 September 2023
Revised: 31 October 2023
Accepted: 1 November 2023
Published: 3 November 2023

1. Introduction

Viticulture and wine-making represent long-standing activities in Greece. The area around Mt Pangaion (Figure 1) in the north-east part of the country provides an example where myth and scientific evidence meet: according to myth, the god of wine-making, Dionysos, lived with his followers, the Maenands on the slopes of Mt Pangaion. Thousands of charred grapevine seeds, together with grapevine pressings and clay cups used for drinking the grape juice or wine, have been recently unearthed from the Neolithic settlement of Dikili Tash located in the valley on the north of Mt Pangaion, making this area the oldest known wine-making site in Europe [1].

The term "Asproudi" comes from the word "Aspo" ("Άσπρο" in Greek means white) and etymologically represents a case of diminution, a phenomenon quite common in the Greek language, through which new words are produced that enrich the original meaning with a delightful and cheerful mood while indicating new properties.

Figure 1. Map of Greece: the toponyms mentioned in the text are shown.

In the wine-related Greek literature, the term "Asproudi" was found as early as 1888 referring to grapevines cultivated in various areas of the country, such as in the central parts of Peloponnese, in Thessaly, and in the north Aegean islands (Figure 1) [2]; a few years later, however, the term was used to define the most dominant grapevine variety in the country [3] pointing out the color of the fruit. In the study "Greek Ampelography", a total of 12 varieties were described under the term "Asproudi" [4,5]. Nearly all of them were identified and discriminated either by the corresponding toponym of the cultivation area or by an adjective referring to a particular and well-observed feature. In recent years, some "heretic" statements, judging on the phenotypic characteristics, referred that "Asproudes" represent groups of white varieties that need to be separated [6], regardless of how difficult the task might be [7]. Therefore, a working hypothesis was emerged, according to which the term "Asproudi", together with its variants (e.g., "Asprouda", "Asproudes"), is a general term that has served as a verbal repository for grouping in the same heterogeneous group different varieties, exclusively on the basis of their light berry color and their time of maturation.

"Asproudi" vines are large with high vividness; it is considered as an early ripening local variety. Ripening occurs from early to mid-August, producing about four to five kilograms of grapes per vine. The wine that is produced by this variety has received many international awards pointing out the high value of this autochthonous grapevine material.

The Hellenic Agricultural Organization DIMITRA (ELGO-DIMITRA) officially maintains the reference collection of the country—more than five hundred autochthonous accessions—in three vineyards: (i) In Lykovrysi (Athens). This is the oldest and largest vineyard established in 1929. During the 1950s and 1960s, clonal experiments were performed for some of the most common autochthonous and international varieties, while in the 1980s numerous expeditions throughout the country ended up in the collection of nearly all of the autochthonous varieties. (ii) In Thermi (Thessaloniki). This is a copy of the Lykovrysi collection enriched with native varieties collected from the northern parts of the country. (iii) In Heraklion (Crete). This is a collection of the Cretan varieties as well as of varieties cultivated in the islands of the southern Aegean Sea.

From an economical/consumer point of view, the world market has started turning to distinct high-quality wines produced from the less-utilized autochthonous grapevine varieties imposing an absolute necessity to revalorize and preserve these varieties. This trend is not limited to specific regions and is observed in countries such as Italy [8], Germany [9], Argentina, Chile, and Bolivia [10]. The preservation and revitalization of these varieties aim to provide distinctive identities to wines and promote biodiversity in vineyards. Additionally, there is a need to adapt to changing environmental conditions, such as climate change, which may require the cultivation of new varieties for sustainable viticulture [11]. However, the market introduction of new resistant selections can be impeded by preconceptions about their wine quality. Overall, the cultivation and promotion of autochthonous grapevine varieties contribute to the diversity and sustainability of the wine market, investing in a novel high-quality wine product of unique organoleptic characteristics.

The pathway towards the valorization and utilization of such cultivars commences with the crucial phase of accurate varietal identification. This, however, could turn out to be a difficult task due to the occurrence of synonyms, homonyms, and misidentifications [12]. Ampelographic identification is based on phenotypic differences. The primary drawback of this approach is that identification solely relies on visual data, necessitating an expert with extensive training to ensure effectiveness. Nevertheless, misidentifications may still occur due to the sheer number of existing cultivars and their resemblance, exacerbated by the impact of external factors on grapevine morphology [13]. At present, grapevine identification is typically augmented through the use of molecular markers. Molecular analysis offers a more precise identification and characterization, as its outcomes are not influenced by environmental factors [14]. Among the various types of DNA markers, microsatellite markers (SSRs) have been widely employed in grapevine identification. These markers are highly informative and their application can yield distinctive profiles that provide unambiguous identification of grapevine cultivars, independent of environmental factors, diseases, or vineyard practices. SSR markers are locus-specific and co-dominant, enabling the inference of familial relationships between different grapevine cultivars [15].

The revival of autochthonous grape varieties emphasizes the necessity of evaluating the produced wine, particularly due to the lack of previous research data. Molecular identification of the plant material that is used for the production of wine or related products together with their chemical analysis are needed to ensure authenticity [16]. A novel and more extensive approach involving molecular identification of the plant material used, together with a chemical analysis of the produced must and the wine, followed by sensory analysis of the final product, have been introduced recently [17] on the study of two minor Greek grapevine varieties, Karnachalades and Bogialamades. Previously, a multidisciplinary approach was used for the accurate ampelographic description of four Albanian varieties ultimately aiming to improve local economies [18].

Herein, we describe our involvement with the "Asproudi" grapevines of a commercial vineyard in Peloponnese. Initially, we were interested in revealing the molecular profile of this autochthonous Greek variety. Interestingly, however, we found out that it is a population of distinct genotypes. To evaluate their oenological potential, we performed targeted micro-vinifications for each of these genotypes; must and wine chemical analyses were carried out, followed by the sensory analysis on the final products. At later stages, when the molecular profile of the grapevine varieties maintained in the Greek reference collection became available, the molecular profiles of the genotypes found in the "Asproudi" population were compared to them. Some genotypes of the "Asproudi" population identified as varieties maintained and conserved in the reference collection while some others still remain unidentified.

2. Materials and Methods

2.1. Monemvasia Winery Management

The Monemvasia Winery vineyard, owned by the Tsimbidis family, is located in Monemvasia (36°40′59.1″ N 22°54′53.6″ E), the eastern part of Laconia facing the Aegean

Sea (Figure 1). It is planted in a homogeneous sandy-clay field in a sloppy hill. Vines are planted in R110 rootstocks, in 2.5 m × 1.1 m blocks, and trained on double cordon. All vine training, irrigation, and spraying operations were regularly performed each year. The pruning system that followed was three spurs in each cordon.

2.2. Molecular Studies

Young leaves had been collected from various plants of the Monemvasia Winery vineyard and were kept on ice until stored at −80 °C for further use. Genomic DNA was extracted from about 100 mg of the frozen tissue using the commercially available NucleoSpin Plant II kit (Macherey-Nagel, Düren, Germany) according to the manufacturer's instructions. The integrity of the extracted genomic DNAs was evaluated by agarose gel electrophoresis, and the concentration was estimated by using a Quawell (Q3000 UV-Vis Spectrophotometer, Quawell Technology Inc., San Jose, CA, USA) spectrophotometer.

Polymerase chain reactions (PCRs) were performed as before [17,19], in a volume of 20 μL using 25 to 30 ng genomic DNA as a template, 200 mM of each dNTP, 10 pmol primers, 2 μL 10× KAPATaq DNA Polymerase buffer, and 1 U KAPATaq DNA Polymerase (KapaBiosystems, Cape Town, South Africa). Ten pairs of primers were used: VVS2 [20], VVMD5 and VVMD7 [21], VVMD25, VVMD27, VVMD28, and VVMD32 [22], and VrZAG62, VrZAG67, and VrZAG79 [23]. Forward primers were 5′–end fluorescently labeled with different fluorophores: FAM, HEX, ROX, and TAMRA. Primers were custom labeled according to (i) each dye's absorption and emission wavelength and (ii) the length of the amplified product to avoid overlapping during gel electrophoresis. PCR amplifications were performed in a 96-well MiniAmp Thermal Cycler (Applied Biosystems, Foster City, CA, USA) as follows: 1 cycle (95 °C, 2 min), 35 cycles [95 °C, 15 s; 52 to 60 °C (depending on the primers), 15 s; 72 °C, 10 s], and 1 cycle (72 °C, 20 min). PCR fragments were separated using capillary electrophoresis in a 3730xl DNA Analyzer (Applied Biosystems, CA, USA). Data analysis, sizing, and genotyping were performed using the GeneMapper (version 4.0) software. The GenAlEx 6.5 program [24] was used for statistical analysis (Genetic Distance—Codom Genotypic). Data were then exported to the MEGA4.1(Beta) program [25] as a Tri-Matrix using the default options. Dendrograms were constructed using the UPGMA method in the MEGA4.1(Beta) program.

2.3. Harvest and Standard Grape Juice Analysis

"Asproudi" vines from the Monemvasia Winery were grown under the same viticultural management until their manual harvest; they were all harvested during the same day, early in the morning in order to avoid oxidations [26].

The exact harvest date (last week of August 2019) was decided considering the same criteria as every year: desirable physiochemical parameters (°Brix, titratable acidity, and pH value) and technological maturity. The grapes were harvested by hand with care to ensure consistency and were pooled in one (or more) basket per group. Groups were defined according to the outcome of the microsatellite analysis. All baskets were transferred to the Laboratory of Enology and Alcoholic Drinks at the premises of the Agricultural University of Athens, and stored overnight in a cooling chamber (at 4 °C), before being moved to the experimental winery for vinification. As a reference, harvest of the "Asprouda" variety, maintained by ELGO-DIMITRA at Lykovrysi (38°04′09.1″ N 23°46′32.9″ E) was also performed.

Prior to fermentation, a set of physiochemical parameters related to maturity were conducted in the must. The analytical methods recommended by the International Organization of Vine and Wine (OIV) were used to determine the sugar concentration and titratable acidity of the grape juices. Thereafter, each group was divided into three small-scale vinifications. Due to lower acidity levels in white musts often causing the polymerization of phenolic compounds and resulting in brown deposits and therefore causing darkening of white wine [27], 1 kg/tn of tartaric acid was added in each vinification.

2.4. Small-Scale Vinification Protocol, and Must Analysis

Grapes were destemmed, crushed, and softly pressed—the grape juice was sulfated with 15 mg/L of sulfur dioxide (SO_2) during the crushing. The pressed juice was placed in 10 L plastic bottles, and 3 mg/L Safizym pectinase enzyme (Safizym Clean, Fermentis, Marquette-lez-Lille, France; endo-polygalacturonase) was added in order to facilitate sedimentation. The bottle headspace was purged with N_2 before the bottles were sealed and left overnight at 4 °C for sedimentation. Clear juice, coming from each group, was racked off the sediment. Conventional analysis (pH, °Brix, and titratable acidity) was conducted in the clear must of each group. Two liters (2 L) of clear must was transferred into clean 3 L tanks, in triplicates for each group. The clear grape must was inoculated with Safoeno GV 107 (Fermentis, Lille, France) yeast, prepared in accordance with the manufacturer's instructions. As for the yeast nutrient (20 g/100 kg), SpringFerm™ (Fermentis, Lille, France) was used—this includes inactivated yeast (rich in growth factors). Fermentations in triplicates of each group were conducted in a temperature-controlled environment (16 to 18 °C).

Alcoholic fermentation showed a regular trend, and was considered finished when the reducing sugar concentration was lower than 2 g/L. At the end of fermentation (approximately after 10 days), wines were racked and they were stored for the stabilization process at a controlled temperature (4 °C). Finally, before sealing the wines, 30 mg/L SO_2 was added for protection.

2.5. Analysis of Conventional Oenological Parameters and Total Phenolic Index

After sealing the wines in the plastic bottles, they were analyzed using the OIV methods [28] on alcohol percentage, reducing sugars, pH value, and total acidity.

The total polyphenol index (TPI) was determined by measuring the 280 nm absorbance of a 1:100 dilution of wine with a spectrophotometer, using a 10 mm quartz cuvette and multiplying the absorbance value by 100 [29]. The total polyphenol concentration was determined by the Folin–Ciocalteu assay, with the micro-scale protocol [30]. The results were expressed as mg/L of gallic acid equivalents (GAE).

A modification of the model described by Singleton and Kramling [31] was used to assess browning development. Wine lots of 30 mL were filtered and placed in a 30 mL, screw-cap glass vial (7.5 cm length, 2.1 cm internal diameter). Samples were subjected to heating at a constant temperature of 55.0 ± 0.2 °C in a heating chamber. Aliquots were withdrawn at 24 h intervals over a period of 13 days, and browning was measured at A420. The samples were then immediately returned to the vials to maintain the initial headspace volume.

2.6. Sensory Evaluation of the Wine

Sensory trials were carried out by twelve trained assessors (recruited by the Laboratory of Enology and Alcoholic Drinks; equal sex distribution) with professional experience in the wine industry, and at least one year experience in tasting white wines. Wine samples used for the sensory evaluation were unfiltered, and the evaluation was performed two months after sealing the tanks.

Samples (25 mL) of wines were presented in a randomized order for each participant. Samples were served in ISO standard glasses 3591 [32] and numbered by a random three-digit number; each glass was covered with a glass cup in order to avoid diffusion of odorants [33,34]. The evaluation consisted of describing the appearance, aroma, taste, and harmony of each wine sample; the taster participants were first asked to describe each wine by a list of seven descriptors ("Color Intensity", "Aroma Intensity", "White Fruits/Flowers", "Vegetal Aroma", "Taste Balance", "Acidity", "Aftertaste"), and then to proceed to a quantitative assessment, using a scale of 0 to 10 (from lowest to highest intensity).

2.7. Statistical Analysis

Small scale vinifications were performed in triplicates. All values are presented as the mean ± standard error. All values are presented as the mean and standard deviation. Statistical analyses were performed using the Statgraphics Centurion application (version 1.0.1.C). The significance of the results was determined with an unpaired *t*-test or one-way ANOVA with Tukey's test. Multivariate statistical data analysis (MVA) of the samples was performed with XLstat (XLSTAT 2017: Data Analysis and Statistical Solution for Microsoft Excel; AddinSoft, Paris, France, 2017).

The sensory evaluation data were analyzed by a non-parametric Kruskal–Wallis one-way analysis of variance using Statgraphics Centrurion. The Kruskal–Wallis Non-Parametric Hypothesis Test is used when a variable does not meet the normality assumptions of a one-way ANOVA. When the *p*-values were <0.05, a post-hoc Mann–Whitney–Wilcoxon test was applied to compare, one by one, the wines for each variable.

3. Results

3.1. Molecular Studies Showed That "Asproudi" of the Monemvasia Winery Is a Population of Different Genotypes

Searching for autochthonous grapevine material in Peloponnese, an initial sampling mission was performed at the early stages of the vegetative period (1 June) of 2018 in the Monemvasia Winery vineyard. A total of sixteen samples were collected from the vineyard; eight of the samples were collected randomly from the vines of the white variety that was known as "Asproudi". Molecular identification on microsatellite loci was performed on these samples and on the autochthonous varieties "Monemvassia" (spelling according to the updated national Catalogue: ya530_57378_020322-2), "Kydonitsa", "Assyrtiko, "Glykerithra", and "Gaidouria" which are commonly cultivated in the wider geographical area of Monemvasia, and in other parts of the Monemvasia Winery vineyard. Samples of these five reference varieties were also collected from the grapevines that are maintained in the reference collection at Lykovrysi. This analysis ended up in the distinction of the eight Monemvasia Winery "Asproudi" samples in four different and discreet groups not related to the five reference varieties analyzed (Figure 2; Groups i, ii, iii, iv); since the empirical names of the Monemvasia Winery working team were considered, the molecular analysis detected misnamings (indicated by asterisks in Figure 2): "Glykerithra-MW-2018" and "Gaidouria-MW-2018" are indeed "Assyrtiko". As no studies, neither ampelographic nor molecular, were performed prior to that analysis, this distinction was the first evidence that the Monemvasia Winery "Asproudi" material is actually a heterogeneous population of different genotypes. The emerging question was whether all genotypes that make up "Asproudi" have been represented in this primary analysis. To answer this question, sampling was repeated the following year, when bunches were mature (the last days of July 2019). Thirty-one samples were collected: phenotypic differences were considered when choosing the vines to sample. In addition, twenty-one samples of white grapevines cultivated in the surrounding vineyards ("Unknown" samples from Estate Loulouda and Estate Sarra) were also collected. The analysis on the same microsatellite loci brought up seven groups (Table 1 and Figure 3; hereafter, the constituting groups are called "Groups": A, B, C, D, E, F, G).

Table 1. Microsatellite analysis at nine loci; the microsatellite loci are designated in the top line. Allele size (in base pairs) are shown; asterisks (*) indicate the varieties that are maintained by ELGO-DIMITRA at the Lykovrysi's grapevine vineyard (Athens); "MW": "Monemvasia_Winery"; "Est_Loulouda" and "Est_Sarra" are estates located in the wider Monemvasia area; "2018" or "2019" indicate the year the samples were collected. Xinomavro, Koundoura lefki, and Cabernet Sauvignon were included as reference material. Data for the VVMD5 are not shown because the data produced were not available for all samples.

	VVS2		VVMD7		VVMD25		VVMD27		VVMD28		VVMD32		VrZAG62		VvZAG67		VvZAG79	
Asproudi-MW-2018-1	141	155	238	246	243	253	183	187	259	279	250	254	195	203	138	160	241	241
Asproudi-MW-2018-2	133	143	246	248	237	239	187	191	249	259	256	270	201	203	138	150	241	249

Table 1. *Cont.*

	VVS2		VVMD7		VVMD25		VVMD27		VVMD28		VVMD32		VrZAG62		VvZAG67		VvZAG79	
Asproudi-MW-2018-3	143	143	238	252	237	239	177	187	255	259	248	254	187	199	130	148	241	249
Asproudi-MW-2018-4	133	143	246	248	237	239	187	191	249	259	256	268	201	203	138	150	241	249
Asproudi-MW-2018-5	143	151	238	252	253	261	179	183	249	261	248	256	187	187	124	130	249	257
Asproudi-MW-2018-6	143	151	238	252	253	261	179	183	249	261	248	256	187	187	124	130	249	257
Asproudi-MW-2018-7	143	143	238	252	237	239	177	187	255	259	248	254	187	199	130	148	241	249
Asproudi-MW-2018-8	133	143	246	248	237	239	187	191	249	259	256	270	201	203	138	150	241	249
Asproudi-MW-2019-1	141	155	238	246	243	253	183	187	259	279	250	254	195	203	138	160	241	241
Asproudi-MW-2019-2	133	143	246	248	237	239	187	191	249	259	256	270	201	203	138	150	241	247
Asproudi-MW-2019-3	143	143	240	252	237	239	177	187	255	259	248	254	187	199	130	148	241	249
Asproudi-MW-2019-4	133	143	246	248	237	239	187	191	249	259	256	270	201	203	138	150	241	247
Asproudi-MW-2019-5	143	143	238	252	237	239	177	187	255	259	248	254	187	199	130	148	241	247
Asproudi-MW-2019-6	131	143	238	242	239	239	179	191	245	261	248	254	187	187	130	138	239	253
Asproudi-MW-2019-7	143	151	240	252	253	261	179	183	251	261	248	256	187	187	124	130	247	255
Asproudi-MW-2019-8	143	149	238	252	253	261	179	183	249	261	248	256	187	187	124	130	247	255
Asproudi-MW-2019-9	143	143	238	252	237	239	177	187	255	259	248	254	187	199	130	148	241	247
Asproudi-MW-2019-11	143	143	240	252	237	239	177	187	255	259	248	254	187	199	130	148	241	249
Asproudi-MW-2019-13	133	143	246	248	237	239	187	191	249	259	256	270	201	203	138	138	241	247
Asproudi-MW-2019-14	133	143	246	248	237	239	187	191	249	259	256	270	201	203	138	138	241	247
Asproudi-MW-2019-15	143	151	240	252	253	261	179	183	249	261	248	256	187	187	124	130	247	255
Asproudi-MW-2019-16	143	143	238	252	237	239	177	187	255	259	248	254	187	199	130	148	241	247
Asproudi-MW-2019-17	143	151	240	252	253	261	179	183	249	261	248	256	187	187	124	130	247	255
Asproudi-MW-2019-18	133	135	238	246	237	239	177	177	255	259	250	270	187	187	148	148	241	247
Asproudi-MW-2019-19	133	143	246	248	237	239	187	191	249	259	256	270	201	203	138	150	241	247
Asproudi-MW-2019-20	143	149	238	252	253	261	179	183	249	261	248	256	187	187	124	130	247	255
Asproudi-MW-2019-21	133	143	246	248	237	239	187	191	249	259	256	270	201	203	138	150	241	247
Asproudi-MW-2019-22	143	149	238	252	253	261	179	183	249	261	248	256	187	187	124	130	247	255
Asproudi-MW-2019-23	143	151	238	252	253	261	179	183	249	261	248	256	187	187	124	130	247	255
Asproudi-MW-2019-24	143	143	240	252	237	239	177	187	255	259	248	254	187	199	130	148	241	247
Asproudi-MW-2019-25	143	149	238	252	253	261	179	183	249	261	248	256	187	187	124	130	247	255
Asproudi-MW-2019-26	143	143	240	252	237	239	177	187	255	259	248	254	187	199	130	148	241	249
Asproudi-MW-2019-27	143	143	238	252	237	239	177	187	255	259	248	254	187	199	130	148	241	247
Asproudi-MW-2019-28	143	143	238	252	237	239	177	187	255	259	248	254	187	199	130	148	241	249
Asproudi-MW-2019-29	133	143	246	248	237	239	187	191	249	259	256	270	201	203	138	150	241	247
Asproudi-MW-2019-30	133	143	246	248	237	239	187	191	259	259	256	270	201	203	138	150	241	249
Unknown-Est_Loulouda-46	143	143	238	252	237	239	177	187	255	259	248	254	187	199	130	148	241	247
Unknown-Est_Loulouda-47	133	143	246	248	237	239	187	191	249	259	256	270	201	203	138	150	241	249
Unknown-Est_Sarra-48	133	143	246	248	237	239	187	191	249	259	256	268	201	203	138	150	239	239
Unknown-Est_Sarra-49	143	151	240	252	253	261	179	183	251	261	248	256	187	187	124	130	247	255
Unknown-Est_Sarra-50	143	143	238	252	237	239	177	187	255	259	248	254	187	199	130	148	241	247
Unknown-Est_Sarra-51	143	151	240	252	253	261	179	183	249	261	248	256	187	187	124	130	247	255
Unknown-Est_Sarra-52	143	149	238	252	253	261	179	183	249	261	248	256	187	187	124	130	247	255
Asprouda_Aitoloakarnanias-1 *	133	143	246	248	237	239	187	191	249	259	256	270	201	203	138	150	241	249
Asprouda_Aitoloakarnanias-2 *	133	143	246	248	237	239	187	191	249	259	256	270	201	203	138	150	241	249
Asprouda-1 *	133	143	246	248	237	239	187	191	249	259	256	270	201	203	138	150	241	249
Asprouda-2 *	133	143	246	248	237	239	187	191	249	259	256	270	201	203	138	150	241	249
Asprouda or Dimitreiko-1 *	133	143	246	248	237	239	187	191	249	259	256	270	201	203	138	150	241	247
Asprouda or Dimitreiko-2 *	133	143	246	248	237	239	187	191	249	259	256	268	201	203	138	150	241	247
Xinomavro *	131	131	248	248	237	239	177	179	229	245	248	250	193	203	122	136	235	247
Koundoura lefki *	139	143	244	248	239	247	183	191	237	259	250	256	195	201	144	154	235	247
Cabernet Sauvignon *	137	151	238	248	237	247	173	187	235	237	238	238	187	193	122	136	243	243

The eight genotypes analyzed in 2018 were identified in these seven groups, together with the samples from the wider area (Estate Loulouda and Estate Sarra). These results demonstrated that the "Asproudi" of the Monemvasia Winery vineyard represents indeed a population of at least seven genotypes. Some of these genotypes are represented in high numbers (Groups A and D) in the "Asproudi" population of the Monemvasia Winery vineyard while some others are represented minimally (Groups B, C, and G consisted only of one vine). At that point, the emerging question was whether the seven genotypes that make up the "Asproudi" population in the Monemvasia Winery vineyard represent unknown autochthonous material or they are registered in the National Catalogue and in the Greek bibliography. To answer this question, the molecular profile of the seven Monemvasia Winery Groups were compared to the molecular profile of many of the varieties maintained in the reference collection by ELGO-DIMITRA; this comparison was performed in summer 2022 (when such molecular data became available for the genetic material maintained in the reference collection). The analysis showed that: Group A possesses a high degree of similarity to the variety "Asprouda" (synonym: Dimitreiko), originally collected from Arkadia (the central part of Peloponnese), and "Asprouda Aitoloakarnanias"; both maintained in the reference collection. Group D is related to "Arkadino", also a variety of Arkadia, while Groups E and F are both highly close to the variety "Proimo aspro", a white variety collected from the area of Serres—a region in the very north of the country. For the single member Groups B, C, and G, only the latter was found to be closely related to the variety "Svarna" collected from Magnesia (central Greece), while the former two remained unidentified. Some of the Monemvasia Winery "Asproudi" genotypes have also been detected in neighboring vineyards (Estate Loulouda and Estate Sarra; "Unknown" samples 46 to 52; Figure 3 and Table 1) supporting the concept that these genotypes represent common local grapevine genetic material.

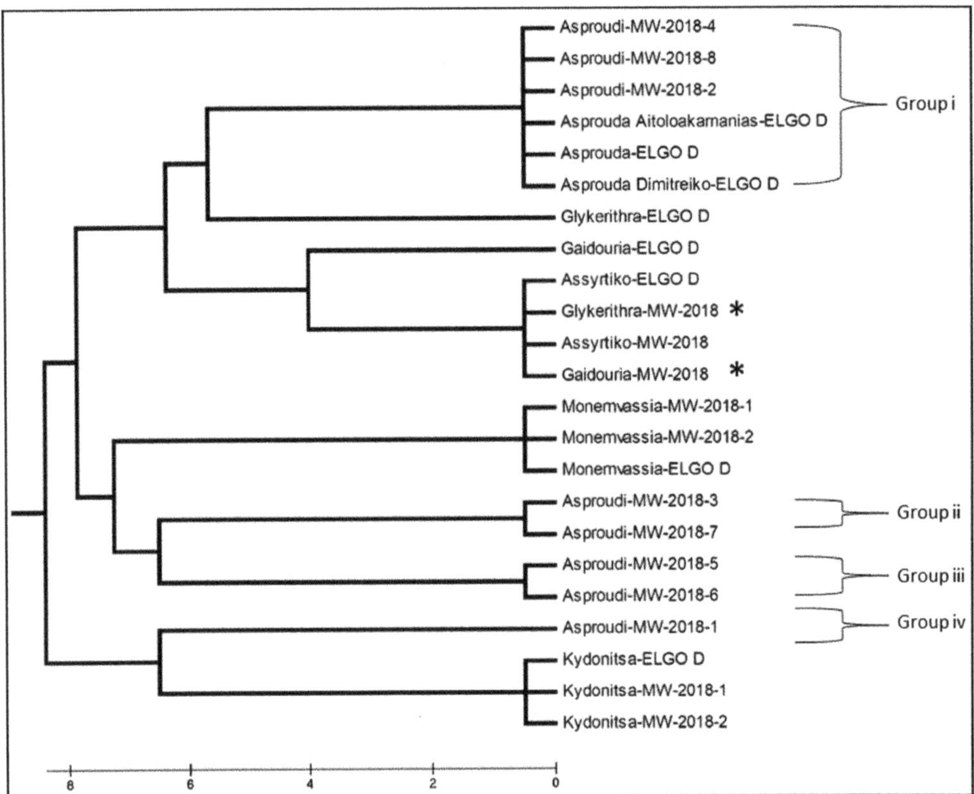

Figure 2. Dendrogram showing the relationship among the "Asproudi" genotypes of the Monemvasia Winery (MW) and the reference samples collected from the reference collection maintained by ELGO-DIMITRA (ELGO D). Since the empirical names of the MW were used, misnamings have been detected (indicated by asterisks): "Glykerithra-MW-2018" and "Gaidouria-MW-2018" are indeed "Assyrtiko".

3.2. Oenological Potential of the Monemvasia Winery "Asproudi" Constituting Genotypes

As long as it was clear that the Monemvasia Winery "Asproudi" was indeed a population of discreet genotypes, the aim of this work was to evaluate the oenological potential of the constituting genotypes.

Harvest time was decided by the experts of the Monemvasia Winery taking into consideration the technological maturity of the grapes; harvest was performed in the same day (27 August 2019; Table 2) for all groups. Group G was heavily infected by fungal infections; therefore, it was excluded from further experimentation on vinification. An additional sample was harvested on 4 September from "Asprouda" (synonym: "Dimitreiko") maintained in the reference ELGO-DIMITRA grapevine collection (hereafter: Group H). Table 2 shows the sugar concentration and titratable acidity of the grape juices of the different groups, confirming that they were at different ripen stages: °Brix values in musts varied from 17.2 (Group C) to 26.3 (Group B). In the oenological field, the different groups provide grape berries with different levels of maturity (Table 2): the pH values varied between 3.47 (Group C) and 3.93 (Group H). Group H showed by far the highest content of total acidity (4.95 g/L) compared to the lowest 3.1 value (Groups B and D).

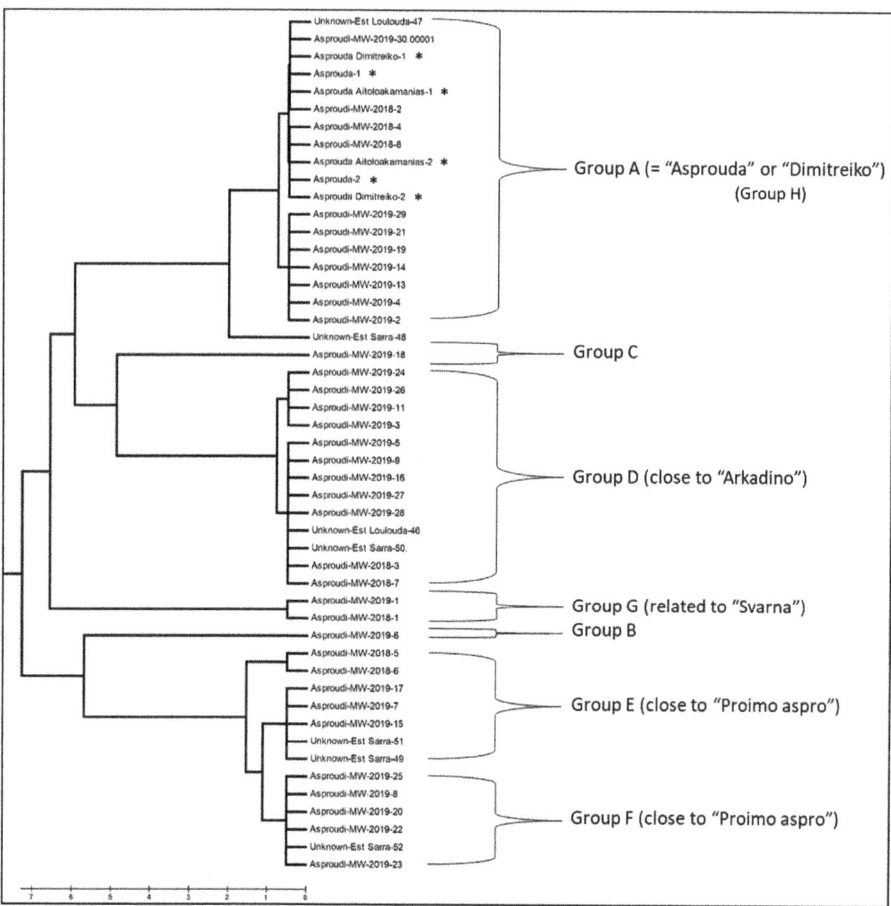

Figure 3. Dendrogram showing the distinction of the "Asproudi" Monemvasia Winery (MW) genotypes to seven groups (A to G). Samples from the area that surrounds the Monemvasia Winery vineyards were also included in this analysis (Unknown-Est_Loulouda and Unknown-Est_Sarra) together with reference material from the reference collection (indicated by an asterisk). "2018" and "2019" indicate the year that sampling occurred.

A key factor to produce wines with high-quality characteristics is the distinctive aroma. The warm and dry summer of the respective vintage of Monemvasia during the ripening period provided must with low levels of total acidity, recorded in all Monemvasia Winery "Asproudi" groups. Therefore, tartaric acid was added to the musts to adjust the pH of the must and ensure a smooth alcoholic fermentation.

Wines were analyzed approximately two months after the end of fermentation to record the evolution of ethanol content, pH, and total acidity in the micro-vinifications (Table 3). The ethanol content of the wines produced by the Groups B, D, A, and G was high; this is in agreement with the corresponding sugar concentration observed in grapes.

An exception is recorded: the wine produced from Group D. Total acidity indicates the freshness of the wines, and the stability of a wine over time; in this respect, the wines from Groups A and B attained the highest values of acidity (7.8 and 7.42 tartaric acid g/L, respectively; Table 3), whereas the lowest values were recorded by wines from Group H and F (5.24 and 5.77 tartaric acid g/L, respectively). Group H had the highest wine pH value (3.6), whereas Group C had the lowest (2.88). All wines can be considered as dry wines because the reducing sugar content was less than 2 g/L (Table 3).

Table 2. Details of harvest and vinification (cultivation area, harvest time, and vinification time), and conventional must analysis (°Brix, pH, and total acidity) of all groups analyzed. Groups A to F refer to the Monemvasia Winery vineyard "Asproudi" plant material, while Group H refers to the "Asprouda" variety that is maintained in the ELGO-DIMITRA vineyard at Lykovrysi. Means ± Standard errors followed by a different letter, in the column, do differ by Tukey's HSD test at 5% probability. Values represent means of triplicate determinations ± standard error.

Grape Group	Cultivation Area	Harvest Date	Vinification Date	Total Soluble Solids (°Brix)	pH	Total Acidity (Tart. Ac. g/L)
A	Monemvasia	27 August 2019	28 August 2019	24.1 ± 0.3 c	3.62 ± 0.04 b	4.30 ± 0.1 b
B	Monemvasia	27 August 2019	28 August 2019	26.3 ± 0.2 a	3.87 ± 0.07 a	3.10 ± 0.16 d
C	Monemvasia	27 August 2019	28 August 2019	17.2 ± 0.3 f	3.47 ± 0.04 c	3.75 ± 0.10 c
D	Monemvasia	27 August 2019	28 August 2019	24.8 ± 0.4 b	3.8 ± 0.09 a	3.10 ± 0.21 d
E	Monemvasia	27 August 2019	28 August 2019	20.9 ± 0.5 d	3.68 ± 0.07 ab	3.75 ± 0.19 c
F	Monemvasia	27 August 2019	28 August 2019	21.4 ± 0.6 de	3.69 ± 0.07 ab	3.70 ± 0.07 c
G	Monemvasia	27 August 2019	---	---	---	---
H	Lykovrysi	4 September 2019	5 September 2019	22.5 ± 0.4 e	3.93 ± 0.06 a	4.95 ± 0.04 a

Table 3. Conventional wine analysis (alcohol volume, pH, and total acidity) of the wines from all seven groups; Groups A to F refer to the Monemvasia Winery "Asproudi" Groups, while Group H refers to the "Asproudi" variety that is maintained in the ELGO-DIMITRA vineyard at Lykovrysi. Means ± Standard errors followed by a different letter, in the column, do differ by Tukey's HSD test at 5% probability. Values represent means of triplicate determinations ± standard error.

Grape Group	Alcohol Volume (V.V.%)	pH	Total Acidity (Tartaric Acid g/L)	Residual Sugars (g/L)
A	13.5 ± 0.01 c	3.093 ± 0.014 bc	7.80 ± 0.07 f	0.76 ± 0.02 c
B	15.4 ± 0.04 e	2.975 ± 0.021 ab	7.42 ± 0.10 e	0.61 ± 0.02 f
C	9.2 ± 0.07 a	2.888 ± 0.016 a	6.82 ± 0.035 d	0.67 ± 0.02 df
D	14.6 ± 0.11 d	2.948 ± 0.044 ab	6.86 ± 0.051 d	1.40 ± 0.02 a
E	12.3 ± 0.11 b	3.150 ± 0.096 c	6.11 ± 0.018	0.93 ± 0.02 b
F	12.2 ± 0.09 b	3.026 ± 0.040 abc	5.77 ± 0.03 b	0.72 ± 0.023 cd
H	13.3 ± 0.07 c	3.601 ± 0.026 f	5.24 ± 0.03 a	1.44 ± 0.023 a

Groups B and D were at a higher degree of maturity (Table 2), thus provided wines with higher values of absorption and phenolics, in contrast to Groups C, E, and F which had not reached comparable maturity levels. Similar results have been reported [35]: wines from early harvests showed a pale yellow-soft color and lower color intensity. The wines produced from grapes with a high degree of phenolic maturation (Groups B and D) had a higher absorption value at 420 nm (Table 4), so they appeared darker in color and with higher values of phenolics (Figure 4). Therefore, the degree of the grapes ripening during harvest is a critical factor for the color of the produced wine as well as for its aromatic potential. Wines from Groups B, D, and A have shown maximum ethanol content, the highest values of absorbance at 420 nm, and degrees in TPI and K factor (Table 4). The results imply that the identified factors are unique to each distinct group, underscoring the significance of scrutinizing these groups more extensively with regards to viticultural and oenological utilization and exploitation. It can be stated that the wines from the genotypes of the Monemvasia Winery "Asproudi" plant material can be distinguished according to their classic parameters.

Table 4. Compositional factors and browning characteristics determined in the wines from all groups analyzed: Groups A to F refer to the Monemvasia Winery "Asproudi" Groups, while Group H refers to the "Asproudi" variety that is maintained in the ELGO-DIMITRA vineyard at Lykovrysi. Means ± Standard errors followed by a different letter, in the column, do differ by Tukey's HSD test at 5% probability. Values represent means of triplicate determinations ± standard error.

Grape Group	420 nm	Folin Ciocalteu (Gallic Acid mg/L)	TPI	k Factor
A	0.0965 ± 0.0016 d	2.8511 ± 0.0303 b	9.5266 ± 0.3407 c	0.0026 ± 0.00026 c
B	0.3485 ± 0.0025 f	3.9944 ± 0.0032 cd	14.1266 ± 0.0878 d	0.0082 ± 0.00016 e
C	0.0525 ± 0.0007 a	2.5947 ± 0.1097 a	7.21330 ± 0.0838 a	0.0013 ± 0.00017 ab
D	0.1780 ± 0.0016 e	3.8696 ± 0.0488 c	13.9866 ± 0.1215 d	0.0068 ± 0.00048 d
E	0.0610 ± 0.0009 b	2.8063 ± 0.0092 b	8.2200 ± 0.07540 b	0.0006 ± 0.00040 a
F	0.0515 ± 0.0002 a	2.5097 ± 0.0013 a	7.8400 ± 0.08210 ab	0.0022 ± 0.00031 bc
H	0.0880 ± 0.0009 c	4.0590 ± 0.0382 f	7.5266 ± 0.2087 a	0.0008 ± 0.000005 a

Figure 4. Visual wine color evaluation—from left to right, the wines of Groups A to H (no wine from Group G).

In the analysis of compositional factors and browning characteristics, a repeated pattern was observed: wines of Groups B, D, and A received the highest values in the absorption at 420 nm, Folin index, TPI factor, and K factor (Table 4) with one exception: wine from Group H in the Folin index. These results might be explained by the high oxidation of those wines and the tendency to be oxidized in a shorter period than the rest of the wines.

Absorbance at 420 nm varied from 0.0515 to 0.3485 (Groups F and B, respectively). Total phenolics showed great variance: the values of equivalent of gallic acid mg/L in the Folin Ciocalteu assay ranged between 2.5097 (Group F) and 4.059 (Group H), whereas the values of the TPI ranged between 7.2133 (Group C) and 14.1266 (Group B). Finally, the accelerated browning test provided significant differences among the k factor (from 0.0008 to 0.0082) (Table 4). The wines that showed significantly lower values regarding the k factor were those made from Groups E and H: practically, these two groups produce wines which would develop a brown color later than the others.

3.3. Sensory Analysis Confirmed That Different Wines Are Produced by Different Genotypes

Sensory analysis was conducted two months after the end of alcoholic fermentation—during this period wines were kept at 4 °C for protein and tartaric stabilization. Quantitative descriptive sensory analysis was assessed for average wine odor and taste intensity scores (Table 5 for statistical data elaboration; Figure 5).

The sensory descriptor "Color Intensity" showed a diverse performance, with the wines of Groups B and D possessing the highest values. In "Aroma Intensity", wine from Group H showed by far the highest score, whereas all the other groups possessed comparable scores. As for the odoriferous "White Fruits/Flowers" character, wine produced by Group B presented the lowest value—all other wines presented a comparable performance. Wine from Group A received the highest score in "Vegetal Aroma". In respect to the descriptor of "Taste Balance", the lower scores were recorded by the Groups B and C. The panelists judged wines B and D with the highest value for the "Acidity" descriptor while wine H received the lowest value. Finally, no significant differences were observed among the wines for the descriptor "Aftertaste".

Table 5. The Kruskal–Wallis test and, when significant, ($p < 0.05$) the Mann–Whitney–Wilcoxon test were applied for multiple comparisons to the results of the sensory scores for the wines produced by the different fertilization treatments. Test statistics: the Kruskal–Wallis test was statistically significant when $p < 0.05$. Different letters in each row indicate significant differences ($p < 0.05$) among different samples.

Sensory Descriptors	Kruskal–Wallis Test p-Value	Post-Hoc Mann–Whitney–Wilcoxon Test						
		A	B	C	D	E	F	H
Color Intensity	2.004×10^{-8}	b	d	ab	c	d	a	ab
Aroma Intensity	0.007	a	a	a	a	a	a	b
White Flowers/Fruits	0.004	a	b	a	a	a	a	a
Vegetal Aroma	0.03	a	ab	b	ab	ab	b	b
Taste Balance	0.0124	a	b	b	ab	a	ab	a
Acidity	0.02	a	a	a	a	ab	ab	b
Aftertaste	0.04	ab	a	b	a	ab	ab	a

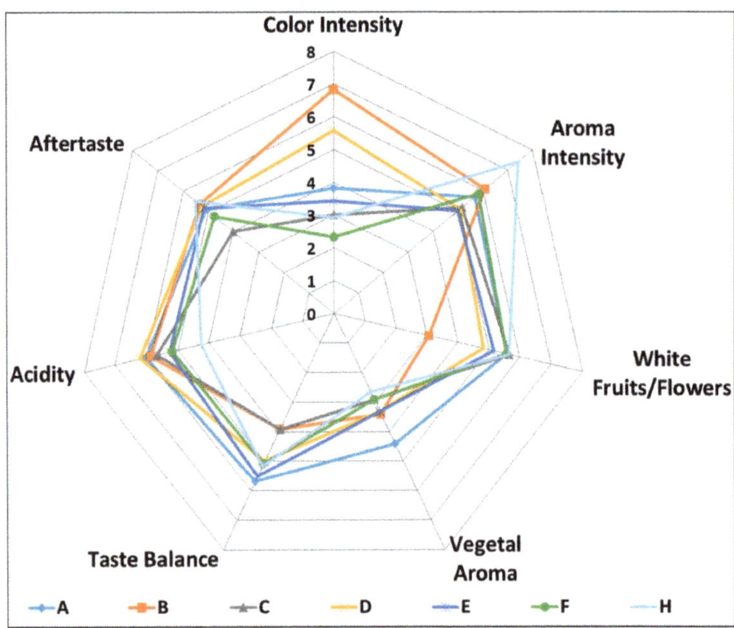

Figure 5. Spider plot of the sensory profile of the experimental wines produced by the various Monemvasia Winery "Asproudi" Groups, as outlined by a group of trained panelists from the Laboratory of Enology and Alcoholic Beverages of the Agricultural University of Athens. Wines were judged using predefined quality attributes on a scale from 1 (absent) to 10 (high).

The Kruskal–Wallis test was applied due to the fact that it is a rank-based test that is similar to the Mann–Whitney U test, but can be applied to one-way data with more than two groups. The Kruskal–Wallis test does not address hypotheses about the medians of the groups. Instead, the test addresses if it is likely that an observation in one group is greater than an observation in the other (Table 5). The outcome of the Kruskal–Wallis test unravels differences among the groups, but does not unravel which groups are different from other groups; this was achieved by performing the post-hoc Mann–Whitney–Wilcoxon testing (Table 5).

Principal component analysis (PCA) showed that the wines produced by the various Monemvasia Winery "Asproudi" genotypes differ in their classic parameters, confirming the assumption that different varieties produce different wines. Projection on the plot clearly separated the samples into two main groups: one on the left and one on the right of the Y axis (Figure 6a). Each group is subdivided into two subgroups: one above and one below the X axis, so as, in the end, an even distribution of the "Asproudi" Groups in the plot is observed providing additional support to the concept that different and distinctive characteristics of each variety lead in the production of wines with a particular sensory profile while applying the same winemaking technique. It is noted that the same genotype cultivated in different terroirs (Group A in Monemvasia and Group H in Lykovrysi) may produce wines with different chemical and sensorial characteristics.

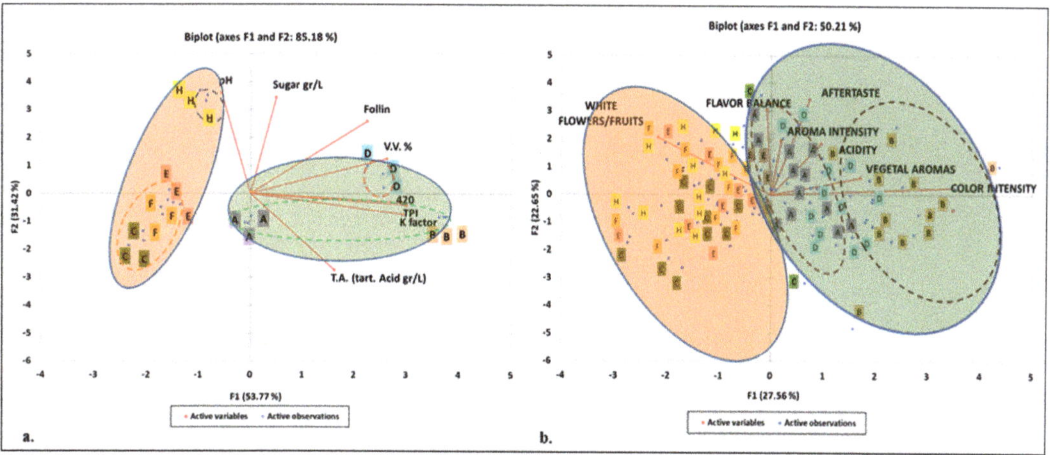

Figure 6. Unsupervised classification using Principal component analysis (PCA). (**a**) PCA applied to chemical parameters of wines of the different groups identified in the Monemvasia Winery "Asproudi" population. (**b**) PCA applied to the seven sensory descriptors from the experimental wines of the different "Asproudi" Groups. Samples in the score plots was colored according to the groups.

To analyze further the relationships of the attributes to the wine samples, PCA was conducted on the covariance matrix using seven additional terms ("White Flowers/Fruits", "Flavor Balance", "Aroma Intensity", "Aftertaste", "Acidity", "Vegetal Aromas", "Color Intensity") (Figure 6b). PC I explained 27.56% of the variance, whereas PC II explained 22.65% of the variance, totaling 50.21% of the variation of the data represented in the biplot. PCA showed that "Color Intensity" and "Vegetal Aroma" were associated mostly with the A, B, and D Groups, whereas "White Flower/Fruits" was associated with the G, C, and F Groups. As a consequence, the wines produced by different groups could be determined after building two PCAs: the wines of the H, F, C, and E groups are localized in the left side and the wines produced by the A, B, and D groups are localized on the right part of the two PCAs (Figure 6a,b).

Besides demonstrating the associations among the descriptors, PCA can also be used to display the relative "locations" of the samples with respect to each other and their characterizing attribute. Furthermore, a significant finding is that the discrimination regarding the chemical and sensorial analysis of the Monemvasia Winery "Asproudi" plant material is closely related with the grouping according to the molecular classification.

4. Discussion

For years, the identification of grapevine varieties was based solely on ampelographic descriptions. This method, however, is time consuming, requires experienced personnel, and depends on the terroir, the cultivation techniques, and the sanitary stage of the vines. During the last two decades, microsatellite analysis, a method based on DNA technology, has been used instead [36]. A complementary combination of the former, traditional ampelography, and the latter, modern ampelography, should be used, to achieve scientific progress.

"Asproudi" has long been cultivated in the central and south-east parts of Peloponnese. It is considered as a discreet grapevine variety, regardless of the recent statements that support it is a population of different genotypes [6,7]; it is noted, however, that in the recently updated National Catalogue of cultivated grapevine varieties in Greece (ya530_57378_020322-2), the term "Asproudi" has been replaced by the term "Asproudes" (plural of "Asproudi"). Our current work demonstrates that the Monemvasia Winery "Asproudi" is comprised of at least seven genotypes. One of these seven genotypes was identified as "Asprouda" (synonym: "Dimitreiko"), a registered variety maintained in the reference grapevine collection at the ELGO-DIMITRA premises. It was also found that the microsatellite profile of "Asprouda" is similar to the profile of "Asprouda_Aitoloakarnanias", another accession in the ELGO-DIMITRA reference collection. Another four of the seven genotypes were found to be genetically related to three registered accessions, whereas two genotypes remained unidentified. Future work could aim towards a comprehensive ampelographic description in combination with the molecular analysis of the four genotypes and also of the two unidentified (and undescribed) ones.

The chemical as well as the polyphenolic profile of a given grapevine cultivar serves as a significant indicator of its inherent genetic and commercial potential, and can be used as a valuable tool for discerning between distinct cultivars. The wines underwent analysis approximately two months subsequent to the cessation of the fermentation process. It is necessary to focus on the differences of total acidity among the "Asproudi" genotypes. This feature has added value due to climate change, which is considered as the main factor for the decreased acidity of wines, as well as the production of unbalanced wines [37].

Moreover, noteworthy variations were detected among the genotypes with regards to the maturity of grape berries. Therefore, it is important to ensure that cultivars do not ripen too early because ripening during the hottest period of the summer results in unbalanced wines that can be high in alcohol, lacking acidity, freshness, and aroma expression features [37–39].

During the grape ripening period, the sugar concentration increases whilst the acidity level declines. Grapes from cooler areas have higher levels of acidity, which is linked to slower grape ripening, compared to grapes from warmer climate areas [40]. It has also been reported that lower acidity levels in white wines are often the cause of the polymerization of phenolic compounds, resulting in brown deposits and therefore causing darkening of white wine [27].

The absorbance at 420 nm exhibits comparability with the previously reported values pertaining to the phenomenon of the browning test observed in Greek white wines over time [41]. According to the determined k factor, the genotypes of the Monemvasia Winery "Asproudi" provide wines with a decreased color intensity, low phenolic compounds, and the tension to brown later than the other Greek autochthonous white grapevine varieties, such as Assyrtiko, Moschofilero, Roditis, Petroulianos, and Malagousia [41,42].

"Color Intensity" is a key descriptor in the sensory evaluation; the wines of Groups B and D displayed the most elevated levels. Consistent with prior investigations on white wines, it is plausible that amplified mouthfeel qualities may be linked to a heightened phenolic compound presence [43]. Previous studies differentiated clones of the white grapevine variety Albarino based on their physiochemical parameters [44].

Our efforts extensively documented the chemical and sensory variability of wines obtained from different "Asproudi" groups that were grafted onto the identical rootstock and cultivated under similar mesoclimatic conditions. It is important to note that altered chemical and sensory characteristics of wine were observed across different groups even when the same vinification protocol was applied. Six groups of the Monemvasia Winery "Asproudi" underwent evaluation, revealing significant differences in the classic parameters among musts and wines, as well as in sensory profiles. The outcome of this examination led to the conclusion that chemical compounds can serve as a valuable tool for the classification and differentiation of various clones, and those wines may possess distinct characteristics not only in relation to varieties but also to groups within a given variety. The application of PCA demonstrated that wines originating from different groups differ in their classic parameters. For instance, Groups B, D, and A exhibited the highest levels of ethanol content. Classic parameters analysis confirmed that distinct groups produce different wines. These findings suggest that the parameters are characteristic to individual groups, highlighting the importance of characterizing these groups for industrial use and consumer preference.

Characterizing autochthonous grapevine germplasm is crucial for preventing the erosion of genetic resources and the loss of local oenological products that are deeply rooted in the traditions of the region. However, in addition to conventional molecular and ampelographic analyses, it is also necessary to consider information about the natural environment in order to assess the optimal growth conditions for the genotype–environment relationship. Moreover, understanding the spatial distribution of local biodiversity within the indigenous territory can provide insights into the degree of erosion risk, which is inversely related to the extent of diffusion, as well as the importance of on-farm conservation. Abandoning vineyards would not only result in landscape loss and land degradation, but also undermine the preservation of the local biodiversity. Therefore, in the current research it is proposed that integrated criteria for evaluating and promoting local varieties are adopted as the means to promote the sustainability of grapevine production in the face of climate-related obstacles.

The multidisciplinary approach—molecular identification, must and wine evaluation, and sensory analysis—indicates that the genotypes found under the name "Asproudi" in the Monemvasia Winery vineyard could satisfactorily be used separately. This practice can satisfy the various grapevine growing and oenological requirements in different environmental conditions, taking advantage of their individualities, such as earliness of ripening, productivity, and ability to influence wine sensory characteristics in terms of body and complexity of aroma. Shaping the final quality features of the Monemvasia Winery, "Asproudi" grapes at harvest can offer valuable support to orient viticultural practices aimed at enhancing the quality of grape production in light of growing site and clone/variety preference. Alternatively, depending on the range of oenological objectives, it may appear advantageous for the simultaneous cultivation of several clones, chosen according to their level of production and aromatic complexity that can be used in mixes.

The objective of a sustainable agriculture/viticulture, as advocated by the new Rural Development Program (2014–2020) in alignment with the European 2020 Horizon objectives, will also entail the preservation of agro-biodiversity and landscape, as well as the mitigation of habitat fragmentation or simplification. In the Monemvasia region, numerous autochthonous grapevine cultivars are currently at risk alongside their respective landscapes. This investigation focuses on a local grapevine genotype, namely "Asproudi," and furnishes an analysis of its molecular characteristics, as well as the chemical and sensory descriptors, with the ultimate aim of utilizing distinct grapevine genotypes for the sustainable exploitation of genetic resources.

5. Conclusions

The multidisciplinary approach incorporated in the current study managed to distinguish a population of varieties that are known as "Asproudi". Each of these genotypes was evaluated in terms of vinification by chemical and sensory analysis.

Therefore, to answer the initial question, "How to improve a successful product", our suggestion would be to use scientific research as an essential tool in everyday practice. This, although it may be considered as a self-evidenced assumption, is still not the main issue in many cases. The application of modern scientific research could be an important advantage to the primary sector representing two-way communication and cooperation between the scientists and the people of the primary sector (viticulturists, farmers, nurseries, etc.): the common goal would be to improve agricultural production, and ultimately the national economy and the private income. Our current work aims to serve as a first step towards this perspective.

Author Contributions: Conceptualization: G.M., D.-E.M. and Y.K.; methodology: G.M., D.-E.M. and Y.K.; software: G.M. and D.-E.M.; formal analysis: G.M., D.-E.M. and Y.K.; investigation: G.M. and D.-E.M.; resources: G.M., Y.K., P.H. and G.T.; data curation: G.M., D.-E.M. and Y.K.; writing—original draft preparation: G.M. and D.-E.M.; writing—review and editing: G.M., D.-E.M., Y.K. and P.H.; visualization: G.M., D.-E.M., Y.K. and P.H.; supervision: G.M., Y.K. and P.H.; project administration: G.M. and D.-E.M.; funding acquisition: G.M., Y.K. and P.H. All authors have read and agreed to the published version of the manuscript.

Funding: This research received no external funding.

Institutional Review Board Statement: This study was conducted according to the guidelines of the Declaration of Helsinki. Ethical review and approval for this study were waived due to the anonymity of the interviews and the request for non-sensitive information.

Informed Consent Statement: Panelists gave informed consent before participating in this study.

Data Availability Statement: The data presented in this study are available on request from the corresponding author (pending privacy and ethical considerations).

Acknowledgments: The authors greatly appreciate the valuable comments and suggestions of the anonymous reviewers that helped improve the quality of this paper.

Conflicts of Interest: The authors declare no conflict of interest.

References

1. Valamoti, S.M. Harvesting the 'wild'? Exploring the context of fruit and nut exploitation at Neolithic Dikili Tash, with special reference to wine. *Veget. Hist. Archaeobot.* **2015**, *24*, 35–46. [CrossRef]
2. Poniropoulos, E.I. *Greek Viticulture and Vinification*; Michalopoulo: Athens, Greece, 1888.
3. Rousopoulos, O.A. *A Practical Guide for the Viticulturist, Winemaker and Distiller*; Viticulture: Athens, Greece, 1894; Volume 1.
4. Kribas, V.D. *Greek Ampelography*; Ministry of Agriculture: Athens, Greece, 1944; Volume 2, pp. 18–25.
5. Kribas, V.D. *Greek Ampelography*; Ministry of Agriculture: Athens, Greece, 1949; Volume 3, pp. 27–70.
6. Stavrakas, D.E. *Ampelography*; Ziti Publications: Athens, Greece, 2010.
7. Spinthiropoulou, C. *Wine Varieties of the Greek Vineyard*; Olive Press Publications: Corfu, Greece, 2000.
8. Crespan, M.; Migliaro, D.; Larger, S.; Pindo, M.; Petrussi, C.; Stocco, M.; Rusjan, D.; Sivilotti, P.; Velasco, R.; Maul, E. Unraveling the genetic origin of 'Glera', 'Ribolla Gialla' and other autochthonous grapevine varieties from Friuli Venezia Giulia (northeastern Italy). *Sci. Rep.* **2020**, *10*, 7206. [CrossRef]
9. Töpfer, R.; Trapp, O. A cool climate perspective on grapevine breeding: Climate change and sustainability are driving forces for changing varieties in a traditional market. *Theor. Appl. Genet.* **2022**, *135*, 3947–3960. [CrossRef]
10. Gutiérrez-Gamboa, G.; Liu, S.Y.; Pszczólkowski, P. Resurgence of minority and autochthonous grapevine varieties in South America: A review of their oenological potential. *J. Sci. Food Agric.* **2020**, *100*, 465–482. [CrossRef]
11. Alsafadi, K.; Bi, S.; Bashir, B.; Alsalman, A.; Srivastava, A.K. Future Scenarios of Bioclimatic Viticulture Indices in the Eastern Mediterranean: Insights into Sustainable Vineyard Management in a Changing Climate. *Sustainability* **2023**, *15*, 11740. [CrossRef]
12. Pastore, C.; Fontana, M.; Raimondi, S.; Ruffa, P.; Filippetti, I.; Schneider, A. Genetic characterization of grapevine varieties from Emilia-Romagna (northern Italy) discloses unexplored genetic resources. *Am. J. Enol. Vitic.* **2020**, *71*, 334–343. [CrossRef]
13. Diago, M.P.; Fernandes, A.M.; Millan, B.; Tardáguila, J.; Melo-Pinto, P. Identification of grapevine varieties using leaf spectroscopy and partial least squares. *Comput. Electron. Agric.* **2013**, *99*, 7–13. [CrossRef]
14. Tomić, L.; Štajner, N.; Javornik, B. Characterization of grapevines by the use of genetic markers. In *The Mediterranean Genetic Code-Grapevine and Olive*; Poljuha, D., Sladonja, B., Eds.; Intech Open Science: Rijeka, Hrvatska, 2013; pp. 1–25. [CrossRef]
15. Cappellini, E.; Gilbert, M.T.P.; Geuna, F.; Fiorentino, G.; Hall, A.; Thomas-Oates, J.; Collins, M.J. A multidisciplinary study of archaeological grape seeds. *Naturwissenschaften* **2010**, *97*, 205–217. [CrossRef]

16. Purwidyantri, A.; Azinheiro, S.; Roldán, A.G.; Jaegerova, T.; Vilaça, A.; Machado, R.; Cerqueira, M.F.; Borme, J.; Domingues, T.; Martins, M.; et al. Integrated Approach from Sample-to-Answer for Grapevine Varietal Identification on a Portable Graphene Sensor Chip. *ACS Sens.* **2023**, *8*, 640–654. [CrossRef]

17. Miliordos, D.E.; Merkouropoulos, G.; Kogkou, C.; Arseniou, S.; Alatzas, A.; Proxenia, N.; Hatzopoulos, P.; Kotseridis, Y. Explore the rare—Molecular identification and wine evaluation of two autochthonous Greek varieties: "Karnachalades" and "Bogialamades". *Plants* **2021**, *10*, 1556. [CrossRef]

18. Kullaj, E.; Bacu, A.; Thomaj, F.; Fiku, H.; Argyriou, A. Albanian grapevine cultivars: Preliminary results of molecular, phenolic and ampelometric profiles and relatedness. *Vitis* **2015**, *54*, 111–113.

19. Merkouropoulos, G.; Michailidou, S.; Alifragkis, A.; Argiriou, A.; Zioziou, E.; Koundouras, S.; Nikolaou, N. A combined approach involving ampelographic description, berry oenological traits and molecular analysis to study native grapevine varieties of Greece. *Vitis* **2015**, *54*, 99–103.

20. Thomas, S.M.; Scott, N.S. Mirosatellite repeats in grapevine reveal DNA polymorphisms when analyzed as sequence–tagged sites (STSs). *Appl. Genet.* **1993**, *86*, 985–990. [CrossRef]

21. Bowers, J.E.; Dangl, G.S.; Vignani, R.; Meredith, C.P. Isolation and characterization of new polymorphic simple sequence repeat loci in grape (*Vitis vinifera* L.). *Genome* **1996**, *39*, 628–633. [CrossRef] [PubMed]

22. Bowers, J.E.; Dangl, G.S.; Meredith, C.P. Development and characterization of additional microsatellite DNA markers for grape. *Am. J. Enol. Vitic.* **1999**, *50*, 243–246. [CrossRef]

23. Sefc, K.M.; Regner, F.; Turetschek, E.; Glossl, J.; Steinkellner, H. Identification of microsatellite sequences in Vitis riparia and their applicability for genotyping of different Vitis species. *Genome* **1999**, *42*, 367–373. [CrossRef]

24. Peakall, R.; Smouse, P.E. GenAlEx 6.5: Genetic analysis in Excel. Population genetic software for teaching and research–an update. *Bioinformatics* **2012**, *28*, 2537–2539. [CrossRef] [PubMed]

25. Tamura, K.; Dudley, J.; Nei, M.; Kumar, S. MEGA4: Molecular Evolutionary Genetics Analysis (MEGA) software version 4.0. *Mol. Biol. Evol.* **2007**, *24*, 1596–1599. [CrossRef]

26. Grose, C.H.; Martin, D.J.; Stuart, L.; Albright, A.; McLachlan, A.R.G. Grape harvest time and processing method can be used to manipulate 'Sauvignon Blanc' wine style. *Acta Hortic.* **2016**, *1115*, 139–146. [CrossRef]

27. Darias-Martín, J.J.; Rodríguez, O.; Díaz, E.; Lamuela-Raventós, R.M. Effect of skin contact on the antioxidant phenolics in white wine. *Food Chem.* **2000**, *71*, 483–487. [CrossRef]

28. OIV. *Compendium of International Methods of Wine and Must Analysis*; International Organisation of Vine and Wine: Paris, France, 2018; Volume 1.

29. Ribéreau-Gayon, P.; Glories, Y.; Maujean, A.; Dubourdier, D. Phenolic compounds. In *Handbook of Enology*; John Willey & Sons Ltd.: New York, NY, USA, 2000.

30. Arnous, A.; Makris, D.P.; Kefalas, P. Effect of principal polyphenolic components in relation to antioxidant characteristics of aged red wines. *J. Agr. Food Chem.* **2001**, *49*, 5736–5742. [CrossRef]

31. Singleton, V.L.; Kramling, T.E. Browning of white wines and an accelerated test for browning capacity. *Am. J. Enol. Viticul.* **1976**, *27*, 157–160. [CrossRef]

32. *ISO 3591*; Sensory Analysis Apparatus: Wine–Tasting Glass. International Standards Organization (ISO): Geneva, Switzerland, 1977.

33. Langlois, J.; Ballester, J.; Campo, E.; Dacremont, C.; Peyron, D. Combining olfactory and gustatory clues in the judgment of aging potential of red wine by wine professionals. *Am. J. Enol. Vitic.* **2010**, *61*, 15–22. [CrossRef]

34. Cartier, R.; Rytz, A.; Lecomte, A.; Poblete, F.; Krystlik, J.; Belin, E.; Martin, N. Sorting procedure as an alternative to quantitative descriptive analysis to obtain a product sensory map. *Food Qual. Prefer.* **2006**, *17*, 562–571. [CrossRef]

35. Gómez-Míguez, M.J.; Gómez-Míguez, M.; Vicario, I.M.; Heredia, F.J. Assessment of colour and aroma in white wines vinifications: Effects of grape maturity and soil type. *J. Food Eng.* **2007**, *79*, 758–764. [CrossRef]

36. This, P.; Jung, A.; Boccacci, P.; Borrego, J.; Botta, R.; Costantini, L.; Crespan, M.; Dangl, G.S.; Eisenheld, C.; Ferreira-Monteiro, F.; et al. Development of a standard set of microsatellite reference alleles for identification of grape cultivars. *Theor. Appl. Genet.* **2004**, *109*, 1448–1458. [CrossRef]

37. Palliotti, A.; Tombesi, S.; Silvestroni, O.; Lanari, V.; Gatti, M.; Poni, S. Changes in vineyard establishment and canopy management urged by earlier climate-related grape ripening: A review. *Sci. Hortic.* **2014**, *178*, 43–54. [CrossRef]

38. Duchêne, E.; Huard, F.; Dumas, V.; Schneider, C.; Merdinoglu, D. The challenge of adapting grapevine varieties to climate change. *Clim. Res.* **2010**, *41*, 193–204. [CrossRef]

39. van Leeuven, C.; Seguin, G. The concept of terroir in viticulture. *J. Wine Res.* **2006**, *17*, 1–10. [CrossRef]

40. Schmit, T.M.; Rickard, B.J.; Taber, J. Consumer valuation of environmentally friendly production practices in wines considering asymmetric information and sensory effects. *J. Agr. Econ.* **2013**, *64*, 483–504. [CrossRef]

41. Kallithraka, S.; Salacha, M.I.; Tzourou, I. Changes in Phenolic Composition and Antioxidant Activity of White Wine during Bottle Storage: Accelerated Browning Test versus Bottle Storage. *Food Chem.* **2009**, *113*, 500–505. [CrossRef]

42. Salacha, I.M.; Kallithraka, S.; Tzourou, I. Browning of white wines: Correlation with antioxidant characteristics, total polyphenolic composition and flavanol content. *Intern. J. Food Sci. Tech.* **2008**, *43*, 1073–1077. [CrossRef]

43. Olejar, K.J.; Fedrizzi, B.; Kilmartin, P.A. Enhancement of Chardonnay antioxidant activity and sensory perception through maceration technique. *LWT—Food Sci. Tech.* **2016**, *65*, 152–157. [CrossRef]
44. Zamuz, S.; Carmen Martínez, M.; Vilanova, M. Primary study of enological variability of wines from different clones of *Vitis vinifera* L cv. Albariño grown in Misión Biológica de Galicia. *J. Food Compos. Anal.* **2007**, *20*, 591–595. [CrossRef]

Article

What Smallholders Want: Effective Strategies for Rural Poverty Reduction

Ruerd Ruben 

Development Economics Group & Wageningen Economic Research, Wageningen University and Research, 6706 KN Wageningen, The Netherlands; ruerd.ruben@wur.nl

Abstract: Since poverty is particularly concentrated amongst smallholder farmers, development programs intend to support rural livelihoods and agricultural entrepreneurship. The final impact of these programs remains, however, rather limited due to insufficient understanding of key challenges that smallholder families are facing. Many well-intended initiatives for reinforcing smallholder production systems and for strengthening their commercial relationships meet conceptual and practical limitations that reduce their effectiveness. Smallholder livelihoods are most constrained because behavioural drivers for adopting innovations and for upgrading value-chain relationships are not well understood and are frequently overlooked. This article discusses the analytical linkages between the key causes of smallholder poverty, the constraints that limit the effectiveness of ongoing rural development initiatives, and the prospects for alternative strategies to support behavioural change. A better understanding of what smallholders want and need may lead to fundamentally new policy propositions. It is argued that technological change in smallholder production or integration into market systems will only take place if embedded in behavioural change mechanisms that are complemented by appropriate institutions and governance regimes. This asks for coordinated structural reforms in farm and community organisation, value chain integration and more effective public-private cooperation.

Keywords: rural poverty; smallholders; Innovation; empowerment; impact; behavioural change

Citation: Ruben, R. What Smallholders Want: Effective Strategies for Rural Poverty Reduction. *Sustainability* **2024**, *16*, 5525. https://doi.org/10.3390/su16135525

Academic Editor: Dimitris Skalkos

Received: 17 April 2024
Revised: 19 June 2024
Accepted: 20 June 2024
Published: 28 June 2024

1. Introduction

There is an ongoing and lively debate on the most appropriate strategies for improving the position of smallholders in tropical commodity chains in Sub-Sahara Africa [1–3]. While many studies focus on opportunities for increasing yields or improving prices, far less attention is usually given to opportunities for reducing risk or strengthening trust in tropical value chains. This divergence is largely due to the fact that the behavioural drivers for smallholder decision-making are little understood, and, therefore, many well-intended initiatives that try to reinforce smallholders' incomes and livelihoods meet major constraints that limit their effectiveness [4].

In this article, we outline an analytical framework to better understand the key objectives that smallholder farmers pursue. Our analysis is grounded in the systematic analysis of empirical field studies on smallholder behaviour, complemented by longstanding expertise in farm-household production, consumption and marketing decisions [5,6]. This provides the basis for a further analysis of different strategic initiatives for improving smallholder welfare, such as fair trade certification and living income benchmarks. We show that many of these market-based interventions were rather ineffective due to the limited consideration given to the required changes in the behavioural drivers for the adoption of innovations and the engagement in market relationships.

The article relies on a conceptual approach (see Figure 1) that starts with an overview of key constraints that smallholder farm-households have to face, followed by an analysis of why many strategic initiatives for 'making markets work for the poor' are delivering such limited impact [7]. This paves the way for a further discussion on the possibilities of providing better incentives for behavioural change that enable smallholder livelihoods to respond to key challenges.

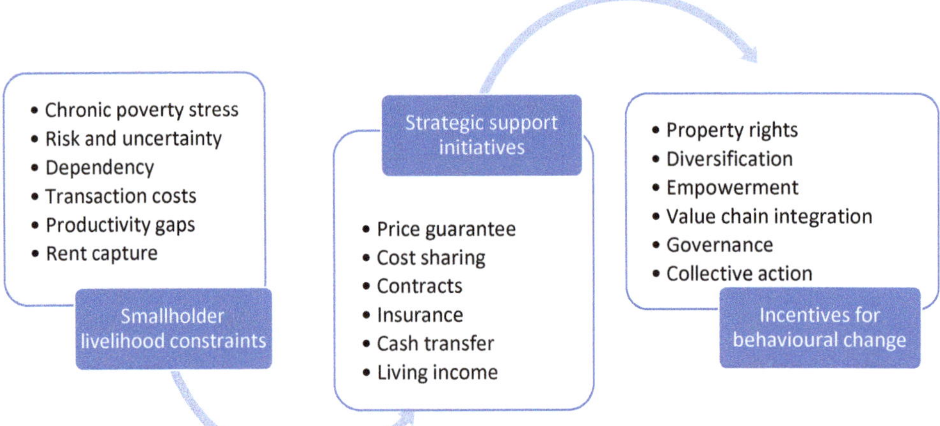

Figure 1. Linkages between smallholder livelihoods, strategic development initiatives, and incentives for behaviour change.

Smallholders are defined as family-operated farms that mainly use family labour for their agricultural activities. They comprise farms up to 2 hectares that make intensive use of scarce land resources and use part of the produce for direct household consumption [8]. Smallholder livelihoods strongly rely on agriculture, usually a mixture of cash crops, livestock rearing and subsistence farming. In addition, they engage in off-farm work and non-farm (self)employment to diversify their household income. Part of the rural labour is moving out of agriculture and absorbed into low-productivity services and informal urban activities, such as retail trade, transportation, food preparation, etc. Smallholders still depend for essential necessities on community exchange networks that provide access to water, food, feed, shelter and services.

Smallholder farmers represent an important segment of global agrifood systems where a significant part of rural poverty is still concentrated. Of the 570 million farms worldwide—83% of which are in sub-Saharan Africa (9%) and Asia (74%)—475 million are smallholder farms (<2 hectares). These farms operate about 24% of the world's agricultural land and provide around 30–34% of the food supply [9]. Some 400 million rural people live in extreme poverty, of which 76% are in sub-Saharan Africa. Yet, many of these poor farmers are systematically neglected in agricultural and rural development programs.

After several decades of gradual reduction of rural poverty, it started to increase again with the COVID-19 pandemic. Currently, around 680 million people (=8.5% of the world population) still face (extreme) poverty. Two out of three people experiencing poverty live in rural areas and much of this poverty is concentrated in sub-Saharan Africa. Rural poverty is most strongly affecting families living in conflict regions and/or facing the effects of climate change [10]. Especially women and children suffer from the consequences of rural poverty in terms of malnutrition, stunting and wasting [11], as well as the negative educational and psychosocial effects.

The current literature on smallholder engagement in tropical value chains shows important limitations. Most studies mention a range of practical challenges that need to be addressed to improve farming systems and household revenues and identify some actions to overcome these problems [12,13]. However, they fail to outline the key drivers and mechanisms to modify this situation. This article argues that—instead of just combining different actions for strengthening the smallholder sector—it is important to better understand the underlying drivers of rural household behaviour for adjusting their livelihoods and engaging in any of these strategies. Therefore, we need to identify which issues shape prospects for adopting innovations into smallholder livelihoods and how they can become engaged in strategies for improved resource management and better governance of agrifood value chains.

These issues are especially important in tropical commodity chains of cocoa, coffee, tea and cotton. Bymolt et al. [14] assess the opportunities for poverty reduction in the cocoa sector in Ghana and Ivory Coast—looking at the role of land tenancy, production systems and marketing regimes—and conclude that more fundamental institutional and market reforms are required for improving farmer' welfare and living wages and to overcome existing structural bottlenecks. In a similar vein, Waarts and Kiewisch [15] study on cocoa farmers in West Africa shows that neither increases in yields nor better farm-gate prices offer sustainable alternatives for reaching minimum living income. The authors, therefore, make a plea to support the poorest cocoa farmers towards income diversification and engagement in non/off-farm and self-employment. Taher [16] reaches a similar conclusion for Indonesian cocoa farmers, showing that adoption of improved production technologies is only possible if appropriate risk management practices are in place. Therefore, major emphasis should be given to better market linkages and improved governance to enable farmers to achieve a decent living standard.

Recent studies on value chains for coffee and cotton also provide evidence that initiatives for improving yields and prices through (fairtrade or organic) certification of cropping systems and upgrading of agro-processing facilities focus too much on the supply-side facilities (credit and technical assistance services, etc). They generally neglect the high investment costs and the intensive labour demands that make adoption less attractive to many smallholders and underestimate the marketing and governance constraints (such as price fluctuations, weak cooperatives, over-certification, etc.) that limit their feasibility [17].

Value chains for staple food crops (maise, wheat, rice) report similar constraints for the adoption of improved practices and the involvement of relevant stakeholders, particularly women. Jain et al. [18] show that promising strategies for sustainable intensification of smallholder farms—such as water harvesting and the system of rice Intensification—have considerable potential to scale to reach more than 50% of smallholder farmers who plant staple crops, but significant barriers to adoption remain related to limited access to land and water, labour shortages and market constraints. Adoption of climate-smart agricultural practices is especially limited due to high risks and insecure revenue streams [19].

Whereas all studies seem to agree on the idea that there is *'no one size fits all'* solution, they tend to disagree on the concrete priorities and policy pathways for implementing a combined set of measures for strengthening farmers' livelihoods. In essence, this is due to their reliance on partial and strictly operational approaches that make rather limited use of systems transformation thinking [20]. Current literature shows a preponderance of studies focussing on the adoption of new technologies for (semi-)commercial farmers but is notably scarce in generating knowledge on adaptive livelihood change in a smallholder context. This is partly due to the fact that behavioural change processes are far more complex, face more resistance and, thus, require a longer time period, which is why structural bottlenecks are identified earlier than institutional constraints. This article intends to fill this gap by focussing on the drivers of smallholder behaviour and identifying policy alternatives to support rural poverty reduction.

2. Key Challenges That Smallholders Face

Agricultural and rural development programs in the global South tend to focus strongly on strategies for improving yields since smallholders only operate small plots of land and are therefore considered to be better off if returns to land improve [21]. While land is certainly scarce and land ownership remains insecure in many settings, livelihood strategies of poor rural households also include a number of other objectives that are equally important—albeit more difficult—to pursue.

Even while local livelihood realities are substantially different between farmers and locations, the empirical literature on smallholder preferences, needs and priorities points towards a coherent set of farm-household objectives that are considered of primary importance to guarantee their resilience and survival and eventually enable them to raise their living standards beyond poverty [11,22,23].

Based on a thorough review of the literature on the determinants of income and revenues for smallholder farmers and the structural causes of poverty amongst rural households, we identify six main challenges that notably influence smallholder behaviour and need to be addressed for combatting rural poverty (see Figure 2).

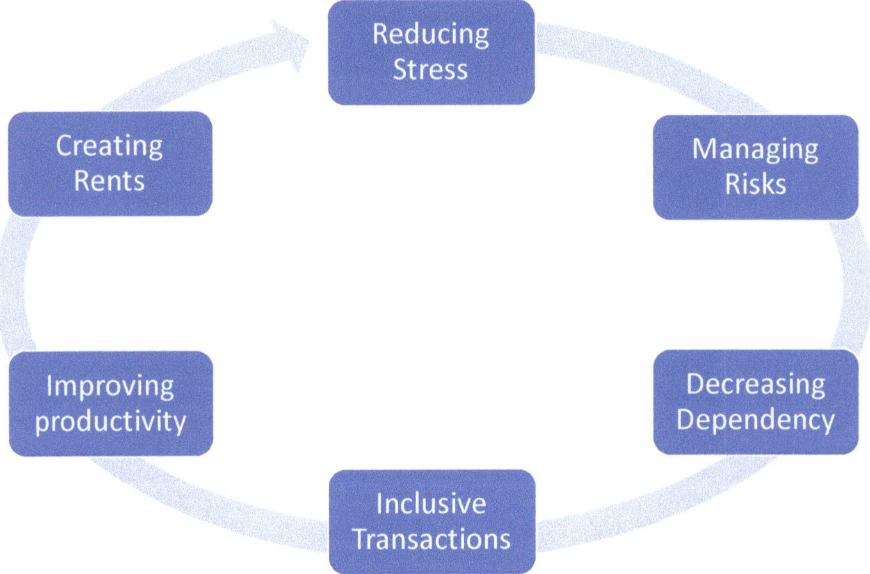

Figure 2. Key livelihood challenges for poverty reduction by smallholder producers.

2.1. Reducing Stress

Smallholder farmers are poor, and this poverty tends to become a chronic feature (that is intergenerationally transmitted) due to the high variability and low predictability of income streams. Consequently, they regularly face critical shortfalls in basic consumption that lead to a context of stress, where mental and physical disparities limit opportunities for coping. Uncertainty about future income streams is further enhanced by strong variations in prices and limited prospects for improving resource productivity.

There is wide evidence that higher incomes do not automatically lead to better nutrition and health [24]. Even while most investments for improving production or trade are controlled by men, income streams steered by women have substantially stronger effects on household food security, child nutrition and the life course of the family.

Considering rural farm households as a unit of production and consumption implies that they are not merely interested in higher net farm incomes but especially in 'consumption smoothing' through stable revenue streams with small fluctuations [25]. Incomes that are less variable and more predictable are appreciated more than just higher incomes. Stable expectations about future income streams enable rural households to invest in education, nutrition, housing and health, in such a way that they can allocate windfall profits to investments in farm household assets. It is therefore advocated to reduce smallholders' stress from the consumer side, using vouchers and cash transfers as incentives for allocating expenditures to better nutrition. This may lead to higher labour productivity and can be paid off by earnings from the production side of the household.

2.2. Managing Risks

Smallholders continuously face chronic poverty traps that are caused by unexpected income shortfalls due to personal disasters (such as illness, death, fire or theft) or local climatic events (drought, floods, storms, plagues and diseases, etc). The management of such risks asks for the maintenance of substantial reserve stocks (cattle, jewellery, etc.) that may be mobilised for risk-coping purposes, but can also lead to compulsory sales of critical assets (land, house, cattle) for overcoming income shortfalls [26].

Building resilience into smallholder households implies that farmers tend to engage more in low-return activities that enable flexible adjustments, usually at the expense of investments with a higher earning capacity. Poor farmers thus become risk-averse and exhibit a low willingness to invest in potential improvements, a trend that can be reversed by providing insurance, deepening education and/or diversifying income-generating activities. Some of these strategies depend on public support.

2.3. Decreasing Dependency

Smallholder farmers usually face insecure land (ownership or use) rights that inhibit making investments in better resource management. More secure and registered land rights offer opportunities for borrowing (using land as collateral) and enlarge the time horizon for expected returns, thus enabling poor farmers to engage in other income-generating activities. In a similar vein, better access to loans and credit may support farmers to increase their returns to land and labour

Efforts for decreasing smallholder dependency usually focus on activity and income diversification through labour engagement in non- and off-farm employment and farm investments for upgrading (financed from off-farm income and remittances), thus creating important spillovers between different activities within the household [27]. Moreover, diversification of household revenues reduces dependence on a single crop and improves farmer's capacity to adapt to shocks and their ability to invest in household nutrition. Simultaneous engagement of farmers with midstream traders in different markets is also helpful in diversifying risks [28,29]. Gender-balanced decision-making on priorities for household resource allocation in production, consumption and trading requires that women's empowerment is reinforced in such a way that lifetime perspectives of livelihoods for all household members are adequately considered.

2.4. Inclusive Transactions

Even while part of smallholder production is likely to be devoted to household food consumption, they also depend on market exchange for generating cash resources to finance non-food expenditures. Farmer's bargaining position in rural markets is heavily affected by high transaction costs that lead to strong price fluctuations and low margins. High transaction costs are caused by low frequency (small trade volumes), limited price information, individual sales and fierce competition.

Smallholder trade transactions on input and output markets are mainly spot exchanges that are characterised by low trust with traders, short-term arrangements and the absence of collective action (cooperative organisation). Many poor farmers operate as net buyers on local markets [30] and are, therefore, more interested in low prices (as consumers) instead of high prices (as producers). This leads to a perverse response to prices and tends to limit opportunities for engaging women in trade as a pathway for supporting changes in intra-household bargaining power.

2.5. Improving Resource Productivity

Smallholder farmers face low and declining crop yields (due to limited input use and progressive land degradation) combined with stagnating labour productivity. Since smallholders' incomes simultaneously depend on the returns to land, labour and capital, attention should be devoted to raising 'Total Factor Productivity' (TFP). While much emphasis is usually given to technical opportunities for improving returns from agricultural production- either through higher yields (relying on better breeds, fertiliser and irrigation) or through better prices—farmers prefer to increase welfare by adding value from resources that are most scarce to them [31]. The key strategy for strengthening rural livelihoods thus starts with investing in the labour capacity of the household workforce. This particularly includes investment in education and human capital that generate prospects for farm innovation and create opportunities for on- and off-farm income generation.

In the smallholder sector, both land and capital are limited, but (family) labour is especially scarce [32,33]. This may sound strange given the high level of local unemployment, but prospects for sustainable intensification and regenerative agriculture require fairly labour-intensive production systems with peak labour requirements that 'ty up' family labour to the farm. It is, therefore, not sufficient to increase only the availability of resources, but incentives and leverage points need to be identified at the institutional and behavioural levels to generate higher returns to production factors.

2.6. Creating Rents

Much of the income growth by smallholder farmers is based on horizontal extension of cultivated area. Far less progress is made with efforts for increasing (physical) resource productivity (i.e., reducing yield gaps) and improving (financial) returns to resources. The value added created by smallholders remains low due to a lack of vital assets and limited access to complementary resources (electricity, storage, transport, etc.).

Therefore, a large part of agricultural production is sold as raw materials at depressed farm gate prices and most of the value added is captured by midstream stakeholders, such as shopkeepers, traders, processors and exporters [29]. Moreover, access to rents is constrained due to limited infrastructure at the farm and village levels and the control over resources dominated by large landowners and predominantly male household members.

It should be noted that there are critical interlinkages between these challenges that need to be addressed through coordinated strategies. Opportunities for reducing risks are related to policies for decreasing dependency and exclusion, whereas prospects for creating rents are based on strategies for enhancing productivity and strengthening competitive markets. These structural causes of poverty amongst rural households are mutually related and cannot be 'solved' with straightforward strategies for improving livelihoods by broadening access to resources, markets, services and information. Instead, they are linked through behavioural relationships characterised by limited confidence and low trust [34,35] that lead to social exclusion from networks and inequality in bargaining power within and outside households [10]. Breaking the vicious circles of rural poverty, therefore, requires that key attention is given to intra-household organisation (i.e., the gender division of work, assets and incomes) and the external governance relationships with value chain stakeholders as well as public and civic institutions.

3. Why Current Rural Development Initiatives Are So Little Effective?

Most investments for smallholder development are focussed on improving farm production systems, while far less attention is usually given to the backwards linkages between the farm and the household, and the forward linkages between the farm and the value chain. Interventions that take the household as an entry point look at family nutrition as a main outcome (that may, in turn, lead to higher labour productivity or reduced losses of working days to health problems) and also focus on the intra-household division of labour and gender bargaining power differences as underlying causes of poverty and malnutrition. In a similar vein, value chain interventions that try to improve the competitive position of smallholders—through better prices and longer-term contractual arrangements—contribute to higher commercial margins and improved security for sales.

Rural development programs face major constraints if they do not adequately consider these interlinkages between farm, household and value chain dimensions. This is mainly caused by structural inequalities in ownership of land and limited access to capital markets and knowledge and information systems [11,36]. In addition, appropriate institutions for farmer organisation and value-chain bargaining are often not in place. Different incentives and instruments can be used to improve the performance of each of these components (see Figure 3). Especially the timing of smallholder support measures—before/during or after the harvest—matters a lot for the generation of revenue streams at critical moments of the family lifecycle. Widening the perspective on farm-household livelihoods enables to target the focus of poverty alleviation programs from 'Investing in farms' to 'investing in people in poor places' [37].

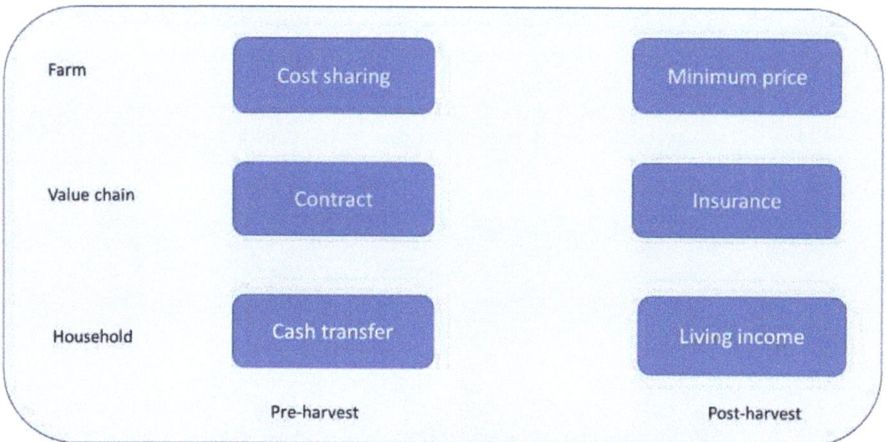

Figure 3. Different instruments for reinforcing smallholder livelihoods.

3.1. Cost Sharing

Cost sharing and Input support programs are used to promote the adoption of Good Agricultural Practices (GAP) that increase the profitability and/or sustainability of farming practices. Most sustainable intensification activities require access to better seed material, (organic) fertilisers and knowledge/training in land- and crop-management practices to enable their adoption by smallholders.

Cost sharing with other upstream partners (input providers, shopkeepers) or with downstream traders and processors is based on pre-finance arrangements that commit part of the harvest as repayment. Even while access to inputs is guaranteed, smallholders might lose a substantial part of their higher productivity by receiving low output prices. Many programs focus on intensification strategies for enhancing smallholder welfare through better resource use efficiency [38,39], but far less attention is usually given to their cost-effectiveness and the increasing labour demands. Therefore, suitable market incentives that support farmers towards the adoption of these practices need to be in place.

3.2. Price Support

Higher and especially more stable (i.e., predictable) farmgate prices are critical to sustaining the minimum livelihoods of many smallholder farmers. Prices will need to rise if 'real costs' are considered and social and environmental externalities are included. Such a true pricing approach enables smallholders to apply more sustainable production practices and to attain household food security [40]. However, consumer's willingness to pay true prices is still unknown.

Minimum prices are the core of several commodity certification programs (such as fair trade, ecological labelling and certificates or origin) that reward smallholders for maintaining certain production standards to gain preferential access to premium markets. Minimum prices can be guaranteed to producers for delivering better and more homogeneous quality or for supporting the use of more sustainable production methods.

In practice, however, minimum prices are only slightly higher than regular prices. They are difficult to enforce and can be easily circumvented. Due to heavy over-certification, farmers can only sell a minor share of their production under premium conditions, and need to pay high costs to obtain the certificate and verify compliance [41].

3.3. Contracts

Contracts between farmers and traders are meant to provide a reliable commitment to deliver inputs or collect output at pre-established conditions. Contracts that include price guarantees still face the risk of non-compliance. Traders sometimes do not honour their obligation of buying the full harvest and smallholders frequently face delayed payments. Otherwise, farmers are tempted by side sales, if market prices become higher than the earlier-agreed price [42,43].

In practice, many contracts are rather 'incomplete' (i.e., do not fully define all price, delivery and quality requirements) and fail to include adequate conflict resolution procedures. Contracts give limited attention to the distribution of value added. Currently, smallholders receive a tiny share of total value added that only represents 4–6% of the final consumer price. More than 80% of the value added is realised in processing, transport and retail activities, often dominated by foreign firms. Local taxes take away another part of the value added. When contracts become more inclusive and embrace stakeholders throughout the value chain, the perspective may eventually be widened from 'fair prices' towards 'fair chain'.

3.4. Insurance

Insurance against the risk of crop losses due to unforeseen weather conditions or natural disasters is an important ex-post device for smallholders' engagement in sustainable intensification [44]. Given their limited internal reserve capacity, farmers cannot afford to take too much risk and, therefore, may prefer to rely on low-input low-productivity production methods. Insurance can offer an alternative that enables smallholders to build resilience and rely on productivity-enhancing external inputs while committing a larger share of production to market exchange.

New inclusive insurance products, such as weather Index-based insurance and health insurance, may also contain a savings component, which can help smallholder farmers build a financial base that can serve multiple purposes. Insurance could also be offered as part of a wider set of business services that support smallholders. Group-based and community-centred rotating savings and borrowing schemes (such as the well-known Savings and Credit Cooperative Societies or SACCOS)—mainly with female members—are particularly effective in enhancing financial inclusion and reducing chronic poverty and malnutrition [45]. Digital and mobile platforms nowadays offer innovative opportunities for widening access to credit and insurance.

3.5. Living Income

Living income benchmarks define a net annual income required for a household in a particular place to afford a decent standard of living for all its members. Some trade arrangements with international companies may include price- or wage criteria that allow farmers and workers to cover minimum living standards and guarantee healthy diets. Living income thus represents the wider dimension of rural livelihoods and connect input- and output markets as well as pre- and post-harvest aspects.

Empirical studies on living income in the cocoa value chain from West Africa, Waarts and Kiewisch [15] show that neither increases in yields nor better farm-gate prices offer sufficient guarantees for reaching the minimum living income benchmark. In a similar vein, Cordes et al. [46] show that most coffee-sourcing programs of roasters and retailers fail to close the living income gap.

Some certification programs deliver periodical lump-sum payments to farmers' cooperatives or rural communities to support investments in collective goods (infrastructure) or social services (education; health care). These investments have a wide coverage and also benefit non-affiliated families. However, local participants are scarcely informed about the governance of these community funds [47] and tend to prefer individual payments instead of collective use of funds [48].

3.6. Cash Transfers

Cash transfers—either through vouchers or as direct payments to smallholders—are increasingly used as a strategy for combatting poverty and supporting a variety of other outcomes. Cash transfers can be made conditional on participation in education or primary health care. They provide social protection to beneficiaries, their households and communities can foster horizontal relationships within communities and vertical relationships with the state through forms of social accountability and citizenship engagement [49].

Cooke et al. [50] show how cash transfers delivered in three (mobile phone) instalments to coffee farmers in Eastern Uganda lead to a significant increase in household welfare (e.g., 40% higher expenditures), as well as improvements in production outcomes (e.g., doubling of coffee revenue, coffee investment). Conditional cash transfers for coffee farmers in Mexico proved to be effective in mitigating the negative effects of falling coffee prices on early childhood development [51]. Similar studies in Zambia find that cash transfers help households cope with agricultural production and price shocks and enable them to substantially increase their food consumption and overall food security [52]. Other experiences with monthly cash transfers to smallholders indicate that engagement of children in hazardous work strongly declines since households are better protected against adverse shocks, such as sickness, bereavement in the family, income fluctuations or loss of agricultural production.

Many studies are available that outline a broad set of policy interventions to support a more inclusive and sustainable smallholder sector. Key strategies include higher output prices, input support, yield improvement, crop diversification, better market linkages and improved value chain governance. Some measures may contradict each other (i.e., improving yields usually requires higher input costs), whereas others lead to perverse results (i.e., higher prices stimulate production; however, additional market supply will exercise downwards pressure on prices and margins). Experts, therefore, agree that a more comprehensive set of measures is needed to support smallholder commodity farmers [12,53]. The 'right' composition of this package and the size and sequence of the related measures are of fundamental importance for being able to influence smallholder behaviour.

4. Strengthening Smallholder Behaviour and Rural Governance

The fundamental weakness of current studies that focus on improving the smallholder position in tropical supply chains is that excessive attention is given to socio-economic conditions (farm size, prices, income) and the possibilities for technical innovations (better inputs and assistance services, options for digitalisation), thus trying to answer the question: *'what can be done to improve smallholder livelihoods?'*. There is, however, a far more relevant question that has to be addressed, namely: *'how to change the dynamics of the smallholder sector?'*. The latter question focuses on the required changes in behaviour and the improvement of the interactions amongst stakeholders in the value chain.

Supporting the dynamics of agricultural system transformation requires thorough insights into what farmers value most and how they can be motivated to make adjustments in the use of their land, labour, knowledge and other production factors. This is not only dependent upon the access and availability of these resources, but also on the underlying motivations for changing the social relationships that determine access to these resources by all household members.

Notwithstanding the importance of promoting a comprehensive set of interventions that make the smallholder sector both inclusive, sustainable and future-proof. Above all, we need better insights into how these interventions can be helpful for improving farmers' responsiveness and strengthening rural governance. Therefore, attention is given to institutional innovations and incentives for behavioural change that create self-enforcing mechanisms for (re)defining the smallholder's relationships with other (public, private and civic) stakeholders within the agricultural sector. Adoption of any of the before-mentioned innovations in land use, input intensity, labour allocation and trade networks will only take place if the underlying behavioural drivers are effectively addressed.

To support our future thinking on these strategic issues, we can derive from the widely available experimental research (based on participatory field labs and Randomized Controlled Trials) six related system interventions for improving smallholder behavioural responses (see Figure 4).

Figure 4. Key system interventions to support smallholder behavioural change.

Effective rural development interventions should trigger these behavioural change principles in order to guarantee that smallholders engage in—and respond to—the incentives provided by public, private or civic promotion policies or outreach programs. Realising structural change in smallholder behaviour should be based on adjustments in the agrifood system conditions that simultaneously influence smallholder production, exchange and reproduction (consumption) decisions. As outlined in Section 2, structural factors like stress, risk, dependency, distrust and exclusion represent critical constraints that need to be overcome before smallholder livelihoods can be improved. This also requires that due attention is given to the underlying beliefs, traditions and governance mechanisms that determine the scope for behavioural change. Therefore, interventions at the system level need to be identified that influence these coordination mechanisms. We briefly outline the importance of six structural drivers of behavioural change:

4.1. Property Rights for Extending the Time Horizon

Registered and acknowledged (ownership or use) rights over land (+water and trees) provide security for investments and strengthen the time perspective for improving rural household livelihoods [54]. Property rights are important as collateral to borrowing, both for access to (bank) credit as well as for loans from informal moneylenders (mainly delivered as advance payments). They also provide a sense of 'identity' to smallholder farmers who depend on nature for their survival. Land is not only a productive resource but also serves as an old-age pension.

Farmers need a larger time horizon to be able to engage in long-term in-depth investment decisions (such as tree renovation or the purchase of machinery) that deliver revenues after several years. In addition to ownership rights also information on future prices and longer-term delivery contracts are important to enlarge the time horizon. On the other hand, frequent weather changes and immanent trends of climate change tend to reduce the time horizon and can only be mitigated through reliable weather information systems, eventually combined with complex index insurance packages.

Land rights are thus conceived as an Incentive for Investment in agricultural intensification. Land rights registration may be cumbersome as it is usually heavily biased against women and in many cases disregards inheritance rules. Land fragmentation and compulsory land sales strongly reduce household income. Loans based on land collateral are perceived as risky but easily lead to forced land sales if debt payments cannot be made in time [55].

4.2. Diversification for Insurance

Major constraints for reinforcing rural livelihood and enabling sustainable intensification are related to market inefficiencies and uncertainties on future returns. This may be due to several factors: deficient input and seed quality, untimely access to input markets, price variation on output markets, low yields due to drought or excessive rainfall, etc. Some parts of these risks are idiosyncratic or personal (diseases, fire, theft), while other risks are stochastic and faced by many farmers at the same time (climate, market prices, political upheaval, etc.).

Interestingly enough, farmers can only insure themselves against the idiosyncratic risks and look for such insurance through the diversification of activities (e.g., 'putting their eggs in different baskets'). Whereas diversification enables better access to different income streams (thus also managing seasonal variation in prices) and usually improves dietary diversity, it may go at the expense of lower revenues due to the loss of specialisation advantages [56]. More formal insurance devices are costly and, therefore, only affordable to better-off farmers.

4.3. Women Entitlement for Empowerment

Recognition and anchoring of women's rights on land, products and revenues is a key strategy towards more balanced household decision-making. Gender equity is important for decisions on production and diets, as well as for reproductive health, housing and schooling. There is wide evidence that investments in women's capacity-building and empowerment deliver high payoffs in terms of labour productivity, sense of security and physical integrity [57]. Efforts for improving rural livelihoods should therefore start with giving a voice to women in the decision-making processes on key farm, family, household and community issues.

Strengthening women's rights on income, assets and family affairs requires substantial improvements in their bargaining power on livelihood strategies [58]. This implies their involvement in key production and trade activities, as well as the recognition of self-determination rights in family and community affairs. Empowerment of rural women can be supported through their engagement in self-help groups that provide better access to credit, education and information [59].

4.4. Value Chain Trust and Reliability

Linkages between farmers, traders and input providers are subject to large inefficiencies, partly due to unequal competition. This is caused by the small volume of transactions, but is also related to insecurities regarding sales and purchase transactions. Farmers are not always considered as 'reliable' by traders (e.g., when side sales occur against earlier agreements), whereas traders are equally mistrusted by farmers because of the manipulation of weights, delayed payments or untransparent quality grading practices [36].

Investments in mutual trust, reliable and enforceable (longer-term) contracts and reputation are therefore considered critical for improving the efficiency and traceability of transactions in the agrifood supply chain. Such higher efficiency—or lower transaction costs—might enable an improvement in trade margins and a redistribution of value-added shares in favour of upstream segments of the supply chain. There is an urgent need to support 'fair chain' practices based on transparency and a substantial redistribution of revenues between supply chain parties. A tiny 1% reduction in the retail share could be easily translated into a 10% increase in farmer's income [15].

4.5. Cooperation for Collective Action

Individual smallholders are usually weak partners in negotiations with traders and input providers. They therefore need some sort of collective organisation that creates economies of scale and improves their bargaining power. Willingness to cooperate does not emerge automatically and requires some sort of confidence in each other capacities. Moreover, a legal framework for establishing cooperatives needs to be in place. Ostrom [60] outlines that clear access boundaries, open exchange of information and transparent rules and sanctions for conflict resolution are key conditions for successful collective action.

Farmer's groups frequently fail if members have similar properties and problems (the so-called *'coalition of the poor'*) and therefore need some degree of heterogeneity where also midsize and larger farmers join the marketing cooperative. In addition, due attention needs to be given to transparent internal decision-making procedures that guarantee compliance with commitments and to equitable decision-making procedures that avoid coercive dominance by a few members.

4.6. Inclusive Governance

Public investments in physical and commercial infrastructure and services are required to guarantee smallholder's access to markets and information and to support more equal exchange conditions. Smallholder participation in public governance institutions is usually limited to the local or sub-regional level, whereas national policies still strongly rely on clientelism. Smallholder's bargaining position in outgrower schemes still heavily depends on the priorities of international firms.

Local communities need to be engaged in decision-making before committing to major investments. Public levies for land, water and trees as well as market fees easily become a disincentive for further investments in value chain integration. Current taxation regimes hinder functional upgrading and tend to discourage local processing [36]. Most indirect taxes on inputs, consumer goods and processed commodities are already strongly anti-poor. A key role for public policy is therefore related to the enforcement of better quality grading practices, more responsive technical assistance schemes and the support for more transparent market governance mechanisms.

Supporting changes in each of these institutional conditions for smallholder development is likely to create opportunities for overcoming the structural livelihood challenges (as outlined in Section 2). Instead of just reducing external constraints, these actions intend to support adaptive smallholder behaviour, enabling a more constructive response to different types of incentives (that are discussed in Section 3). While each of these system interventions is relevant for reinforcing the smallholder decision-making environment, it is even more important to put into action an interactive framework that combines public interventions (e.g., land right registration; taxation; public investments) with private sector activities (credit and insurance facilities; market transparency) and civic engagement (community-led organisation; women's groups; accountability mechanisms).

5. Conclusions and Outlook

Rural development programs and agricultural development policies have devoted much attention to strategic actions aimed either at providing better access to social and collective services or at investments for improving productivity or raising incomes. Much of these efforts meet major constraints in low-income economies, where high land fragmentation and limited value chain coordination tend to limit prospects for structural change. Whereas structural reforms are certainly important and relevant, we still face insufficient insights into the behavioural drivers for engaging smallholder farmers, families and communities into these transition pathways. This may be the reason why many well-intended initiatives turn out to have little effectiveness in strengthening smallholder livelihoods.

We, therefore, developed an analytical framework that could be helpful in overcoming these challenges and may contribute to more responsive and inclusive smallholder-oriented policies and programs. This is mainly based on a systematic analysis of secondary information sources that provide insights into major trends and patterns in smallholder decision-making strategies and outline possible leverage points for pathways for agrifood system transformation. This analysis provides comprehensive insights into the intrinsic driving forces for smallholder behaviour (Section 2) and explains the disappointing impact of many current agricultural development programs for rural poverty alleviation (see Section 3). We, therefore, highlighted (in Section 4) some alternative strategies for more inclusive and sustainable pathways towards agrifood system transformation that more adequately capture the possibilities and needs of smallholder producers.

Such an approach requires that due attention is given to the behavioural constraints that define the reactions of resource-poor households to different kinds of external incentives. In addition, the focus of interventions on farm-level, household-level or value chain-level activities and their adequate timing throughout the lifecycle matters a lot for the effective uptake of transformative livelihood strategies. Finally, some fundamental institutional changes are necessary to enable smallholder farmers to take advantage of the range of rural development incentives.

The analysis and examples presented in this article point to a few underestimated complexities in the dynamics and performance of smallholder livelihoods. First, scarcity in resource endowments is subject to much confusion. Whereas smallholder farmers—by definition—meet major land constraints, on many occasions—along the season and the lifecycle—family labour is the most limiting factor (especially for land preparation and harvesting activities). Promotion of labour-intensive sustainable innovations of agricultural production systems—such as agroforestry or conservation agriculture—therefore frequently fails.

Second, intra-household relationships and decision-making arrangements are hardly addressed. Many studies still rely implicitly on a unitary farm-household approach that overlooks the specific interests of women and children in innovations that render results in terms of nutrition (instead of only production) and in terms of professional careers (instead of just education services). Gender equity is frequently the stepping stone towards more inclusive smallholder development.

Third, linkages of farm-households with external governance networks are of critical importance for creating trust and confidence in durable and sustainable development pathways. Both horizontal (spatial) cooperation and vertical (value chain) linkages can contribute to the engagement and commitment of smallholders with rural development initiatives, and provide longer-term perspectives for rural transformation programs. Structural changes in the governance framework for agricultural development programs should therefore precede any other—more focussed—incentives that are aimed at the reduction of poverty and malnutrition in rural areas.

Finally, two important issues still deserve attention. First, there is considerable variation amongst different types of smallholder farmers, both in terms of farm size and the degree of specialisation, as well as in terms of resource-use intensity, labour productivity and crop yields. This implies that agrarian policies and rural development programs need to be differentiated to consider the demands and opportunities of different types of farmers. Second, for the selection of the appropriate set of policies and incentives, we need identify the key constraints for improving farm-household income, revenues and welfare (the so-called 'most limiting factor'). Policy-making is based on the selection of instruments and it is inefficient just to combine many different incentives without fully understanding their mutual relationships. Making choices is an intrinsic part of policy making!

In summary, the structural transformation of smallholder livelihoods is pursued, priority attention should be given to the reform of governance mechanisms, institutional innovation and reinforcement of ownership structures. This may pave the way to support behavioural responses by smallholder farmers. We need 'clever incentives' to enable smallholders to escape from poverty. Hopefully future discussions are based on careful listening to what drives smallholder resource use decisions and thus may support more fundamental changes in their behaviour.

Funding: This research received no external funding.

Data Availability Statement: The data presented in this study are available on request from the corresponding author.

Conflicts of Interest: The author declares no conflict of interest.

References

1. Cramb, R.; Manivong, V.; Newby, J.C.; Sothorn, K.; Sibat, P.S. Alternatives to land grabbing: Exploring conditions for smallholder inclusion in agricultural commodity chains in Southeast Asia. *J. Peasant Stud.* **2017**, *44*, 939–967. [CrossRef]
2. Neimark, B.; Osterhoudt, S.; Alter, H. A new sustainability model for measuring changes in power and access in global commodity chains: Through a smallholder lens. *Palgrave Commun* **2019**, *5*, 1. [CrossRef]
3. Fischer, D. (Ed.) *Working with Smallholders: A Handbook for Firms Building Sustainable Supply Chains*; International Finance Corporation: Washington, DC, USA, 2013.
4. Binswanger, H.; Rosenzweig, M. Behavioral and Material Determinants of Production Relations in Agriculture. *J. Dev. Stud.* **1986**, *22*, 503–539. [CrossRef]
5. Ruben, R. Impact assessment of commodity standards: Towards inclusive value chains. *Enterp. Dev. Microfinance* **2017**, *28*, 81–97. [CrossRef]
6. Ruben, R.; Slingerland, M.; Nijhoff, H. *Agro-Food Chains and Networks for Development*; Springer Verlag: Berlin, Germany, 2006.
7. Vorley, B.; Fearne, A.; Ray, D. *Regoverning Markets: A Place for Small-Scale Producers in Modern Agrifood Chains?* Routledge: London, UK, 2016.
8. Lowder, S.K.; Skoet, J.; Raney, T. The Number, Size, and Distribution of Farms, Smallholder Farms, and Family Farms Worldwide. *World Dev.* **2016**, *87*, 16–29. [CrossRef]
9. Ricciardi, V.; Ramankutty, N.; Mehrabi, Z.; Jarvis, L.; Chookolingo, B. How much of the world's food do smallholders produce? *Glob. Food Secur.* **2017**, *17*, 64–72. [CrossRef]
10. Shepherd, A.; Brunt, J. (Eds.) *Chronic Poverty: Concepts, Causes and Policy*; Palgrave Macmillan: Basingstoke, UK, 2013; 328p, ISBN 9780230579347.
11. Barrett, C.; Little, P.; Carter, M. Understanding and reducing persistent poverty in Africa. Introduction to a special issue. *J. Dev. Stud.* **2013**, *42*, 167–177. [CrossRef]
12. Bartol, T. Smallholders and small-scale agriculture: Mapping and visualization of knowledge domains and research trends. *Cogent Soc. Sci.* **2023**, *9*, 2161778. [CrossRef]
13. Wortmann, C.; Amede, T.; Bekunda, M.; Ndung'u-Magiroi, K.; Masikati, P.; Snapp, S.; Stewart, Z.P.; Westgate, M.; Zida, Z.; Kome, C.E. Improvement of smallholder farming systems in Africa. *Agron. J.* **2020**, *112*, 5325–5333. [CrossRef]
14. Bymolt, R.; Laven, A.; Tyszler, M. *Demystifying the Cocoa Sector in Ghana and Côte D'ivoire*; Royal Tropical Institute (KIT): Amsterdam, The Netherlands, 2018.
15. Waarts, Y.; Kiewisch, M. *Balancing the Living Income Challenge: Towards a Multi-Actor Approach to Achieving a Living Income for Cocoa Farmers*; Wageningen Economic Research and Mondelēz International: Wageningen/Chicago, The Netherlands, 2021.
16. Taher, S. Factors Influencing Smallholder Cocoa Production: A Management Analysis of Behavioural Decision-Making Processes of Technology Adoption and Application. Ph.D. Thesis, Wageningen University, Wageningen, The Netherlands, 1996.
17. Chelkeba, S.D.; Erko, B.; Fikirie, K. Production and adoption constraints of coffee improved varieties in Jimma zone; Southwest Ethiopia. *J. Sci. Agric.* **2019**, *3*, 33–40.
18. Jain, M.; Barrett, C.B.; Solomon, D.; Ghezzi-Kopel, K. Surveying the Evidence on Sustainable Intensification Strategies for Smallholder Agricultural Systems. *Annu. Rev. Environ. Resour.* **2023**, *48*, 6.1–6.23. [CrossRef]
19. Kangogo, D.; Dentoni, D.; Bijman, J. Adoption of climate-smart agriculture among smallholder farmers: Does farmer entrepreneurship matter? *Land Use Policy* **2021**, *109*, 105666. [CrossRef]
20. Acevedo, M.; Pixley, K.; Zinyengere, N.; Meng, S.; Tufan, H.; Cichy, K.; Bizikova, L.; Issacs, K.; Ghezzi-Kopel, K.; Porciello, J. A scoping review of adoption of climate resilient crops by small-scale producers in low-and middle-income countries. *Nat. Plants.* **2020**, *6*, 1231–1241. [CrossRef] [PubMed]

21. Larson, D.F.; Muraoka, R.; Otsuka, K. Rural Development Strategies and Africa's Small Farms. In *The Role of Smallholder Farms in Food and Nutrition Security*; Gomez y Paloma, S., Riesgo, L., Louhichi, K., Eds.; Springer: Cham, Switzerland, 2020. [CrossRef]
22. Abraham, M.; Pingali, P. Transforming smallholder agriculture to achieve the SDGs. In *The Role of Small Farms in Food and Nutrition Security*; Riesgo, L., Gomez-Y-Paloma, S., Louhichi, K., Eds.; Springer: New York, NY, USA, 2020.
23. Terlau, W.; Hirsch, D.; Blanke, M. Smallholder farmers as a backbone for the implementation of the Sustainable Development Goals. *Sustain. Dev.* **2018**, *27*, 523–529. [CrossRef]
24. Carletto, G.; Ruel, M.; Winters, P.; Zezza, A. Farm-level Pathways to Improved Nutritional Status: Introduction to the Special Issue. *J. Dev. Stud.* **2015**, *51*, 945–957. [CrossRef]
25. Molitor, K.; Braun, B.; Pritchard, B. The effects of food price changes on smallholder production and consumption decision-making: Evidence from Bangladesh. *Geogr. Res.* **2017**, *55*, 206–216. [CrossRef]
26. Rahut, D.B.; Aryal, J.P.; Marenya, P. Understanding climate-risk coping strategies among farm households: Evidence from five countries in Eastern and Southern Africa. *Sci. Total Environ.* **2021**, *769*, 145236. [CrossRef] [PubMed]
27. Sibhatu, K.T.; Krishna, V.V.; Qaim, M. Production diversity and dietary diversity in smallholder farm households. *Proc. Natl. Acad. Sci. USA* **2015**, *112*, 10657–10662. [CrossRef]
28. Reardon, T. The hidden middle: The quiet revolution in the midstream of agrifood value chains in developing countries. *Oxf. Rev. Econ. Policy* **2015**, *31*, 45–63. [CrossRef]
29. Ruben, R.; Kuijpers, R.; Dijkxhoorn, Y. Mobilizing the Midstream for Supporting Smallholder Intensification. *Land* **2022**, *11*, 2319. [CrossRef]
30. de Janvry, A.; Fafchamps, M.; Sadoulet, E. Peasant Household Behaviour with Missing Markets: Some Paradoxes Explained. *Econ. J.* **1991**, *101*, 1400–1417. [CrossRef]
31. Mumba, M.; Edriss, A.-K. Determinants and Change in Total Factor Productivity of Smallholder Maize Production in Southern Zambia. *J. Sustain. Dev.* **2018**, *11*, 170. [CrossRef]
32. Alwang, J.; Siegel, P.B. Labor Shortages on Small Landholdings in Malawi: Implications for Policy Reforms. Labor Shortages on Small Landholdings in Malawi: Implications for Policy Reforms. *World Dev.* **1999**, *27*, 1461–1475. [CrossRef]
33. Dorward, A. Farm size and productivity in Malawian smallholder agriculture. *J. Dev. Stud.* **1999**, *35*, 141–161. [CrossRef]
34. Alemayehu, M.; Beuving, J.; Ruben, R. Risk Preferences and Farmers' Livelihood Strategies: A Case Study from Eastern Ethiopia. *J. Int. Dev.* **2018**, *30*, 1369–1391. [CrossRef]
35. de Vries, J.R.; Turner, J.A.; Finlay-Smith, S.; Ryan, A.; Klerkx, J. Trust in agri-food value chains: A systematic review. *Int. Food Agribus. Manag. Rev.* **2022**, *26*, 175–197. [CrossRef]
36. De Brauw, A.; Bulte, E. *African Farmers, Value Chains and Agricultural Development: An Economic and Institutional Perspective*; Palgrave Studies in Agricultural Economics and Food Policy; Springer International Publishing: Cham, Switzerland, 2021.
37. Ravallion, M. *The Economics of Poverty: History, Measurement and Policy*; Oxford University Press: New York, NY, USA, 2016.
38. Konja, D.T.; Mabe, F.N.; Alhassan, H. Technical and resource-use-efficiency among smallholder rice farmers in Northern Ghana. *Cogent Food Agric.* **2019**, *5*, 1. [CrossRef]
39. Mezgebo, G.K.; Mekonen, D.G.; Gebrezgiabher, K.T. Do smallholder farmers ensure resource use efficiency in developing countries? Technical efficiency of sesame production in Western Tigrai, Ethiopia. *Heliyon* **2021**, *7*, e07315. [CrossRef]
40. True Price. *A Roadmap for True Pricing—Vision Paper*; True Price Foundation: Amsterdam, The Netherlands, 2019.
41. de Janvry, A.; McIntosh, C.; Sadoulet, E. Fair trade and free entry: Can a disequilibrium market serve as a development tool? *Rev. Econ. Stat.* **2015**, *97*, 567–573. [CrossRef]
42. Benos, T.; Sergaki, P.; Kalogeras, N.; Tzinalas, D. Coping With Side-Selling in Cooperatives: A Members' Perspective. *Ann. Public Coop. Econ.* **2022**, *95*, 177–199. [CrossRef]
43. Keenan, M.; Fort, R.; Vargas, R. *Liquidity, Scale Economies, and Shocks: Theoretical and Empirical Determinants of Side-Selling in Peruvian Specialty Coffee Cooperatives*; Wageningen University: Mimeo, The Netherlands, 2021.
44. Tadesse, M.A.; Shiferaw, B.A.; Erenstein, O. Weather Index Insurance for Managing Drought Risk in Smallholder Agriculture: Lessons and Policy Implications for Sub-Saharan Africa. *Agric. Food Econ.* **2015**, *3*, 26. [CrossRef]
45. Sedai, A.K.; Vasudevan, R.; Pena, A.A. Friends and benefits? Endogenous rotating savings and credit associations as alternative for women's empowerment in India. *World Dev.* **2021**, *145*, 105515. [CrossRef]
46. Cordes, K.; Sagan, M.; Kennedy, S. Responsible Coffee Sourcing: Towards a Living Income for Producers. 2021. Available online: https://ssrn.com/abstract=3894124 (accessed on 10 August 2021).
47. Sellare, J. New insights on the use of the Fairtrade social premium and its implications for child education. *J. Rural Stud.* **2022**, *94*, 418–428. [CrossRef]
48. Loconto, A.M.; Arnold, N.; Silva-Castañeda, L.; Jimenez, A. Responsibilising the Fairtrade Premium: Imagining better decision-making. *J. Rural Stud.* **2021**, *86*, 711–723. [CrossRef]
49. Molyneux, M.; Jones, N.; Samuels, F. Can Cash Transfer Programmes Have 'Transformative' Effects? *J. Dev. Stud.* **2016**, *52*, 1087–1098. [CrossRef]
50. Cooke, M.; Mukhopadhyay, P.; Stein, D. Cash Crop: Evaluating Large Cash Transfers to Coffee Farming Communities in Uganda; GiveDirectly & IDinsight for Benckiser Stiftung Zukunft (BSZ): 2019. Available online: https://www.givedirectly.org/wp-content/uploads/2019/06/Cash_Crop_Ugandan_CoffeeRCT.pdf (accessed on 10 May 2023).

51. Gitter, S.R.; Monley, J.; Barham, B. *The Coffee Crisis, Early Childhood Development, and Conditional Cash Transfers. IDB Working Paper # 245*; Inter-American Development Bank: Washington, DC, USA, 2011.

52. Lawlor, K.; Handa, S.; Seidenfeld, D.; Zambia Cash Transfer Evaluation Team. Cash transfers enable households to cope with agricultural production and price shocks: Evidence from Zambia. *J. Dev. Stud.* **2019**, *55*, 209–226. [CrossRef] [PubMed]

53. Adolph, B.; Allen, M.; Beyuo, E.; Banuoku, D.; Barrett, S.; Bourgou, T.; Bwanausi, N.; Dakyaga, F.; Derbile, E.K.; Gubbels, P.; et al. Supporting smallholders' decision making: Managing trade-offs and synergies for sustainable agricultural intensification. *Int. J. Agric. Sustain.* **2021**, *19*, 456–473. [CrossRef]

54. Place, F.; Otsuka, K. Land Tenure Systems and Their Impacts on Agricultural Investments and Productivity in Uganda. *J. Dev. Stud.* **2002**, *38*, 105–128. [CrossRef]

55. Tulone, A.; Galati, A.; Pecoraro, S.; Carroccio, A.; Siggia, D.; Virzì, M.; Crescimanno, M. Main intrinsic factors driving land grabbing in the African countries' agro-food industry. *Land Use Policy* **2022**, *120*, 106225. [CrossRef]

56. de Roest, K.; Ferrari, P.; Knickel, K. Specialisation and economies of scale or diversification and economies of scope? Assessing different agricultural development pathways. *J. Rural Stud.* **2018**, *59*, 222–231. [CrossRef]

57. Anderson, C.; Leigh, T.W.; Reynolds, P.B.; Vedavati, P.; Carly, S. Economic benefits of empowering women in agriculture: Assumptions and evidence. *J. Dev. Stud.* **2021**, *57*, 193–208. [CrossRef]

58. Quisumbing, A.R.; Ahmed, A.; Hoddinott, J.F.; Pereira, A.; Roy, S. *Designing for Empowerment Impact in Agricultural Development Projects: Experimental Evidence from the Agriculture, Nutrition, and Gender Linkages (Angel) Project in Bangladesh*; IFPRI DP # 1957; International Food Policy Research Institute: Washington, DC, USA, 2020. [CrossRef]

59. Alemu, S.H.; Van Kempen, L.; Ruben, R. Women Empowerment Through Self-Help Groups: The Bittersweet Fruits of Collective Apple Cultivation in Highland Ethiopia. *J. Hum. Dev. Capab.* **2018**, *19*, 308–330. [CrossRef]

60. Ostrom, E. *Governing the Commons: The Evolution of Institutions for Collective Action*; Cambridge University Press: Cambridge, UK, 1990.

sustainability

MDPI

Article

Innovative Intelligent Cheese Packaging with Whey Protein-Based Edible Films Containing Spirulina

Vasiliki G. Kontogianni [1,*], Ioanna Kosma [2], Marios Mataragas [1], Eleni Pappa [1], Anastasia V. Badeka [2] and Loulouda Bosnea [1]

[1] Dairy Research Department, Hellenic Agricultural Organization-DEMETER, 45221 Ioannina, Greece; mmatster@gmail.com (M.M.); pappa.eleni@yahoo.gr (E.P.); louloudabosnea@gmail.com (L.B.)
[2] Laboratory of Food Chemistry, Department of Chemistry, University of Ioannina, 45110 Ioannina, Greece; i.kosma@uoi.gr (I.K.); abadeka@uoi.gr (A.V.B.)
* Correspondence: vgkontog@gmail.com

Abstract: The use of edible and biodegradable films and coatings as active packaging for cheese has recently attracted great attention as it meets the concept of sustainability and ensures safety. *Spirulina* is a rich source of high-added-value biocompounds, which could be used as functional ingredients. In the present study, *spirulina* was added in different concentrations (0.5; 1; 2; 4% *w/w*) to the edible films produced from whey protein concentrate-based solutions. The films were characterized according to their optical parameters (color); they were studied for their total phenolic content, and the viability of the films in simulated gastric juice was investigated. The possible use of the developed films for intelligent food packaging, as colorimetric pH indicators, was also investigated. Finally, a preliminary evaluation of selected films containing *spirulina* (WPC-based films containing 2% *spirulina*) as packaging for "kefalotyri" cheese was also assessed. The effect of these films, applied as packaging for "kefalotyri" cheese during two months of refrigerated storage, was evaluated. GC-MS analysis was used to evaluate the effect of the *spirulina* odor of the film with *spirulina* incorporated and the cheese products where the film was applied.

Keywords: whey protein concentrate; edible film; *spirulina*; cheese packaging; functional food

Citation: Kontogianni, V.G.; Kosma, I.; Mataragas, M.; Pappa, E.; Badeka, A.V.; Bosnea, L. Innovative Intelligent Cheese Packaging with Whey Protein-Based Edible Films Containing Spirulina. *Sustainability* **2023**, *15*, 13909. https://doi.org/10.3390/su151813909

Academic Editor: Filippo Giarratana

Received: 7 August 2023
Revised: 7 September 2023
Accepted: 14 September 2023
Published: 19 September 2023

1. Introduction

Since cheeses are considered biologically and biochemically to be in the unstable food category, there has been an increase in the demand for innovative proper packaging systems to promote their safety [1]. The need to design bio-based food materials has increased, and the use of biodegradable biopolymers for food packaging has attracted great attention given the lack of sustainability of conventional food packaging [2]. Active packaging such as edible and biodegradable films is suggested in order to avoid undesired microbial growth and lipid oxidation on cheese [3]. Edible films and coatings can enhance the organoleptic and nutritional properties of cheese, either due to their film/coating composition, which may have beneficial properties by itself, or due to their capacity to incorporate active ingredients, which are eaten with the coating [4]. Whey protein (WP), a nutrient-rich byproduct of the cheese industry, has been applied to edible film formulations, which could be consumed with the cheese, generate no waste, lengthen its shelf life, and increase its quality [5]. Active packaging could provide the potential for bioactive compounds to be introduced both into packaged foods [6,7] and into the consumer's gastrointestinal system, at a controlled rate [8]. Another concept, called intelligent or smart packaging materials, is also trending in the same area. Such packaging can detect, monitor, and transmit information about changes occurring inside the food to the outside world [9]. Therefore, consumers can visualize changes in the state of the material and avoid wasting food and resources to a certain extent. This kind of food packaging material is more in line with the requirements of sustainable development.

WP edible films with antimicrobial and antioxidant agents with a natural origin incorporated can have a functional effect on the surface of cheese, as detailed below. Pluta-Kubica et al. [10] developed films produced from furcellaran–whey protein isolate solutions with the addition of yerba mate and white tea extract, which were successfully applied in order to improve the shelf life of soft rennet-curd cheese. Recently, Robalo et al. [11] prepared films produced from WP-based solutions enriched with green tea extracts and implemented them as a packaging material for Latin-style fresh cheeses, which efficiently protected them. In one of our previous works [12], the application of whey protein concentrate (WPC)-based films enriched with rosemary and sage infusions as a packaging material for soft cheese was able to protect the soft cheese from spoilage or pathogenic bacteria.

Spirulina (sp.) is abundant in important human health constituents, such as proteins, vitamins, amino acids, and minerals, and is a natural polyphenol source with many biological functions [13]. The safety and effectiveness of sp. biocompounds in the treatment of many human diseases has led scientists to apply sp. or its bioactive compounds to foods [14,15]. However, its inclusion, owing to the peculiar flavor and aroma associated with sp., has a negative impact on the final product's sensory attributes. A recent review covered studies on the use of sp. biomass and its blue extract for the enrichment of edible films and coatings (gelatin-based, chitosan-based, etc.) and their application to fruits, meat, fish, etc. [16]. Our research group used WPC, a troublesome byproduct of the cheese industry, to develop edible films with the addition of a commercial sp. powder in different concentrations, applying different treatments [17].As far as we know, this research constitutes the first approach toward the production of WPC-based films enriched with sp. powder.

Therefore, the present study expands this research, in which sp. was added in different concentrations (0.5; 1; 2; 4% w/w) to the edible films produced from WPC-based solutions, and the physicochemical, mechanical, and antioxidant properties, along with the optical parameters of the prepared films were evaluated [17].In the current investigation, the above films were characterized according to their optical parameters (color), and they were studied for their total phenolic content. The possible use of the developed films as an intelligent pH change indicator was also investigated along with the viability of the films in simulated gastric juice. The primary goal of this research was to assess the possible application of the WPC films with the best properties as a packaging material for "kefalotyri" cheese. WPC-based films containing 2% sp. were applied as packaging to "kefalotyri" cheese stored in refrigerated conditions to determine their influence on the cheese's microbiological stability. GC-MS analysis was used to evaluate the effect of sp. odor in the film with sp. incorporated And the cheese products where the film was applied.

2. Materials and Methods

2.1. Chemicals and Standards

All chemical reagents and solvents used in the current study and the ingredients used for the production of the film systems were the same as previously described [17].

Ringer's solution tablets, potato dextrose agar, and plate count agar were purchased from Neogen Culture Media (Heywood, UK).

2.2. Preparation of Film Systems

The films were prepared according to Kontogianni et al. [17]. The film-forming solution was prepared by dissolving 10 g WPC and 5 g glycerol in 100 mL water. The pH of each solution was adjusted to 5.95 using 3% w/v CH_3COOH. Then, the sp. powder at different concentrations (0.5; 1; 2; 4; 6; 8% w/w) and 1 drop of Tween 80 were added. The obtained solutions were stirred magnetically and homogenized with an Ultra Turrax homogenizer. The dispersion was placed in a common ultrasonic bath for 15 min at 70 °C, homogenized with the Ultra Turrax, and degassed by sonication for 30 min. After cooling the solutions at room temperature, their pH was adjusted to 5.5, and they were filtrated with a cheese cloth. Finally, the films were dried and stabilized.

2.3. Color Measurements

The Hunter color (L*, lightness; a*, redness; b*, yellowness) values were measured using a HunterLab, Model D25 L optical sensor (Hunter Associates, Reston, VA, USA). Each sample was analyzed using three independent repetitions with four to six measurements taken on each repetition (n = 3 × 4). The results reported are the mean of the above determinations.

2.4. Total Phenolic Content

For the preparation of the film solutions, we used the procedure followed for the antioxidant properties' evaluation as previously described [17]. The total phenolic content (TPC) was determined according to the method described by Jamróz et al. [18], expressed as mg of gallic acid/g of film. The analysis of each sample consisted of two independent measurements.

2.5. Viability of Films in Simulated Gastric Juice and Intelligent Material Analysis

For simulated gastric juice (SGJ), we dissolved pepsin in a 0.5% NaCl (w/v) solution to a concentration of 3 g/L. The pH was adjusted to 2.0 with a 1.0 M HCl solution. The gastric juice was freshly made for use on the same day. Next, 1 g of Control films and the WPC-based edible films incorporating six different concentrations of sp. powder were suspended in 9.0 mL of SGJ solution, placed in glass Petri dishes, and incubated at 37 °C for 3 h under orbital shaking in a formal orbital shaker (Thermo Electron Corporation, Waltham, MA, USA). Optical observation was made at regular time intervals, every 5 min for the first hour of observation and every 30 min for the 2 h left until the films were fully disintegrated.

The trial of the intelligent material assay was according to Jamróz et al. [18].

2.6. Application as Packaging for "kefalotyri" Cheese Pieces

The WPC-based films containing 2% sp. powder were selected as the best concentration for the evaluation of their potential as a packaging material for "kefalotyri" cheese portions. "Kefalotyri" is a traditional hard Greek cheese, manufactured from a mixture of sheep and goat milk. It has a flat cylindrical shape, firm texture, salty taste, and strong flavor. The cheese composition according to the label was: maximum moisture 38.0%; minimum dry fat 40%.; protein 26%; and salt 2.7% (the "kefalotyri" cheese was a commercial product purchased from local supermarket).Two cheese pieces in a trapezium shape (~5 cm long, 2.5 and 3.5 cm wide, 0.5 cm thickness) were layered with one film (Ø 90 mm), placed in a glass Petri dish, and vacuum packaged and sealed in sterile LDPE granule extruded bags (18 × 30 cm size). Furthermore, Control films were used to cover the cheese pieces in the same way for comparison purposes. The packages were stored at 4 °C for 7 days. Then, they were left at room temperature (23 °C) for 2 h before opening and evaluation. The sensory evaluation was performed by three analysts individually. The experiments were repeated three times. Each judge evaluated three packages with edible films as packaging for "kefalotyri" cheese pieces. The observations about the wholeness of the films after the storage and their removal from the cheese, about changes of color or odor in the cheese pieces after storage, and the smell and appearance of the film were recorded by questionnaires based on the respective analysis performed by [19]. Actually, they recorded the film wholeness after opening the package of the LDPE bag, expressed as the percentage of broken films to the total number of films, and easiness of the separation of the film from the cheese, expressed as the percentage of each evaluation related to the total evaluated surfaces. The samples used for microbiological analyses were prepared as follows. Pieces of cheese,12–14 g, in a trapezium shape were placed in sterile Petri dishes and distributed into three groups. The first group was kept without a surface film (Control, containing only "kefalotyri" cheese); the second group was layered with the edible film without sp., while the third group was layered with edible films containing 2% (w/w) sp., respectively. Films before use were sterilized as previously described [12]. The above samples were kept for two months of refrigerated storage (stored at 4 °C), and proper cheese samples were collected for analysis at 0, 30, and 60 d of storage.

2.7. Microbiological Analyses

For the microbiological analyses of the cheese samples, we used the procedure as previously described [12].Viable counts for total mesophilic counts (TVCs), molds, and yeasts were performed in duplicate. More specifically, total mesophilic counts were enumerated on plate count agar (PCA) (30 °C for 72 h), and yeasts and molds were enumerated on potato dextrose agar (PDA) (30 °C for 72 h). All counts were recorded as logcfu/g.

2.8. Analysis of Volatile Compounds: Identification and Percentage Ratio of Volatile Compounds Using SPME–GC/MS

For the volatile analysis, the method described by Vatavali et al. [20] was used with minor modifications. Briefly, three grams of sample ("kefalotyri", film, or sp. powder) were used for analysis. SPME was performed with the fiber as previously described [20]. The samples were placed in an 80 °C water bath and stirred at 800 rpm. There were 30 min used for the sample to equilibrate and 15 min for the exposure of the fiber. The column was initially maintained at 40 °C for 2 min, heated to 170 °C at a rate of 5 °Cmin^{-1}, heated to 260 °C at a rate of 10 °Cmin^{-1}, and held for 2 min. For the volatile compounds of sp., the method according to Ozogul et al. [21] was used. Peak identification was performed by the comparison of the mass spectra of the eluting compounds to those of the Wiley library [22]. The retention indices (RIs) of the volatile compounds were calculated using-n-alkane (C8–C20) standard solution (Fluka, Buchs, Switzerland).

All determinations were carried out in triplicate.

2.9. NIR Spectroscopy: Spectra Acquisition

The homogenized "kefalotyri" sample and the "kefalotyri" cheese sample layered with film containing 2% sp. after storage at 4 °C for 7 days were subjected to NIR spectroscopy using a FoodScan 2 spectrophotometer (FOSS Analytical, Hillerod, Denmark).

2.10. Statistical Analysis

ANOVA was applied to the results using SPSS software [23]. Tukey's test was used to assess differences between means, and differences were considered significant at the level of $p < 0.05$.

3. Results

3.1. Color Measurements

One of the most-important factors of edible films developed for food products is color. It can influence both the product appearance and consumer acceptance. The color properties of WPC-based films with sp. incorporated are shown in Table 1. As was expected, the L* value, which gives the lightness, was higher in Control films, and its value decreased as the sp. powder concentrations increased from 0.5 to 4% ($p < 0.05$). The reduced L* value with increased sp. concentrations indicated that the films with sp. became darker. The color value between red and green is expressed by the a* value; the highest was found in the film with 1% sp. (-21.57 ± 1.04); the lowest was found in film with 4% sp. (0.90 ± 0.03), recording statistically significant differences. The color value between yellow and blue, which is expressed by the b* value, was found to be the highest in the film with 0.5% sp. (24.86 ± 0.45) and the lowest in the film with 4% sp. (-0.27 ± 0.05), also statistically significant ($p < 0.05$). Yellowness, which is expressed by the b* value, increased with the addition of 0.5% sp. and, then, with increasing sp. concentrations, decreased. In accordance with the findings of Balti et al. [24], the addition of sp. had a remarkable impact on the color of the resulting WPC films and could be attributed to the presence of phenolic compounds and colored substances.

Table 1. Color of WPC films with the addition of sp. and total phenolic content of the films.

Film	L *	A *	B *	TPC (mg Gallic/g Film)
Control	75.17 ± 0.44 [e,*]	−3.36 ± 0.39 [d]	15.91 ± 0.43 [c]	62.00 ± 4.35 [b]
0.5% sp.	51.88 ± 0.29 [d]	−18.80 ± 1.44 [b]	24.86 ± 0.45 [e]	41.00 ± 2.22 [a]
1% sp.	29.99 ± 0.23 [c]	−21.57 ± 1.04 [a]	21.73 ± 0.47 [d]	86.75 ± 5.11 [d]
2% sp.	14.75 ± 0.29 [b]	−8.95 ± 0.66 [c]	6.22 ± 0.44 [b]	100.75 ± 6.21 [e]
4% sp.	7.08 ± 0.13 [a]	−0.90 ± 0.03 [e]	−0.27 ± 0.05 [a]	74.75 ± 3.52 [c]

* Values with different letters in the same column indicate statistically significant differences; $p < 0.05$.

3.2. Total Phenolic Content

Film containing 0.5% sp. showed the lowest total phenolic content (TPC) (41.00 ± 2.22 mg gallic acid/g), even lower than the Control film (62.00 ± 4.35 mg gallic acid/g). In general, the TPC of films with sp. incorporated increased with an increasing amount of sp. powder until the concentration of 2% sp.($p < 0.05$).The DPPH assay results were in accordance with these results [17], where the edible film containing 2% sp. powder showed the highest percentage of radical scavenging activity(higher than the films containing 4, 6, and 8% sp.).

Miranda et al. [25] reported that the major phenolic compounds identified in sp. were salicylic, trans-cinnamic, synaptic, chlorogenic, quinic, and caffeic acids. Our results are in agreement with those of Balti et al. [24], who found that the TPC in crab chitosan films made with sp. extract increased with increasing sp. extract concentration. However, they reported that the TPC in crab chitosan films made with sp. ranged from 5.62 to 21.85 mg GAE/g film, quite lower values compared to our films.

3.3. Simulated Gastric Juice Test and Intelligent Material Analysis

The incubation of WPC-based films containing 0.5; 1; 2; and 4% sp. powder and the incubation of the Control films in SGJ imitating the low-pH and highly digestive conditions in the stomach resulted in the digestion of the films (Figures 1 and 2). The above films can be hydrolyzed by pepsin in the simulated gastric juice. According to optical observations, the WPC-based films containing 0.5; 1; 2; and 4% sp. powder and the Control films disintegrated within 1 h (60 min), and only small fragments of the films were still detectable. Especially, the WPC-based films containing 4% sp. disintegrated faster than the others and almost disappeared entirely within 60 min. On the contrary, the WPC-based films containing 6 and 8% sp. remained intact throughout the treatment with simulated gastric juice, even after 2 h (Figure 2).

The food transit time in the stomach has been found to be between 2 and 4 h, while the emptying time of liquid foods from the stomach is about 20 min. Therefore, WPC-based films containing 0.5; 1; 2; and 4% sp. and the Control films could be used as a primary packaging (the packaging that is in direct contact with the food), which is contained in a secondary or tertiary packaging and can probably be consumed safely with the food as a food additive and work as a functional food. According to Nourmohammadi et al. [26], who encapsulated sp. using alginate and whey protein concentrate (WPC) by the emulsification method and investigated its addition to non-fat stirred yogurt during storage, the total release of the sp. from the microcapsules was observed in simulated intestinal fluid. Sp. as a prebiotic source enhances the population of *Lactobacillus* in the human intestine by producing growth factors such as extracellular carbohydrates. The *Lactobacillus* population in the human intestinal tract is enhanced using sp. as a source of prebiotic substances by liberating growth substances such as extracellular carbohydrates [27].

Intelligent packaging is a crucial aspect of food technology that can help to improve food quality control. Considering that the change in the acidity of foods is common when they begin to deteriorate, this simplistic, optical measurement can be utilized to detect pH changes in food products. The color response of the pH indicator of the WPC films with sp. and the Control films was tested by immersing them in different pH solutions (pH 3.0 and pH 12.0). The color of the WPC films with the addition of sp. in all different concentrations did not change after being placed either solution.

Recently, a research group developed films with C-phycocyanin as a pH indicator for food quality control and confirmed that, at an alkaline pH above 8, the films lost their blue and green tones, giving a color inclining toward white [28]. Such a phenomenon was attributed to some natural pigments present in sp. that are susceptible to pH changes, such as β-carotene, tocopherols, phycoerythrin, and chlorophylls. Furthermore, according to Kuntzler et al. [29], who incorporated sp. biomass into nanofibers, color changes occurred between pH 5 and 7, a pH value range that includes an extensive diversity of fresh foods.

3.4. Evaluation of Films as Packaging Material of "kefalotyri" Cheese Pieces

In order to select a film containing sp. in a concentration that showed the best characteristics as a packaging material for "kefalotyri", we compared all the parameters evaluated in our previous work [17] and in the present study. The ultrasound-treated WPC films with 2% sp. had the best uniform appearance and homogeneous texture. The incorporation of sp. had a beneficial impact on the mechanical properties of the edible films; the film with 2% sp. presented the highest TS value (TS for Control: 0.69 ± 0.07; for 2% sp.: 2.55 ± 0.36; for 4% sp.: 1.43 ± 0.27), exhibiting a strong structure and being more flexible than the film with 4% (presenting a higher E% value, E% for Control: 51.09 ± 5.69; for 2% sp.: 12.10 ± 1.40; for 4% sp.: 9.44 ± 1.38). It also had the highest percentage of radical scavenging activity. Based on the results of the current study, the film with 2% sp. showed good color properties and the highest total phenolic content and could be hydrolyzed by pepsin in the simulated gastric juice, which disintegrated within 1 h. Therefore, films with 2% sp. were selected to be applied as packaging for "kefalotyri" cheese surfaces.

Figure 1. Photographs of WPC-based films (control, 0.5% sp., and 1% sp.) containing different concentrations of sp. immersed in SGJ for different times.

Figure 2. PPhotographs of WPC-based films (2% sp., 4% sp, and 6% sp.) containing different concentrations of sp. immersed in SGJ for different times.

In the evaluation of the films' wholeness after storage with cheese pieces at 4 °C for 7 days in LDPE bags (after opening the bag), the Control films were stuck to the bag when the package was opened, and small pieces of the film stayed on the surface of the cheese (Figures 3a and S1). The color of the Control film was almost the same as that of the cheese, so it is necessary to focus on Figure S1a in order to detect the broken parts of the Control film material that stayed stuck to the bag. In the evaluation of the film wholeness after opening the bag, 98.71% of the Control films broke during opening the package and stayed stuck to the bag. On the contrary, all the WPC-based films containing 2% sp. stayed intact on the surface of the cheese after 7 days of storage at 4 °C (Figure 3b). Regarding the ease of the film separation from the cheese both for the Control films and the films with sp. added, it was not possible to separate them from the cheese surface. The packaging material did not affect the smell of the cheese, neither for the Control film nor the film with sp. The evaluators did not find any changes to the odor of the cheese caused by the packaging material laid on the surface of the cheese. As they indicated, the odor of the Control film and the film with sp. added was the same and did not resemble the unwanted aroma of sp. The evaluators reported a slight change in the tonality on the surface of the cheese after the removal of the WPC-based films containing 2% sp. It turned darker, but it did not stain the cheese surface green (Figure 3c; the same piece after removing the film containing 2% sp. (on the right side) with a freshly cut piece of cheese (on the left side) for comparison).

Figure 3. (**a**) "Kefalotyri" cheese layered with Control film after storage at 4 °C for 7 days; (**b**) cheese layered with WPC-based films containing 2% sp. after storage at 4 °C for 7 days; (**c**) the same piece after removing the film containing 2% sp. (on the right side) with a freshly cut piece of cheese (on the left side) for comparison.

The above packaging material (film with sp.) was unable to be separated from the cheese surface after a storage time of 7 days at 4 °C intact, so probably, we should consider and evaluate the possibility of the film being consumed with the cheese and constituting a functional food consumed together with the cheese. WP is an animal protein with great nutritional value that closely matches the human body's requirements for all eight essential amino acids; thus, WP-based films have nutritional properties themselves [4]. WP- based films are non-toxic, biodegradable, safe to use, tasteless, odorless, and biocompatible with cheese products [30]. Since all the film ingredients are safe for consumption, this does not threaten the health of the final consumer, but the consumer could benefit from the health-promoting effects of sp. and its great nutritional value. *Spirulina* has received Generally Recognized As Safe(GRAS) certification from the Food and Drug Administration (FDA), allowing it to be used as a food or food supplement [31]. Both the WPC and sp. powder used in this study were commercial products sold as dietary supplements, accompanied by toxicological tests and all the necessary tests by national organizations. This packaging material possesses trustworthy characteristics. It is non-toxic, has minimal manufacturing expenses, and is completely compatible with the packaged product. Utilizing a film with sp. incorporated as a packaging material for a food that could be consumed with it could prove an effective method to suppress the unwanted aroma and even taste of the core material. It could be used probably to cover the undesirable taste of sp. and enhance the customer's satisfaction. Further analysis of the odor of the cheese after the removal of the film containing 2% sp. and the odor of the film on its own using instrumental analysis would prove this speculation. Furthermore, the organoleptic evaluation of the cheese after the removal of the film with sp. and the combination of the films and the cheese is recommended.

Cruz-Diaz et al. [19] evaluated films from whey proteins as a separation material for cheese slices; the films exhibited similar results regarding slice separation ease and slice wholeness after separation with those of the commercial material, without modifying the cheese color and odor. Pluta-Kubica et al. [10] assessed the organoleptic quality of soft rennet-curd cheese wrapped with films produced from furcellaran–whey protein isolate solutions with the addition of yerba mate and white tea extract. They concluded that the application of the films as packaging positively influenced the consistency and the overall quality of the cheese samples, although it had a negative impact on the appearance of the film, the appearance after removing the film, and the smell of the cheese samples.

3.5. Microbiological Changes of "kefalotyri" Cheese with Edible Films Applied

The results of the total mesophilic counts, molds, and yeasts are shown in Table 2. Until the 30th day of storage, no spoilage bacteria were found in any of our cheese samples. In the Control samples (cheeses without film), 5.3 logcfu/g of yeasts were counted on the 30th day of storage and persisted to rise until they amount to 7.6 logcfu/g on Day 60. Yeast spoilage was also obvious from Day 30 on those samples as colorful stains were observed on the surface of the cheeses without films. No spoilage bacteria were evident until the end of the storage days in cheese samples layered with films containing 2% sp. (Table 2).

Table 2. Total TVC and yeast/mold levels of "kefalotyri" cheese during two months of refrigerated storage.

Time (days) of Storage	Kefalotyri log cfu/g		Kefalotyri + Film Control log cfu/g		Kefalotyri + Film *spirulina* log cfu/g	
	TVC	Yeasts/Molds	TVC	Yeasts/Molds	TCV	Yeasts/Molds
0	7.2 ± 0.0 [a,*]	3.2 ± 1.1 [A,**]	7.2 ± 0.1 [b]	3.1 ± 0.0 [A]	7.4 ± 0.1 [b]	2.6 ± 0.5 [A]
30	8.0 ± 0.1 [b]	5.3 ± 1.6 [A]	7.2 ± 0.1 [a]	3.6 ± 1.9 [A]	7.8 ± 0.3 [b]	3.2 ± 1.4 [A]
60	9.2 ± 1.0 [b]	7.6 ± 1.3 [B]	8.2 ± 0.4 [b]	5.6 ± 1.0 [B]	7.7 ± 0.1 [a]	3.6 ± 0.2 [A]

* Values with different lowercase letters in the same row indicate statistically significant differences, $p < 0.05$; ** values with different capital letters in the same row indicate statistically significant differences, $p < 0.05$.

Balti et al. [24] showed that chitosan edible films with sp. extract incorporated exerted good antibacterial activity, exhibiting better results against Gram-positive bacteria (*L. monocytogenes*) than Gram-negative bacteria (*E. coli, P. aeruginosa,* and *S. typhimurium*). The chlorogenic acid contained in sp. extract was the bioactive substance to which the antimicrobial properties of the above films could be attributed. Closely related to this study, hake fillets packaged in gelatin films containing *Spirulina platensis* protein concentrate [32] showed enhanced storage stability. According to the results, the aerobe mesophiles, psychrotrophs, proteolytics, lipolytics, and *Enterobacteriaceae* counts were significantly decreased. Phycocyanin exhibited a significant in vitro antibacterial activity against food-borne pathogens such as *S. aureus, M. luteus, E. coli,* and *Pseudomonas* spp. [33].

Recently, Stejskal et al. [34] observed a remarkable antimicrobial effect of a gelatine-based packaging film with a protein concentrate from sp. applied on the quality of refrigerated Atlantic mackerel. According to Martelli et al. [35], the antimicrobial activity of the crude sp. extract was attributed to the synergistic effect between the wide diversity of bioactive phenolic compounds and other compounds found in sp., such as carotenoids, C-phycocyanin, and chlorophyll (a and b).

3.6. Analysis of Volatile Compounds of Films, sp. Powde r, etc.

The volatile compounds of the films are shown in Table 3. A total of 43 compounds were identified in the WPC-based film containing 2% sp. and 30 in the Control film. There were 28 common compounds among these films. The same major compounds were identified in the films, namely glycerin (44.21% for Control film and 33.25% for 2% sp.); acetic acid (18.10% for Control film and 11.55% for 2% sp.); nonanoic acid (8.92% for Control film and 11.51% for 2% sp.); octanoic acid (5.83% for Control film and 6.35% for 2% sp.); nonanal (3.84% for Control film and 2.96% for 2% sp.); hexanal (3.29% for Control film and 3.31% for 2% sp.); decanoic acid (3.01% for Control film and 2.12% for 2% sp.); heptadecane (2.51% for Control film and 11.15% for 2% sp.); and hexanoic acid (2.04% for Control film and 2.84% for 2% sp.). The 2% sp.film had 15 compounds more than the Control film (pentanal; disulfide dimethyl;(E)-2-hexenal;(E)-2-heptenal; heptanoic acid; 1-octanol; cyclohexanol, 2,6-dimethyl-;9H-pyrrolo[3',4':3,4]pyrrolo[2,1-a]phthalazine-9,11(10H)-dione,10-ethyl-8-phenyl; 1,3-cyclohexadiene-1-carboxaldehyde, 2,6,6-trimethyl-; 1-cyclohexene-1-carboxaldehyde; 2,6,6-trimethyl-;alpha-ionone; 4,5,6,7-tetrahydro-7,7-dimethyl-1(3H)-isobenzofuranone; beta-ionone epoxide; 6(Z),9(E)-heptadecadiene; 4-oxo-beta-ionone), and the Control film had 2 more compounds than the 2% sp. film (butanoic acid and 2,6-di(t-butyl)-4-hydroxy-4-methyl-2,5-cyclohexadien-1-one).

Table 3. Percentage ratio of volatile compounds of the films.

A/A	RI Exp *	RI Lit **	Compound	Control (%)	2% sp. (%)
1	571	606	Acetic acid	18.1	11.55
2	690	695	Pentanal	-	0.16
3	780	785	Disulfide, dimethyl	-	0.03
4	790	784	Butanoic acid	0.43	-
5	793	810	Hexanal	3.29	3.31
6	848	854	(E)- 2-hexenal,	-	0.07
7	890		Oxime-, methoxy-phenyl-	0.33	0.27
8	895	893	Styrene;	0.3	0.23
9	895	899	n- heptanal	0.39	0.34
10	925	915	Pyrazine, 2,5-dimethyl-	0.27	0.43
11	933	963	2(E)-heptenal	-	0.88
12	955	970	Hexanoic acid	2.04	2.84
13	989		Glycerin	44.21	33.25
14	1067	1083	Heptanoic acid	-	0.74
15	1067	1062	2 octenal	0.41	1
16	1070	1070	1-octanol	-	0.36
17	1100	1099	Nonanal	3.84	2.96
18	1110	1088	Maltol	0.62	0.4
19	1110	1110	Cyclohexanol, 2,6-dimethyl-	-	0.75
20	1184	1193	Octanoic acid	5.83	6.35
21	1200		9H-pyrrolo[3',4':3,4]pyrrolo[2,1-a]phthalazine-9,11(10H)-dione,10-ethyl-8-phenyl	-	0.1
22	1210	1203	Decanal	0.57	0.37
23	1218	1198	1,3-cyclohexadiene-1-carboxaldehyde, 2,6,6-trimethyl-	-	0.13
24	1230	1280	1H-pyrrole-2,5-dione, 3-ethyl-4-methyl-	0.18	0.78
25	1233	1208	1-cyclohexene-1-carboxaldehyde, 2,6,6-trimethyl-	-	0.13
26	1238	1237	Nonanoic acid	8.92	11.51
27	1289	1300	Tridecane	0.22	0.14
28	1356	1371	Decanoic acid	3.01	2.12
29	1405	1400	Tetradecane	0.15	0.13
30	1425	1617	Tetradecanal	1.35	0.94
31	1475	1437	alpha-ionone	-	0.08
32	1485	1454	(E)-geranylacetone	0.32	0.24
33	1493	1478	2,6-di(t-butyl)-4-hydroxy-4-methyl-2,5-cyclohexadien-1-one	0.06	-
34	1498		Cyclododecane	0.38	0.18
35	1501		4,5,6,7-tetrahydro-7,7-dimethyl-1(3H)-isobenzofuranone	-	0.08
36	1510	1493	.beta.-Ionone	0.85	2.23
37	1515		beta.-Ionone epoxide	-	0.73
38	1585	1538	(2,6,6-trimethyl-2-hydroxycyclohexylidene)acetic acid lactone	0.43	1.54
39	1595	1600	Hexadecane	0.48	0.95
40	1620	1811	Hexadecanal	0.26	0.15
41	1685	1648	Methyl dihydrojasmonate	0.16	0.11
42	1691	1668	6(Z),9(E)-heptadecadiene	-	0.04
43	1694	1694	1-heptadecene	0.1	0.19
44	1695	1659	4-oxo-beta-ionone	-	0.05
45	1699	1700	Heptadecane	2.51	11.15

* Experimental retention indices' values based on the calculations using the standard mixture of alkanes; ** retention indices of the identified compounds according to the literature data cited in the NIST MS library.

The volatile compounds for the sp. powder are shown in Table 4. A total of 83 compounds were identified. The major compounds identified were heptadecane (39.32%), pentadecane (14.27%), hexadecane (9.04%), beta-ionone epoxide (5.10%), 2(4H)-benzofuranone, 5,6,7,7a-tetrahydro-4,4,7a-trimethyl- (3.46%), 4(5)-acetyl-2-(2-propyl)-1H-imidazole (2.96%), and 8-heptadecane (2.88%).

Table 4. Percentage ratio of volatile compounds of the sp. powder.

A/A	RI Exp *	RI Lit **	Compound	sp. Powder (%)
1	<500	<500	Acetaldehyde	0.30
2	<500	<500	2-propanone	0.26
3	569	606	Acetic acid	0.02
4	610	605	Furan, 2-methyl-	0.02
5	656	650	Butanal, 3-methyl-	0.02
6	659	661	1-butanol	0.02
7	690	695	Pentanal	0.07
8	735	736	1-butanol, 3-methyl-	0.04
9	738	740	1-butanol, 2-methyl-	0.05
10	750	766	1-pentanol	0.40
11	785	798	Hexanal	0.87
12	809	827	Pyrazine, methyl-	0.21
13	825	920	Formic acid, hexylester	1.68
14	901	890	Pyridine, 2,6-dimethyl-	0.09
15	903	899	2-heptanone	0.17
16	910	896	2-heptanol	0.09
17	911	899	n- heptanal	0.16
18	928	915	Pyrazine, 2,5-dimethyl-	1.29
19	959	954	2-heptanone, 6-methyl-	0.10
20	965	963	(E)-2-heptenal,	0.03
21	980	963	Benzaldehyde	0.33
22	981	983	1-octen-3-ol	1.12
23	984	985	6-methyl-5-hepten-2-one	0.40
24	987	992	2-Octanone	0.12
25	987	998	Furan, 2-pentyl-	0.54
26	989		2-heptanol, 6-methyl-	0.06
27	990	1006	Pyrazine, 2-ethyl-6-methyl-	0.04
28	991	976	Benzene, 1,3,5-trimethyl-	0.03
29	1000	1192	Phenol, 2-methoxy-4-methyl-	0.03
30	1025	1035	Benzenemethanol	0.15
31	1030	1047	Cyclohexanone, 2,2,6-trimethyl-	0.59
32	1039		2,3,4,5-tetramethyl-2-cyclopenten-1-one B	0.43
33	1045	1054	(Z)-2-octen-1-ol	0.73
34	1047	1120	2-cyclohexen-1-one, 3,5,5-trimethyl-	0.48
35	1075	1091	Pyrazine, 3-ethyl-2,5-dimethyl-	0.22
36	1079	1198	2-cyclohexen-1-one, 3,4,4-trimethyl-	0.18
37	1085	1110	6-methyl-3,5-heptadien-2-one	0.86
38	1091	1282	Phenol, 5-methyl-2-(1-methylethyl)-	0.08
39	1094		2-(t-butyl)-3-methylthiophene	0.39
40	1100	1110	Cyclohexanol, 2,6-dimethyl-	0.06
41	1115	1123	2-cyclohexene-1-carboxaldehyde, 2,6,6-trimethyl-	0.18
42	1130		4(5)-acetyl-2-(2-propyl)-1H-imidazole	2.96
43	1201	1208	Benzaldehyde, 2,5-dimethyl-	0.34
44	12010	1200	Dodecane	0.39
45	1210	1139	Ethanone, 1-(2-methylphenyl)-	0.25
46	1215	1202	Pyridine, 2-pentyl-	0.49
47	1220	1213	Undecane, 2,6-dimethyl-	0.09
48	1225	1206	1,3-cyclohexadiene-1-carboxaldehyde, 2,6,6-trimethyl-	0.77
49	1240	1280	1H-pyrrole-2,5-dione, 3-ethyl-4-methyl-	0.53

Table 4. *Cont.*

				sp. Powder (%)
A/A	RI Exp *	RI Lit **	Compound	
50	1249	1226	beta-cyclocitral	0.89
51	1259		Cyclododecane	0.03
52	1289	1275	4,8-dimethyl-nona-3,8-dien-2-one	0.08
53	1290	1251	1-cyclohexene-1-acetaldehyde, 2,6,6-trimethyl-	0.08
54	1299		3-cyano-2,4,4-trimethyl-2-cyclohexenone	0.16
55	1301	1300	Tridecane	0.17
56	1303		2-cyclopenten-1-one, 2-pentyl-	0.30
57	1310		2-butyl-3-methylpyrazine	0.14
58	1325		m-cresol, 6-tert-butyl-	0.07
59	1390	1355	2-octenal, 2-butyl-	0.17
60	1394		alpha-ionene	0.27
61	1401		Thiosulfuric acid ($H_2S_2O_3$), S-(2-aminoethyl) ester	0.08
62	1405	1400	Tetradecane	0.64
63	1440	1408	2-undecanone, 6,10-dimethyl-	0.21
64	1452	1437	alpha-ionone	0.54
65	1465		5,9-undecadien-2-one, (E)-6,10-dimethyl-,	0.6
66	1476	1800	Octadecane	0.12
67	1481		4,5,6,7-tetrahydro-7,7-dimethyl-1(3H)-isobenzofuranone	0.39
68	1486	1587	1-hexadecene	0.22
69	1492	1485	4-(2,6,6-trimethylcyclohexa-1,3-dienyl)but-3-en-2-one	0.47
70	1495	1500	Pentadecane	14.27
71	1499		beta-ionone epoxide	5.10
72	1569	1539	2(4H)-benzofuranone, 5,6,7,7a-tetrahydro-4,4,7a-trimethyl-	3.46
73	1575		(Z)-7-hexadecene	0.28
74	1581		(E)-2-nonadecene	0.05
75	1592	1600	Hexadecane	9.04
76	1599	2000	Eicosane	0.02
77	1671	1668	6(Z),9(E)-heptadecadiene	0.62
78	1679	1677	8-heptadecene	2.88
79	1681		(Z)-3-heptadecene	0.2
80	1692	1700	Heptadecane	39.32
81	1798	1800	Octadecane	0.15
82	1813	1846	2-pentadecanone, 6,10,14-trimethyl-	0.17
83	1890	1900	Nonadecane	0.04

* Experimental retention indices' values based on the calculations using the standard mixture of alkanes;
** retention indices of the identified compounds according to the literature data cited in the NIST MS library.

The sp. powder and WPC-based film containing 2% sp. had seven common compounds not found in the Control film. Six of them were minor compounds of sp. powder, namely pentanal;(E)-2-heptanal; cyclohexanol,2,6-dimethyl-; 1,3-cyclohexadiene-1-carboxaldehyde, 2,6,6-trimethyl-; alpha-ionone; 4,5,6,7-tetrahydro-7,7-dimethyl-1(3H)-isobenzofuranone; and 6(Z),9(E)-heptadecadiene (0.03–0.77%). Only beta-ionone epoxide was a major compound of sp. powder (5.10%).

The main components of the *S. platensis* extract that Ozogul et al. [21] determined were dioctylamine (67.64%), monoterpene (7.32%, pentadecane), terpene (6.33%, hexadecane), dodecane (2.34%), jasmololone (2.05%), isopulegone (1.63%), and cymarin (0.79%). In another research, the main volatile constituents of *S. platensis* were heptadecane (39.70%) and tetradecane (34.61%) based on the findings of Ozdemir et al. [36]. Finally, the volatile compounds of a "kefalotyri" cheese sample and of a "kefalotyri" cheese sample layered with film containing 2% sp. after storage at 4 °C for 7 days (after the removal of the film) are shown in Table S1. A total of 12 compounds were identified in the "kefalotyri" cheese sample, which were also identified in the cheese sample layered with the film. Two more compounds were found in this sample, namely 2,3-butanedione and 2-octene, not being compounds of the sp. powder. The above findings support that the WPC-based film containing 2% sp. does not sustain the characteristic odor of sp.

3.7. NIR Spectroscopy: Chemical Composition of "kefalotyri" Cheese Sample with Film

The chemical composition of the "kefalotyri" cheese sample layered with film containing 2% sp. after storage at 4 °C for 7 days after the removal of the film and the cheese with the film homogenized together, as determined by the application of NIR spectroscopy, was fat % 27.73 ± 0.08, moisture % 36.22 ± 0.06, protein % 27.79 ± 0.13, salt % 3.00 ± 0.01, SFA % 18.46 ± 0.92, fat in dry matter % 43.5 and total solids % 63.78 ± 0.06 and fat % 27.83 ± 0.09, moisture % 36.07 ± 0.19, protein % 29.66 ± 0.07, salt % 2.08 ± 0.12, SFA % 17.92 ± 0.90, and fat in dry matter % 43.5 and total solids % 63.93 ± 0.19, respectively. Therefore, consuming cheese together with the WPC-based film containing 2% sp. can increase the protein value and intake of the product, and in general, such packaging can maintain the cheese's nutritional value.

4. Conclusions

The film with 2% sp. showed good color properties and the highest total phenolic content and could be hydrolyzed by pepsin in the simulated gastric juice, disintegrating within 1 h. In conclusion, films with 2% sp. were selected to be applied as packaging of "kefalotyri" cheese surfaces. The spoilage from yeasts was optically evident from Day 30 on the surface of the cheeses without films. Until the end of the storage days, no spoilage bacteria were found in cheese samples layered with films containing 2% sp. The antimicrobial activity of sp. is potentially attributed to the synergistic effect between the wide diversity of bioactive phenolic compounds, such as chlorogenic acid, and other compounds found insp., such as carotenoids, C-phycocyanin, and chlorophyll (a and b). Such a cheese packaging material can be an important and sustainable alternative in the preservation of cheese quality and safety. GC-MS analysis of WPC-based film containing 2% sp. showed that the film does not sustain the characteristic odor of sp. The same major compounds were identified in the Control films and films containing 2% sp. The sp. powder and WPC-based film containing 2% sp. had seven common compounds, not found in the Control film; six of them were minor compounds of sp. powder, and only one was among its major compounds. The WPC-based films containing 2% sp. could be consumed with "kefalotyri" cheese and constitute a functional food. Consuming a film containing sp. can improve consumer acceptance due to the flavor- and color-masking effect of films and could probably prove a good alternative for the wider acceptance of functional products enriched with edible microalgae. Finally, it can increase the protein value and intake of the product, and in general, such packaging can maintain the cheese's nutritional value. However, further research needs to be conducted, and the organoleptic evaluation of the cheese after the removal of the film with sp. and the combination of the films and the cheese is recommended to obtain consumers' acceptance.

Supplementary Materials: The following Supporting Information can be downloaded at: https://www.mdpi.com/article/10.3390/su151813909/s1, Table S1: Volatile compounds of "kefalotyri" cheese and "kefalotyri" cheese covered with 2% sp. film after storage at 4 °C for 7 days after the removal of the film, Figure S1: (**a**) "Kefalotyri" cheese layered with Control film after storage at 4 °C for 7 days; (**b**) cheese layered with WPC-based films containing 2% sp. after storage at 4 °C for 7 days; (**c**) the same piece after removing the film containing 2% sp. (on the right side) with a freshly cut piece of cheese (on the left side) for comparison.

Author Contributions: Conceptualization, V.G.K., I.K., and L.B.; methodology, V.G.K., I.K., and L.B.; formal analysis, V.G.K. and L.B; investigation, V.G.K. and L.B; data curation, V.G.K. and L.B; writing—original draft preparation, V.G.K. and L.B; writing—review and editing, M.M., E.P., and A.V.B.; project administration, L.B.; funding acquisition, L.B. All authors have read and agreed to the published version of the manuscript.

Funding: This research was funded by the research program entitled "Innovative utilization approaches and comparative advantages of cheese whey of ovine/caprine origin from the region of Epirus" (MIS number: 5033108), supported by the action "Strengthening of small and medium-sized enterprises for research programs in the fields of agro-nutrition, health and biotechnology", co-financed by the European Union (European Regional Development Fund) and Greece, under the "Operational Program Epirus 2014–2020" of the National Strategic Reference Framework.

Institutional Review Board Statement: Not applicable.

Informed Consent Statement: Not applicable.

Data Availability Statement: The data are contained within the article.

Conflicts of Interest: The authors declare no conflict of interest.

References

1. Ali, A.M.M.; Sant'Ana, A.S.; Bavisetty, S.C.B. Sustainable preservation of cheese: Advanced technologies, physicochemical properties and sensory attributes. *Trends Food Sci. Technol.* **2022**, *129*, 306–326. [CrossRef]
2. Rodríguez-Félix, F.; Corte-Tarazón, J.A.; Rochín-Wong, S.; Fernández-Quiroz, J.D.; Garzón-García, A.M.; Santos-Sauceda, I.; Plascencia-Martínez, D.F.; Chan-Chan, L.H.; Vásquez-López, C.; Barreras-Urbina, C.G.; et al. Physicochemical, structural, mechanical and antioxidant properties of zein films incorporated with no-ultrafiltered and ultrafiltered betalains extract from the beetroot (*Beta vulgaris*) bagasse with potential application as active food packaging. *J. Food Eng.* **2022**, *334*, 111153. [CrossRef]
3. Jafarzadeh, S.; Salehabadi, A.; Nafchi, A.M.; Oladzadabbasabadi, N.; Jafari, S.M. Cheese packaging by edible coatings and biodegradable nanocomposites; improvement in shelf life, physicochemical and sensory properties. *Trends Food Sci. Technol.* **2021**, *116*, 218–231. [CrossRef]
4. Costa, M.J.; Maciel, L.C.; Teixeira, J.A.; Vicente, A.A.; Cerqueira, M.A. Use of edible films and coatings in cheese preservation: Opportunities and challenges. *Food Res. Int.* **2018**, *107*, 84–92. [CrossRef] [PubMed]
5. De Castro, R.J.S.; Domingues, M.A.F.; Ohara, A.; Okuro, P.K.; dos Santos, J.G.; Brexó, R.P.; Sato, H.H. Whey protein as a key component in food systems: Physicochemical properties, production technologies and applications. *Food Struct.* **2017**, *14*, 17–29. [CrossRef]
6. Daniloski, D.; Petkoska, A.T.; Lee, N.A.; Bekhit, A.E.-D.; Carne, A.; Vaskoska, R.; Vasiljevic, T. Active edible packaging based on milk proteins: A route to carry and deliver nutraceuticals. *Trends Food Sci. Technol.* **2021**, *111*, 688–705. [CrossRef]
7. Mellinas, C.; Valdés, A.; Ramos, M.; Burgos, N.; Del Carmen Garrigós, M.; Jiménez, A. Active edible films: Current state and future trends. *J. Appl. Polym. Sci.* **2016**, *133*, 42631. [CrossRef]
8. Karača, S.; Trifković, K.; Martinić, A.; Đorđević, V.; Šeremet, D.; Cebin, A.V.; Bugarski, B.; Komes, D. Development and characterisation of functional cocoa (*Theobroma cacao* L.)-based edible films. *Int. J. Food Sci. Technol.* **2020**, *55*, 1326–1335. [CrossRef]
9. Ma, Y.; Yang, W.; Xia, Y.; Xue, W.; Wu, H.; Li, Z.; Zhang, F.; Qiu, B.; Fu, C. Properties and Applications of Intelligent Packaging Indicators for Food Spoilage. *Membranes* **2022**, *12*, 477. [CrossRef]
10. Pluta-Kubica, A.; Jamróz, E.; Kawecka, A.; Juszczak, L.; Krzyściak, P. Active edible furcellaran/whey protein films with yerba mate and white tea extracts: Preparation, characterization and its application to fresh soft rennet-curd cheese. *Int. J. Biol. Macromol.* **2020**, *155*, 1307–1316. [CrossRef]
11. Robalo, J.; Lopes, M.; Cardoso, O.; Silva, A.S.; Ramos, F. Efficacy of Whey Protein Film Incorporated with Portuguese Green Tea (*Camellia sinensis* L.) Extract for the Preservation of Latin-Style Fresh Cheese. *Foods* **2022**, *11*, 1158. [CrossRef] [PubMed]
12. Kontogianni, V.G.; Kasapidou, E.; Mitlianga, P.; Mataragas, M.; Pappa, E.; Kondyli, E.; Bosnea, L. Production, characteristics and application of whey protein films activated with rosemary and sage extract in preserving soft cheese. *Lwt* **2022**, *155*, 112996. [CrossRef]
13. Spolaore, P.; Joannis-Cassan, C.; Duran, E.; Isambert, A. Commercial applications of microalgae. *J. Biosci. Bioeng.* **2006**, *101*, 87–96. [CrossRef] [PubMed]
14. Beheshtipour, H.; Mortazavian, A.M.; Mohammadi, R.; Sohrabvandi, S.; Khosravi-Darani, K. Supplementation of *Spirulina platensis* and *Chlorella vulgaris* Algae into Probiotic Fermented Milks. *Compr. Rev. Food Sci. Food Saf.* **2013**, *12*, 144–154. [CrossRef]
15. Wells, M.L.; Potin, P.; Craigie, J.S.; Raven, J.A.; Merchant, S.S.; Helliwell, K.E.; Smith, A.G.; Camire, M.E.; Brawley, S.H. Algae as nutritional and functional food sources: Revisiting our understanding. *J. Appl. Phycol.* **2017**, *29*, 949–982. [CrossRef]
16. Nakamoto, M.M.; Assis, M.; de Oliveira Filho, J.G.; Braga, A.R.C. Spirulina application in food packaging: Gaps of knowledge and future trends. *Trends Food Sci. Technol.* **2023**, *133*, 138–147. [CrossRef]
17. Kontogianni, V.G.; Chatzikonstantinou, A.V.; Mataragas, M.; Kondyli, E.; Stamatis, H.; Bosnea, L. Evaluation of the Antioxidant and Physicochemical Properties of Microalgae/Whey Protein-Based Edible Films. *Biol. Life Sci. Forum* **2021**, *6*, 97. [CrossRef]
18. Jamróz, E.; Kulawik, P.; Krzyściak, P.; Talaga-Ćwiertnia, K.; Juszczak, L. Intelligent and active furcellaran-gelatin films containing green or pu-erh tea extracts: Characterization, antioxidant and antimicrobial potential. *Int. J. Biol. Macromol.* **2019**, *122*, 745–757. [CrossRef]
19. Cruz-Diaz, K.; Cobos, Á.; Fernández-Valle, M.E.; Díaz, O.; Cambero, M.I. Characterization of edible films from whey proteins treated with heat, ultrasounds and/or transglutaminase. Application in cheese slices packaging. *Food Packag. Shelf Life* **2019**, *22*, 100397. [CrossRef]
20. Vatavali, K.; Kosma, I.; Louppis, A.; Gatzias, I.; Badeka, A.V.; Kontominas, M.G. Characterisation and differentiation of geographical origin of Graviera cheeses produced in Greece based on physico-chemical, chromatographic and spectroscopic analyses, in combination with chemometrics. *Int. Dairy J.* **2020**, *110*, 104799. [CrossRef]

21. Özogul, I.; Kuley, E.; Durmus, M.; Özogul, Y.; Polat, A. The effects of microalgae (*Spirulina platensis* and *Chlorella vulgaris*) extracts on the quality of vacuum packaged sardine during chilled storage. *J. Food Meas. Charact.* **2021**, *15*, 1327–1340. [CrossRef]

22. NIST05 National Institute of Standards and Technology. *Mass Spectral Library*; Wiley: West Sussex, UK, 2005.

23. SPSS. *IBM SPSS Statistics for Windows, Version 23.0 (Computer Software)*; IBM Corp.: Armonk, NY, USA, 2014.

24. Balti, R.; Ben Mansour, M.; Sayari, N.; Yacoubi, L.; Rabaoui, L.; Brodu, N.; Massé, A. Development and characterization of bioactive edible films from spider crab (*Maja crispata*) chitosan incorporated with Spirulina extract. *Int. J. Biol. Macromol.* **2017**, *105*, 1464–1472. [CrossRef] [PubMed]

25. Miranda, M.S.; Cintra, R.G.; Barros, S.B.M.; Mancini-Filho, J. Antioxidant activity of the microalga Spirulina maxima. *Braz. J. Med. Biol. Res.* **1998**, *31*, 1075–1079. [CrossRef] [PubMed]

26. Nourmohammadi, N.; Soleimanian-Zad, S.; Shekarchizadeh, H. Effect of Spirulina (*Arthrospira platensis*) microencapsulated in alginate and whey protein concentrate addition on physicochemical and organoleptic properties of functional stirred yogurt. *J. Sci. Food Agric.* **2020**, *100*, 5260–5268. [CrossRef]

27. Mishra, P.; Singh, V.P.; Mohan, S. Spirulina and its Nutritional Importance: A Possible Approach for Development of Functional Food. *Biochem. Pharmacol. Open Access* **2014**, *3*, 1000e171. [CrossRef]

28. Terra, A.L.M.; Moreira, J.B.; Costa, J.A.V.; de Morais, M.G. Development of pH indicators from nanofibers containing microalgal pigment for monitoring of food quality. *Food Biosci.* **2021**, *44*, 101387. [CrossRef]

29. Kuntzler, S.G.; Costa, J.A.V.; Brizio, A.P.D.R.; de Morais, M.G. Development of a colorimetric pH indicator using nanofibers containing Spirulina sp. LEB 18. *Food Chem.* **2020**, *328*, 126768. [CrossRef]

30. Chaudhary, V.; Kajla, P.; Kumari, P.; Bangar, S.P.; Rusu, A.; Trif, M.; Lorenzo, J.M. Milk protein-based active edible packaging for food applications: An eco-friendly approach. *Front. Nutr.* **2022**, *9*, 942524. [CrossRef]

31. FDA GRAS Notification: "Spirulina" microalgae. Available online: https://www.algbiotek.com/spirulina-sertifikalar/Spirulina-GRAS-grn_101.pdf (accessed on 2 September 2023).

32. Stejskal, N.; Miranda, J.M.; Martucci, J.F.; Ruseckaite, R.A.; Barros-Velázquez, J.; Aubourg, S.P. Quality Enhancement of Refrigerated Hake Muscle by Active Packaging with a Protein Concentrate from Spirulina platensis. *Food Bioprocess Technol.* **2020**, *13*, 1110–1118. [CrossRef]

33. Chentir, I.; Hamdi, M.; Li, S.; Doumandji, A.; Markou, G.; Nasri, M. Stability, bio-functionality and bio-activity of crude phycocyanin from a two-phase cultured Saharian *Arthrospira* sp. strain. *Algal Res.* **2018**, *35*, 395–406. [CrossRef]

34. Stejskal, N.; Miranda, J.M.; Martucci, J.F.; Ruseckaite, R.A.; Aubourg, S.P.; Barros-Velázquez, J. The Effect of Gelatine Packaging Film Containing a *Spirulina platensis* Protein Concentrate on Atlantic Mackerel Shelf Life. *Molecules* **2020**, *25*, 3209. [CrossRef] [PubMed]

35. Martelli, F.; Cirlini, M.; Lazzi, C.; Neviani, E.; Bernini, V. Edible Seaweeds and Spirulina Extracts for Food Application: In Vitro and In Situ Evaluation of Antimicrobial Activity towards Foodborne Pathogenic Bacteria. *Foods* **2020**, *9*, 1442. [CrossRef] [PubMed]

36. Ozdemir, G.; Karabay, N.U.; Dalay, M.C.; Pazarbasi, B. Antibacterial activity of volatile component and various extracts ofSpirulina platensis. *Phytother. Res.* **2004**, *18*, 754–757. [CrossRef] [PubMed]

 *sustainability*

Article

Utilization of Spent Coffee Grounds as a Feed Additive for Enhancing the Nutritional Value of *Tenebrio molitor* Larvae

Konstantina Kotsou [1], Theodoros Chatzimitakos [1], Vassilis Athanasiadis [1], Eleni Bozinou [1], Christos G. Athanassiou [2] and Stavros I. Lalas [1,*]

[1] Department of Food Science and Nutrition, University of Thessaly, Terma N. Temponera Str., 43100 Karditsa, Greece; kkotsou@agr.uth.gr (K.K.); tchatzimitakos@uth.gr (T.C.); vaathanasiadis@uth.gr (V.A.); empozinou@uth.gr (E.B.)

[2] Laboratory of Entomology and Agricultural Zoology, Department of Agriculture, Crop Production and Rural Environment, School of Agricultural Sciences, University of Thessaly, Phytokou Str., 38446 Volos, Greece; athanassiou@uth.gr

* Correspondence: slalas@uth.gr; Tel.: +30-24410-64783

Abstract: Increasing demand for sustainable protein sources has spurred interest in the exploration of alternative protein sources with a reduced environmental impact. This study investigates the use of spent coffee grounds (SCG), a widely available by-product, as a feed additive for *Tenebrio molitor* larvae, aiming to contribute to the circular economy and enhance the nutritional quality of the insects. The larvae were fed with a mixture of bran (the conventional feed) and SCGs (10 and 25% w/w). Larval viability, growth, and nutritional composition, including protein, fat, carbohydrates, ash, carotenoids, vitamins A and C, and polyphenols, were evaluated. Increasing the proportion of SCGs in the larvae's feed led to an enhanced nutritional value of the larvae. In particular, crude protein increased by 45.26%, vitamin C showed an increase of 81.28%, and vitamin A showed an increase of 822.79%, while polyphenol content increased by 29.01%. In addition, the oil extracted from these larvae showed enhanced nutritional value and greater resistance to oxidation. The results highlight the promising use of SCGs as a feed additive for *T. molitor* larvae, offering a sustainable approach to enhance their nutritional value. Delving deeper into the results, the addition of 10% SCGs resulted in a 45.26% increase in crude protein compared to the SCG0 sample. Concurrently, increasing SCGs in the dietary substrate led to an increase in vitamin content; in sample SCG25, vitamin C content increased by 81.28% while vitamin A content increased by 822.79% compared to the control sample. Moreover, there was a large increase in polyphenol content with the SCG25 sample showing the highest value, which was a 29.01% increase over the control sample.

Keywords: carotenoids; coffee by-products; edible insects; fat; fatty acids; proteins; proximate composition; vitamin A; vitamin C; yellow mealworm larvae

Citation: Kotsou, K.; Chatzimitakos, T.; Athanasiadis, V.; Bozinou, E.; Athanassiou, C.G.; Lalas, S.I. Utilization of Spent Coffee Grounds as a Feed Additive for Enhancing the Nutritional Value of *Tenebrio molitor* Larvae. *Sustainability* **2023**, *15*, 16224. https://doi.org/10.3390/su152316224

Academic Editor: Dimitris Skalkos

Received: 5 October 2023
Revised: 12 November 2023
Accepted: 21 November 2023
Published: 23 November 2023

1. Introduction

In light of the Earth's increasing population, the widespread consumption of natural resources, and the resulting production of significant amounts of agricultural by-products, there is a continuous and urgent concern for sustainable practices [1,2]. In the scientific and industrial sectors, ongoing efforts are made to establish sustainable methods for managing these discarded by-products in order to reduce the consumption of natural resources [3]. Among the various by-products generated and discarded annually, one noteworthy example is the residue produced after the roasting and brewing of coffee beans [4]. Coffee is the most widely consumed beverage globally and ranks as the second-largest commercial product, following petroleum [5]. According to the study by Getachew et al. [6], the improper disposal of coffee grounds, due to their high tannin and caffeine content, could pose environmental risks. Consequently, research efforts have increased to investigate the coffee grounds' properties and potential applications.

According to previous research, it has been elucidated that spent coffee grounds (SCG) contain bioactive compounds, notably polyphenols, which are well-documented for their antioxidant and antimicrobial attributes [7–9]. These by-products also encompass a variety of phenolic compounds, including caffeic acid, chlorogenic acid, and its isomer, neochlorogenic acid, known for their diverse biological activities, including anticancer, antilipidemic, antidiabetic, antiviral, and antipyretic effects, among others [10–12]. Despite their unsuitability for human consumption, coffee by-products have found utility as an ingredient in poultry feed, demonstrating their compatibility with animal consumption [13]. This multifaceted application of coffee by-products underscores their pivotal role in promoting a circular economy. Consequently, it prompts further exploitation regarding potential additional avenues for their sustainable utilization.

The expanding global population and the aforementioned patterns of resource consumption have led the scientific community toward the exploration of novel, innovative food sources [14]. Among the most promising solutions, the rearing and utilization of insects have garnered significant attention [15]. One promising insect species in this context is the larvae of *Tenebrio molitor* (TM) (Coleoptera: Tenebrionidae), commonly referred to as yellow mealworm larvae, which is frequently used as feed for fish, poultry, and pigs [16]. These larvae have gained recognition for their exceptional nutritional profile, characterized by high protein content, essential amino acids, fats, and essential fatty acids [17,18]. Notably, the TM insect participates inherently in a circular economy model by being reared on substrates derived from various by-products, thus contributing to a sustainable ecosystem. Various by-products have been subjected to examination in order to assess their effects on TM larvae growth. Among these substrates are rice bran, corncob, and potato peels [19] along with a range of industrial by-products, including chicken feed, wheat and rye bran, rapeseed meal and cake, flax cake, and *Silybum marianum* cake [20,21].

The digestive capabilities of mealworm larvae have garnered attention for their ability to process a diverse array of substrates and by-products [22]. Furthermore, these larvae have been documented as having negligible or limited quantities of specific essential nutrients, including vitamins A and C, as well as other bioactive compounds such as polyphenols [23]. Building upon these observations, the aim of our study was the examination of the feasibility of utilizing SCGs as a rearing substrate for TM insects. This investigation serves a dual purpose: firstly, to evaluate the influence of the substrate on the growth and development of the larvae, particularly up to the pupae stage, which follows the larval stage and precedes adulthood in insects [24], and, secondly, to assess the impact of the substrate on the nutritional composition of the larvae. The ultimate objective of this research is the enhancement of the nutritional value of a food source (TM larvae), while simultaneously adhering to principles of cost-effectiveness (by utilizing SCGs, a by-product that is abundantly available, to contribute to the economic viability of insect farming) and ecological sustainability (by repurposing a waste product to contribute to waste reduction and promote a circular economy, aligning with broader sustainability goals). By adopting such an approach, we aspire to contribute to the development of an economically viable and environmentally friendly method for producing a high-quality food resource. This approach represents a novel and environmentally conscious strategy for repurposing waste materials and aligns with the growing emphasis on circular economies in agriculture. Moreover, our dual focus on larval growth and development, coupled with considerations of cost-effectiveness and ecological sustainability, sets our approach apart, contributing to a more comprehensive understanding of the potential of alternative feed sources and their broader implications for sustainable and environmentally friendly agricultural practices. Based on the above, this study aligns with several Sustainable Development Goals (SDGs) by contributing to the global agenda for sustainable development. Firstly, the investigation into utilizing SCGs as a feed additive for TM larvae aligns with SDG 12 (Responsible Consumption and Production) by promoting the efficient and sustainable use of resources, specifically by repurposing a widely available by-product to enhance nutritional value. The emphasis on circular economy principles and waste reduction corresponds to SDG

12 targets. Additionally, the study aligns with SDG 2 (Zero Hunger) by exploring alternative protein sources to meet the increasing demand for sustainable food resources. The enhancement of nutritional content in TM larvae, particularly the substantial increase in crude protein and vitamins, contributes to SDG 3 (Good Health and Well-being) by addressing the nutritional aspect of food security. Furthermore, the research aligns with SDG 15 (Life on Land) by promoting sustainable practices in insect rearing, offering a potential eco-friendly solution to enhance food production while mitigating environmental impact. In summary, this study actively supports the interconnected goals of responsible consumption, zero hunger, good health, and environmental sustainability outlined in the United Nations' Sustainable Development Goals.

2. Materials and Methods

2.1. Chemicals and Reagents

High-performance liquid chromatography (HPLC) grade solvents, specifically, hexane, acetone, ethanol, and methanol, were employed in this study and were supplied from Carlo Erba (Val de Reuil, France). Bradford reagent, ascorbic acid, β-carotene, trichloroacetic acid (163.69 M), hydrochloric acid (6.00 N), and iron (III) chloride (162.20 M) were sourced from Sigma-Aldrich (Steinheim, Germany). Gallic acid, Folin–Ciocalteu reagent, sodium anhydrous carbonate, 2,2-diphenyl-1-picryl-hydrazyl (DPPH), and 2,4,6-tri-2-pyridinyl-1,3,5-triazine (TPTZ) were obtained from Penta (Prague, Czech Republic).

2.2. Insects and Spent Coffee Grounds (SCG) Material

The experiments were performed using newly hatched *Tenebrio molitor* (TM) larvae (7 days old). To obtain the above population, a plastic tray (24 × 29.5 × 10 cm) with an opening in the top cover for adequate ventilation was filled with 2 kg of white flour (as an oviposition substrate), and about 500 adult TM were added over a mesh (to avoid cannibalization of the eggs). In addition, agar was also added as a source of moisture to the insect substrate. Insects were maintained under constant conditions, i.e., 26 ± 1 °C, 55 ± 5% relative humidity (RH), and continuous darkness, for 7 days in order to oviposit. The eggs were left to hatch and removed for further use when the larvae reached 7 days old.

The SCGs were obtained from a local coffee store in Karditsa city, Greece. The coffee grounds (60% Arabica and 40% Robusta) were transferred to the laboratory and underwent lyophilization in order to remove water.

2.3. Feeding Trial

In a series of feeding trials, the growth of TM larvae on wheat bran–based feeding substrates with different percentages of dried SCG, namely, 10% (SCG10), 25% (SCG25), 50% (SCG50), 75% (SCG75), and 100% (SCG100), was evaluated, while wheat bran alone (SCG0) was used as a control. The bran was procured from a local shop in Volos, Greece, and had a particle size <2 mm.

In the preliminary experiments, 50 TM larvae were transferred into plastic, cylindrical vials (diameter 6.5 cm, height 8.8 cm, Carl Roth GmbH & Kg, Karlsruhe, Germany) with a shielded opening in the top cover allowing air circulation, together with 8 g of each substrate. As with the oviposition process, agar was provided as a source of moisture, three times a week while rearing conditions were 26 ± 1 °C, 55 ± 5% RH, and continuous darkness. To avoid depletion and inability of larvae to develop and grow, 2 g of food was added to each vial (the food nearly ran out). Each feed substrate was examined in nine replicates. In order to determine larval survival and development, larvae were separated from the substrate and counted, and their total weight was determined on a precision scale (Equinox EAB125i, Adam Equipment Inc., FoxHollow Road, Oxford, UK). This procedure was repeated every week until the first pupa appeared in each vial.

In order to produce sufficient TM population for further experiments, TM larvae were reared on a larger scale. The substrates that showed satisfactory survival and growth results (*vide infra*) were used (i.e., SCG0, SCG10, and SCG25). Into plastic trays (24 × 29.5 × 10 cm),

1500 larvae were introduced along with 300 g of each substrate and reared for eight weeks under the same conditions while using agar as a moisture source. Larvae were separated by sieving from the rearing substrate, fasted for 24 h, weighed, and euthanized by freezing ($-20\,°C$). Larvae were then placed in a freeze-dryer for 24 h and crushed into a fine powder stored in sterile glass vials at $-40\,°C$ for further analysis.

2.4. Larval Composition Analysis

2.4.1. Water Content

A Biobase BK-FD10P (Jinan, China) lyophilizer was used to remove the moisture content from all larvae samples. Subsequently, the quantification of moisture content within samples was conducted using a gravimetric method referred to by May et al. [25].

2.4.2. Crude Protein Content

In order to extract the proteins from the sample, 1 g of the sample was immersed in 10 mL of distilled water adjusted to pH 12 with 1 M NaOH. The extraction process was carried out for 60 min at 500 rpm and at room temperature. Subsequently, the mixture underwent centrifugation for 5 min at $4500\times g$, and the supernatant was retracted, so as to be further analyzed. To ensure the complete extraction of proteins, the extraction procedure was repeated two more times on the solid residue.

Quantification of protein content, expressed as a percentage, was conducted using the Bradford method [26]. For protein quantification, 900 µL of Bradford reagent was combined with 100 µL of the sample, and this mixture was allowed to react for a duration of 10 min in the absence of light. Ultimately, the absorbance of the samples was quantified at 595 nm using a spectrophotometer (Shimadzu UV-1700 Pharma Spectrophotometer, Kyoto, Japan).

2.4.3. Carbohydrates Content

For the extraction of the carbohydrates, 1 g of the larval sample was added to 10 mL of distilled water. The mixture was stirred at 500 rpm for 1 h at $50\,°C$. The mixture was then centrifuged for 5 min at $4500\times g$ and the supernatant was retracted and further analyzed. The phenol/sulfuric acid method was used to quantify the amount of carbohydrates [27]. In brief, 0.22 mL of the supernatant was transferred to a plastic tube and 0.65 mL of concentrated sulfuric acid and 0.13 mL of 5% w/v aqueous phenol solution were immediately added. The mixture was placed in a water bath at $90\,°C$ for 5 min and then allowed to cool to room temperature for 5 min. The absorbance of the solution was measured at 495 nm using a spectrophotometer while a calibration curve was prepared using D(+)-glucose as a standard.

2.4.4. Ash Content

The ash content of the samples was calculated gravimetrically [26]. Approximately 5 g of the sample was placed in porcelain crucibles and placed in an oven. The temperature of the oven was increased to $550\,°C$ with a rate of $5\,°C/min$, and the samples were heated until no black residues were visible. The crucibles were then placed in a desiccator and left to cool to room temperature. Then the weight of the crucibles was recorded and the ash content was determined.

2.4.5. Total Fat, Fatty Acids, and Calculated Oxidizability Value (COX)

For the determination of the fat content of the larvae, a defatting process was carried out. More specifically, 1 g of sample was mixed with 10 mL of *n*-hexane in a glass vial with a screw cap. Extraction was performed by stirring at $40\,°C$ and 600 rpm for 60 min. To isolate the supernatant, the solution was centrifuged at $4500\times g$ for 10 min. The supernatant was transferred to a pre-weighed flask. The extraction process was carried out two more times on the solid residue and the supernatants were pooled. Finally, the solvent was removed using a rotary evaporator at $40\,°C$. For the identification and quantification of the fatty acids of the samples as well as the calculation of the various indices, the method used in our previous research [26] was used.

2.4.6. Energy Content

The determination of the energy of TM larvae was based on the standard Equation (1):

$$\text{Energy (kcal/100 g)} = (9 \times \% \text{ crude fat}) + (4 \times \% \text{ crude protein}) + (4 \times \% \text{ carbohydrates}) \tag{1}$$

2.4.7. Vitamin C Content

The screening and quantification of the amount of vitamin C samples was performed using a modified colorimetric analysis [28]. An accurately weighed amount of 5 g of powder from each sample of *T. molitor* was mixed with 27 mL of a distilled water:methanol mixture (60:40, *v/v*) and 3 mL of a 10% *w/v* trichloroacetic acid solution. The mixture was stirred vigorously for 60 sec and then 20 mL of *n*-hexane was added. The final solution was stirred at room temperature for 30 min and then centrifuged for 10 min at 10,000× *g*. The lower aqueous phase was withdrawn and used as a sample. Next, 1 mL of the sample was mixed with 0.5 mL of Folin–Ciocalteu reagent (10% *v/v*) and allowed to incubate for 10 min. Absorbance was measured at 760 nm and quantification was performed with an ascorbic acid calibration curve.

2.4.8. β-Carotene–Vitamin A Content

A previously described technique was used to estimate the β-carotene and Vitamin A content [29]. First, 1 g of each sample was added to the extraction stage along with 10 mL of ethanol, and the mixture was stirred at room temperature for 30 min at 300 rpm. The mixture was then centrifuged for 5 min at 3600× *g* while being periodically shaken in an ice bath. Using a standard β-carotene calibration curve and the resulting extract's absorbance at 450 nm, the amount of carotenoid concentration was calculated (range: 0–50 mg/L, equation: y = 0.0182x + 0.0119, R^2 = 0.9982).

2.4.9. Ferric-Reducing Antioxidant Power (FRAP) Assay

The FRAP method was used to determine the antioxidant capacity of the samples according to Makris and Kefalas [30]. The extracted solution as described in 2.4.7 was used in this assay. 50 µL of the extracts were mixed with 50 µL of $FeCl_3$ solution (4 mM in 0.05 M HCl) and incubated at 37 °C for 30 min. Then 900 µL of the TPTZ solution (1 mM in 0.05 M HCl) was added and the absorbance was measured at 620 nm. A calibration curve (concentration range: 50–500 µmol/L in 0.05 M HCl) was generated using ascorbic acid as a standard compound.

2.4.10. Total Polyphenol Content (TPC) Determination

A previously documented technique was used to calculate the extract's total polyphenol content (TPC) [31]. The extracted solution as described in 2.4.7 was used in this assay. A total of 100 µL of the extract was mixed with an equal amount of Folin–Ciocalteu reagent and left for 2 min. Following the addition of 800 mL of Na_2CO_3 solution (5% *w/v*), the mixture was incubated for 20 min at 40 °C without light. Using a standard calibration curve, the absorbance was measured at 740 nm to calculate the TPC, which was represented as mg of gallic acid equivalents (GAE) per g of dry weight (dw) (range: 0–100 mg/L, equation: y = 0.0138x − 0.0044, R^2 = 0.9996) with gallic acid, and the concentration of total polyphenol (C_{TP}) was determined. To calculate the extraction yield of total polyphenol (Y_{TP}), the following Equation (2) was used:

$$Y_{TP}(\text{mg GAE/g dw}) = \frac{C_{TP} \times V}{w} \tag{2}$$

where *V* represents the volume of the extraction medium (in L) and *w* represents the dry weight of the sample (in g).

2.5. Statistical Analysis

For the determination of larval survival and growth, a total of nine measurements were taken. For the proximate composition analysis, three samples were used, each analyzed in triplicate, resulting in a total of nine measurements. Results were expressed as the mean values of the nine measurements using standard deviation (SD). The data was examined to see if they were frequently distributed using the Kolmogorov–Smirnov test. The presence of statistically significant differences ($p < 0.05$) among the samples was assessed using one-way ANOVA (Analysis of Variance) with a post-hoc Tukey HSD (Honestly Significant Difference) Test Calculator (Tukey HSD used with Tukey–Kramer formula). For the statistical analysis, SPSS (version 29) (SPSS Inc., Chicago, IL, USA) software was used.

3. Results and Discussion

The primary focus of the study was to investigate the impact of SCGs as a feed additive on the nutritional value of *Tenebrio molitor* larvae. While acknowledging the importance of chemical composition analysis of the feed, the scope of the study was more centered on *T. molitor* and on the nutritional aspects, such as protein, vitamins, and polyphenols, rather than an exhaustive examination of all chemical components of the feed. Moreover, SCGs have been explored in various studies for their nutritional value, as well as toxic compounds such as ochratoxin A. While the presence of ochratoxin A is a legitimate concern in coffee-related products, existing safety standards may provide insights into the likelihood of contamination. The SCGs were obtained from coffee stores, intended for human consumption. As such, it was taken for granted that the raw materials were thoroughly and carefully examined, so as to be available in the market.

3.1. Survival and Growth of TM Larvae

A number of factors need to be considered in order to evaluate a feed substrate for its suitability for insect rearing [32]. One of the main factors that needs to be examined is the survival of the insects. The selection of bran as a dietary control was based on the fact that it is a widely used substrate that promotes rapid growth and ensures high survival and high larval weight gain [33]. From preliminary experiments, it was learned that larvae in the SCG50, SCG75, and SCG100 samples did not survive more than one week after SCG addition, while those in the SCG0, SC10, and SCG25 samples survived to constitute the examined samples. The % survival rates of the larvae in the three remaining substrates (i.e., SCG0, SCG10, and SCG25) at each week are reported in Table 1. As can be seen, as the SCG content in the larval feeding substrate increases, the number of larvae decreases, albeit to a small extent (i.e., up to 10% in the eighth week). This may be due to the fact that in SCGs several compounds are present, some of which may induce toxicity in the larvae, such as ochratoxin A [34]. However, it is also known that TM larvae can self-select their diet preferences [35]. Therefore, the reduced survival may also be due to reduced feed consumption from their dietary preferences.

Table 1. Survival (%) of *T. molitor* larvae ($\pm$SD) fed for eight weeks with wheat bran (control) (SCG0) and bran fortified with different rates of spent coffee grounds, 10% (SCG10) and 25% (SCG25)) ($n = 9$).

Diets	2nd Week	3rd Week	4th Week	5th Week	6th Week	7th Week	8th Week
SCG0	97 ± 2	95 ± 3	94 ± 3	93 ± 4	91 ± 3	90 ± 2 [a]	89 ± 2 [a]
SCG10	96 ± 4	95 ± 4	93 ± 4	92 ± 3	92 ± 3	87 ± 5 [a,b]	86 ± 5 [a]
SCG25	96 ± 2	95 ± 2	93 ± 3	91 ± 4	90 ± 3	83 ± 3 [b]	80 ± 3 [b]

In all cases, values represent means ± SD. Within each evaluation interval (Week 2–8), means followed by the same superscript letter are not significantly different ($p > 0.05$). Where no letters exist, no statistically significant differences were noted.

The second parameter that must be examined for new feed substrates is the weight gain of the larvae. Results from the weight of the larvae fed with SCGs are shown in Figure 1. As can be seen, a statistically significant increase ($p < 0.0.5$) in the weight gain of the larvae was recorded at weeks 5 and 6 for both SCG percentages, compared to the control larvae. However, in the next two weeks, although a 7.8% increase in the weight gain of the larvae fed with SCG25 was recorded, this was not statistically significantly different ($p > 0.0.5$) compared to the control sample. Although no significant increase in the weight of the larvae was recorded, it is noteworthy that the weight of the larvae did not decline due to selective feeding [35]. This result opens new avenues in insect rearing, given that both domestically and scientifically SCGs and their various extracts act as a repellent for insects such as mosquitoes [36,37].

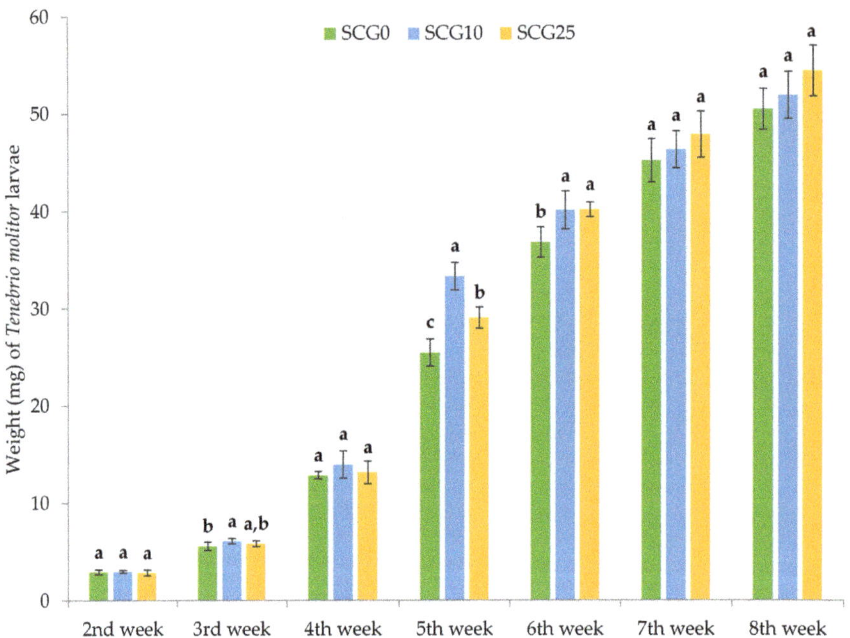

Figure 1. Individual weight (mg) of *T. molitor* larvae fed for eight weeks with wheat bran (control) (SCG0) and bran fortified with different rates of spent coffee grounds, 10% (SCG10) and 25% (SCG25)) ($n = 9$). Statistically significant differences ($p < 0.05$) are denoted with different letters (e.g., a–c).

Although the microbiological safety of products is often studied for food products, this analysis was not carried out in this case. As such, any potential risk to larvae or subsequent consumers would need to be addressed in further research or product development. The present study does not present a ready-to-eat food product but a potential additional ingredient, TM flour, for the preparation of a new product. Our focus on nutritional improvements through SCG supplementation does not negate the significance of safety considerations in food product development. However, we believe that addressing these concerns falls within the purview of future research endeavors that specifically aim to develop ready-to-eat food products using TM flour.

3.2. Evaluation of the Nutritional Value of the Larvae

The first step was to determine the water content of the samples. The larvae fed with bran, considered as the control sample, exhibited a moisture content of approximately 30%, while the larvae fed with varying proportions of SCGs displayed a moisture content of approximately 17.7%. TM larvae present a high water content [38], sometimes ranging from 60.80 to 72.60%. Nevertheless, when the water content is so high, very low percentages of crude protein, in the range of 23.50–16.50%, respectively, also occur [39]. Consequently, the water content in the present study may not reach such high values and this fact may be attributed to their rich protein content, which is presented afterwards.

3.2.1. Proximate Composition

The high crude protein content has instigated substantial interest in yellow mealworm larvae within both the scientific community and the food industry. Consequently, the impact of SCGs on the crude protein content of the larvae was investigated. Detailed proximate composition data for TM larvae, fed with different SCG-based diets, is presented in Table 2. Commonly, TM larvae reared on bran, their primary dietary substrate, typically exhibit a crude protein content ranging from 13.68% to 27.60% [40,41]. However, it is crucial to acknowledge that the variability in crude protein content among larvae is influenced by geographic and rearing conditions, as evident in the diverse results. Although SCGs contain a small amount of crude protein [42], their use as a feed additive resulted in a significant boost in the crude protein levels of larvae. Specifically, the SCG10 sample exhibited a crude protein content of 47.34%, which was a 45.26% increase over the control sample, while the SCG25 sample exhibited a substantial 30.86% increase over the SCG0 group, as its amount of crude protein was 42.65%; both samples' differences were statistically significant ($p < 0.05$). Last but not least, it is worth mentioning that SCG10 was 10.99% more enhanced in crude protein content than SCG25.

Table 2. Proximate composition of *T. molitor* larvae fed for eight weeks with wheat bran (control) (SCG0) and bran fortified with different rates of spent coffee grounds, 10% (SCG10) and 25% (SCG25)) ($n = 9$).

% Composition of Dry Weight	SCG0	SCG10	SCG25
Crude protein	32.59 ± 0.43 [c,*]	47.34 ± 0.17 [a]	42.65 ± 0.26 [b]
Crude fat	22.77 ± 1.29 [a]	20.2 ± 1.07 [a]	21.93 ± 1.55 [a]
Carbohydrates	25.72 ± 0.04 [a]	13.41 ± 0.77 [c]	19.42 ± 0.21 [b]
Crude ash	1.89 ± 0.07 [c]	3.51 ± 0.04 [a]	2.44 ± 0.03 [b]
Energy (kcal/100 g)	438.17 ± 13.49 [a]	424.8 ± 13.39 [a]	445.65 ± 15.83 [a]

* Within each line, statistically significant differences ($p < 0.05$) are denoted with different superscript letters (e.g., a–c).

Given the ongoing exploration of TM larvae as potential substitutes for conventional sources of crude protein like meat, it is important to underscore the magnitude of this increase. For context, beef typically contains ~21.35 g/100 g of crude protein content, while chicken contains ~19.40 g/100 g [43]. In comparison, these contents are 52.64% and 67.99% lower compared to the control sample, and 121.73% and 144.02% lower compared to the SCG10 sample. This underscores the substantial nutritional advantage of TM larvae over conventional protein sources. Notably, the high (47.34%) crude protein content in SCG10 makes TM larvae close to the highest recorded percentage of crude protein in these larvae (50%) [44]. While amino acid profiling provides valuable insights into the specific constituents of proteins, emphasis was placed on the overall protein content, which serves as a pertinent and meaningful indicator of the nutritional enhancement achieved through SCG supplementation. The fundamental premise lies in recognizing that protein content itself is a critical determinant of nutritional value. As such, since our study aimed to contribute to the discourse on sustainable protein sources, with a focus on the circular economy and reduced environmental impact, the demonstrated increase in protein content, independent of the amino acid profile, underscores the potential of SCGs as a viable and sustainable feed additive.

Table 2 shows a noticeable reduction in the percentage of carbohydrates across various samples, which is in line with the observed increment in protein percentages. Specifically, samples SCG10 and SCG25 displayed a substantial decrease of 47.86% and 24.49% (statistically significant at $p < 0.05$) in carbohydrate content when compared to the SCG0 group.

It is noteworthy that the nutritional evaluation of mealworms primarily emphasizes their crude protein content, often omitting essential components like carbohydrates. Nevertheless, in this study, we examined the carbohydrate content as it constitutes a fundamental component of a balanced diet [45]. Beyond their significance in human nutrition, carbohydrates play a pivotal role in the growth and development of insects. A substantial portion of carbohydrates in insects is attributed to chitin [27], which serves as the main material composing the exoskeleton of TM throughout all developmental stages [46]. However, it is observed that as protein levels in the samples increase, there is a concomitant and significant decrease in carbohydrate content. This observation was somewhat expected, as carbohydrates provide the larvae with the requisite energy for their development and the execution of their metamorphosis [35]. Although chitin is undoubtedly a significant component in insect exoskeletons and plays a crucial role in growth and development, its quantification was not carried out since it was not aligned with the primary objectives of the study. In this study, the emphasis was on assessing the overall nutritional changes in TM larvae when supplemented with SCGs. Focus extended beyond individual components to encompass macronutrients, micronutrients, and the holistic nutritional value of the larvae. As far as the human diet is concerned, the consumption of limited carbohydrates can help reduce overall caloric intake without affecting the intake of essential nutrients, e.g., proteins and minerals [47]. Finally, low-carbohydrate diets improve cardiovascular risk factors and prevent or treat diabetes [48].

The determination of crude ash content is of paramount importance as it provides insights into the mineral composition of a food product, a key aspect of its nutritional profile. As elucidated in Table 2, the percentage of crude ash in the SCG-fed samples exhibits a wide variation, from 1.89% in SCG0 to 3.51% in SCG10. Notably, the inclusion of SCGs in the larval diet resulted in a statistically significant ($p < 0.05$) elevation in mineral content. Specifically, SCG10 displayed an 85.14% increase, while SCG25 demonstrated a 29.10% rise in comparison to the control sample. Furthermore, SCG10 emerged as the leading sample in terms of crude ash content, aligning with its prominent position in relation to crude protein levels. It is noteworthy that the highest recorded crude ash value in previous studies on TM larvae was found to be 3.81% [41], a content comparable to that of SCG10. Conversely, the control sample was found to contain 1.89%, which is consistent with prior studies reporting values within the range of 1.23% to 2.20% [40,49]. Collectively, these findings underscore the substantial increase in mineral content attributed to the inclusion of SCGs in the larval diet, with SCG10 achieving or surpassing values reported in prior research.

Another characteristic of TM larvae is their notably high fat content [50]. Extensive investigations have been conducted to elucidate their fat content, reporting a fat content between 20% and 45% [51,52]. According to our results, the observed fat percentages ranged between 20.20% (SCG10) and 22.77% (SCG0). These results were somewhat expected, given the inverse relationship between crude protein and fat percentages, whereby higher crude protein content typically corresponds to lower fat content. The fat percentage exhibited an 11.29% reduction in SCG10 and a 3.69% decrease in SCG25 compared to SCG0. It is important to highlight that while dietary fat serves as a crucial component of a balanced diet, it necessitates prudent consumption. This is underscored by research associating the consumption of high-fat diets with the development of severe health conditions [53].

The percentage of fat in a food product is of paramount significance, but equally critical is the assessment of its oil content and value, which is determined via relevant indicators [54]. Fatty acids encompass diverse health-promoting effects, including the prevention of cardiovascular disease, exhibited by polyunsaturated and monounsaturated fatty acids [55]. The composition of fatty acids in the samples is presented in Table 3, revealing substantial quantities of important fatty acids such as palmitic acid (C16:0), oleic acid (C18:1), and linoleic acid (C18:2 ω-6). Concerning C16:0, higher values were observed in SCG25, which exhibited increases of 8.76% and 10.82% over SCG0 and SCG10, respectively. Additionally, C18:1 is known for its anti-cancer properties [56], and our data in Table 3 demonstrates a significant ($p < 0.05$) increase in C18:1 content as SCG content rises in the TM larval feeding substrates, reaching a peak of 45.87% in SCG25. Similarly noteworthy is the fatty acid C18:2, known for its protective effects against cardiovascular diseases [57]. While the C18:2 content of the samples is high, it is noteworthy that an increase in SCGs in the feed does not appear to increase C18:2 levels. Instead, our findings indicate a significant ($p < 0.05$) decrease of 31.77% between the SCG0 and SCG25 samples. These results collectively underscore the diversity of fatty acids present in TM larvae and their potential health benefits, while shedding light on the nuanced impact of coffee content on specific fatty acid compositions. Nevertheless, it should be pointed out that a high saturated fat intake is associated with atherosclerosis and coronary artery diseases [58] and, as can been seen in Table 3, TM larvae are quite rich in saturated fatty acids.

Table 3. Fatty acid composition of *T. molitor* larvae fed for eight weeks with wheat bran (control) (SCG0) and bran fortified with different rates of spent coffee grounds, 10% (SCG10) and 25% (SCG25)) ($n = 9$).

Fatty Acid (%)	Diets		
	SCG0	SCG10	SCG25
C12:0	0.08 ± 0.00 [b,*]	0.10 ± 0.00 [a]	nd [**]
C14:0	2.06 ± 0.08 [b]	2.77 ± 0.19 [a]	2.36 ± 0.06 [b]
C16:0	19.3 ± 1.37 [a]	18.94 ± 0.53 [a]	20.99 ± 1.26 [a]
C18:0	0.21 ± 0.01 [b]	0.23 ± 0.01 [a]	nd
C18:1	23.98 ± 1.39 [b]	24.83 ± 1.86 [b]	34.98 ± 1.64 [a]
C18:2 (ω-6)	53.87 ± 4.04 [a]	52.67 ± 3.53 [a]	40.88 ± 1.92 [b]
C20:0	0.50 ± 0.03 [b]	0.45 ± 0.03 [b]	0.78 ± 0.02 [a]
∑ SFA [1]	22.15 ± 1.49 [a]	22.5 ± 0.76 [a]	24.13 ± 1.34 [a]
∑ MUFA [2]	23.98 ± 1.39 [b]	24.83 ± 1.86 [b]	34.98 ± 1.64 [a]
∑ PUFA [3]	53.87 ± 4.04 [a]	52.67 ± 3.53 [a]	40.88 ± 1.92 [b]
∑ UFA [4]	77.85 ± 5.43 [a]	77.5 ± 5.39 [a]	75.87 ± 3.57 [a]
PUFA:SFA ratio	2.43 ± 0.02 [a]	2.34 ± 0.08 [a]	1.69 ± 0.01 [b]
MUFA:PUFA ratio	0.45 ± 0.01 [c]	0.47 ± 0.00 [b]	0.86 ± 0.00 [a]
COX [5]	5.79 ± 0.43 [a]	5.67 ± 0.38 [a]	4.56 ± 0.21 [b]
IA [6]	0.35 ± 0.00 [b]	0.39 ± 0.01 [a]	0.4 ± 0.00 [a]
IT [7]	0.55 ± 0.00 [b]	0.57 ± 0.02 [b]	0.62 ± 0.01 [a]
HH [8]	3.63 ± 0.01 [a]	3.55 ± 0.13 [a]	3.25 ± 0.03 [b]
HPI [9]	2.82 ± 0.02 [a]	2.57 ± 0.07 [b]	2.49 ± 0.01 [b]

* Within each line, statistically significant differences ($p < 0.05$) are denoted with different superscript letters (e.g., a–c). ** nd: not detected. [1] SFAs, saturated fatty acids (%): SUM of C12:0, lauric acid; C14:0, myristic acid; C16:0, palmitic acid; C18:0, stearic acid; C20:0, arachidic acid. [2] MUFAs, monounsaturated fatty acids (%): SUM of C18:1, oleic acid. [3] PUFAs, polyunsaturated fatty acids (%): SUM of C18:2, ω-6, linoleic acid. [4] UFAs, unsaturated fatty acids (%): SUM of MUFAs and PUFAs. [5] COX, calculated oxidizability value. [6] IA, Index of atherogenicity. [7] IT, Index of thrombogenicity. [8] HH, Hypocholesterolemic/hypercholesterolemic ratio. [9] HPI, Health-promoting index.

Furthermore, the COX value, an important indicator in assessing the oxidative stability of oil, was calculated. A lower COX value signifies enhanced oxidative stability and, consequently, an extended shelf life of the oil product [59]. In our case, we observed a decrease in the COX value as the SCG content increased, underscoring the fact that SCG consumption by TM larvae promotes a higher shelf life of their oil. Next, other indicators pertinent to the overall quality of oils were also examined. These indicators encompassed the atherogenicity index (IA), thrombogenicity index (IT), and health promotion index (HPI). A lower IA value is indicative of a healthier food product. For instance, high IA values, such as 4.08, are reported for milk [60], a value which is ten times higher than the values recorded for TM larvae oil. In our case, no statistically significant differences ($p < 0.05$) were recorded for SCG-fed larvae. A similar principle applies to the IT index, where lower values are more favorable for human health. For instance, in a study focused on seaweed, another innovative food product alongside insects, the IT value ranged from

0.04 to 2.94 [61]. In contrast, our samples exhibited a maximum value of 0.62 (SCG25). The above comparison further contextualizes the health benefits, highlighting the potential of SCG-fed larvae as a nutritionally favorable and innovative food source. Last but not least, HPI is relevant as the consumption of foods with high values of this index has a positive effect on cardiovascular diseases [62]. Comparing the present HPI data with the corresponding HPI value of meat (2.91) [63], it is obvious that the data are fully comparable while proving that the nutritional value of TM larvae does not differ significantly from conventional food products. This not only supports the larvae's potential as a sustainable alternative but also underscores its role in promoting cardiovascular health. In essence, the analysis of these indices provides a nuanced and comprehensive assessment of the nutritional quality and health implications of TM larvae reared on SCGs. The results not only contribute to the broader understanding of insect-based nutrition but also position SCG-fed larvae as a promising and innovative source of nutrition with potential benefits for human health.

3.2.2. Vitamin C and A Content of TM Larvae

Vitamins C and A belong to the group of essential vitamins, each serving distinct roles crucial for proper human physiological function [64]. TM larvae inherently exhibit vitamin deficiencies [65]. Hence, this study endeavors to naturally enhance their vitamin content via dietary modifications. Results are presented in Table 4. A substantial increase in vitamin content within the examined larvae was recorded as the quantity of SCGs in their dietary substrates increased. While SCGs themselves lack increased ascorbic acid content, they do contain a quantity of up to 8.18 ± 0.39 mg of vitamin C equivalent (VCE)/g SCG [66]. Evidently, this is efficiently absorbed by the larvae during their developmental stages. The elevation in ascorbic acid content was high, exhibiting a statistically significant ($p < 0.05$) rise of 43.63% with the inclusion of 10% SCGs and an even more pronounced increase of 81.28% with the addition of 25% SCGs. This underscores the capacity of the larvae to increase their vitamin C levels, despite prior deficiencies, merely by consuming food waste. Equally noteworthy is the enhancement in β-carotene content, a vital nutrient for humans because it contributes to various therapeutic effects in managing numerous diseases, such as cancer, cardiovascular disorders, COVID-19, cystic fibrosis, and bronchiectasis [67,68]. Larvae fed with this diet can increase their content up to 8.42 µg/g, marking a substantial increase from the control sample containing 0.91 µg/g. The same applies for vitamin A. Larvae fed on bran exhibit a vitamin A content of approximately 2.72 µg RAE/100 g. Intriguingly, the addition of up to 25% SCGs increases this content by 822.79%. Comparatively, chicken contains a mere 6.00 µg/100 g of vitamin A [69]. This underscores the nutritional potential of mealworm larvae, revealing their richness in proteins and essential nutrients, including vitamins, in comparison to conventional protein sources such as chicken.

Table 4. Content of *T. molitor* larvae fed for eight weeks with wheat bran (control) (SCG0) and bran fortified with different rates of spent coffee grounds, 10% (SCG10) and 25% (SCG25)), in vitamin C, β-carotene, and vitamin A ($n = 9$).

Diets	Vitamin C (µg/g)	β-Carotene (µg/g)	Vitamin A (µg RAE/100 g)
SCG0	218.09 ± 2.09 [c,*]	0.91 ± 0.06 [c]	2.72 ± 0.19 [c]
SCG10	313.25 ± 3.24 [b]	4.36 ± 0.56 [b]	13.02 ± 1.68 [b]
SCG25	395.35 ± 1.98 [a]	8.42 ± 0.34 [a]	25.1 ± 1 [a]

* Within each column, statistically significant differences ($p < 0.05$) are denoted with different superscript letters (e.g., a–c).

3.2.3. Antioxidant Properties of the Larvae's Extracts

Table 5 provides a summary of the results of the assessment of the antioxidant activity of mealworm larvae. The results unveil a consistent pattern. As the amount of SCGs in the dietary substrate increases, both the antioxidant activity and polyphenol concentration exhibit an increase. More specifically, the addition of 10% SCGs results in a statistically significant ($p < 0.05$) increase of 9.5% in antioxidant activity, and this effect becomes even more pronounced with the addition of 25% SCGs, where the increase rises to 91.86%. Furthermore, statistically significant differences ($p < 0.05$) are evident in polyphenol content among the samples. A comparative analysis between SCG0 and SCG10 reveals a 23.34% increase, while comparing SCG0 and SCG25 demonstrates a more substantial increase (29.01%). These findings are comparable with prior studies, proposing that SCGs exhibit antioxidant activity and contain polyphenols (approximately 34.43 mg GAE/g) [7,8,70]. Hence, it can be inferred from these results that the nutrients present in the larvae feed substrate are readily absorbed by the larvae, resulting in additive nutritional resources. TM larvae with enhanced antioxidant properties could be used to develop dietary supplements targeting specific health problems, such as immune system support, anti-ageing, or cardiovascular health. So, these larvae can be processed into various food products like protein powders, protein bars, and snacks giving them extra significant properties. However, attention should be paid to the European Union's indications on the consumption of TM larvae. In particular, the Commission states that primary sensitization and allergic reactions to flour proteins from TM larvae may be induced in people with allergies to crustaceans and dust mites [71].

Table 5. Antioxidant properties (FRAP assay) and total polyphenols (TPC) of *T. molitor* larvae fed with different rates of spent coffee grounds, 10% (SCG10) and 25% (SCG25)) ($n = 9$).

Diets	FRAP (µmol AAE/g)	TPC (mg GAE/g)
SCG0	130.8 ± 2.48 [c,*]	43.22 ± 1.8 [b]
SCG10	143.23 ± 1.87 [b]	53.31 ± 1.48 [a]
SCG25	250.95 ± 3.53 [a]	55.76 ± 0.24 [a]

* Within each column, statistically significant differences ($p < 0.05$) are denoted with different superscript letters (e.g., a–c).

4. Conclusions

Coffee consumption generates substantial waste. Our present study reaffirms the suitability of these by-products as animal feed, while opening new avenues for their use in rearing TM larvae. Although, a minor decrease in their survivability was recorded, without having a toll on the weight gain of the larvae, the substrate comprising 75% wheat bran and 25% SCGs proved highly beneficial for TM larvae, enhancing their protein, vitamin, and polyphenol contents, as well as their antioxidant capacity. Our findings underscore the utilization potential of SCGs, promoting the sustainability and ecological impact of coffee consumption. In fact, by adding SCGs in the feed substrate of TM larvae, four goals of the United Nations Sustainable Development Goals (SDGs) can be achieved. Specifically, goal 2 can be achieved through the production of alternative animal feed, combined with the achievement of goal 12 for reducing food industry by-products and promoting a circular economy. TM larvae can be raised using food industry waste, and no negative effects were observed in their growth, development, and metamorphosis. Additionally, goal 2, producing new protein-rich foods, can be achieved with a 45.26% increase in crude protein content in the larvae after their rearing with 10% SCGs. Furthermore, with the increase in protein content, there was an 81.28% increase in vitamin C and an 822.79% increase in vitamin A in the samples that were reared with 25% SCGs. This study further substantiates the notion that insects can effectively consume a wide array of products and efficiently absorb the nutrients offered in their diet. Moreover, it is important to highlight that TM larvae possess the potential to become one of the most nutritionally-rich sources of animal-derived food. These points warrant increased consideration for their consumption either in their natural form or as additives in low–nutritional value products like bread, cakes, and biscuits. Moreover, it must be stated that the present study does not present a ready-to-eat food product but a potential additional ingredient, TM flour, for the preparation of a new product. Therefore, since the utilization of SCGs was found suitable for insect rearing, we urge other researchers to use it for the creation of more nutritious larvae and add it to food, whereas future work could address concerns regarding the microbiological and chemical safety of the new food products.

Author Contributions: Conceptualization, S.I.L. and K.K.; methodology, S.I.L. and K.K.; software, T.C. and V.A.; validation, K.K., E.B., T.C. and V.A.; formal analysis, K.K., T.C., V.A. and E.B.; investigation, K.K. and E.B.; resources, S.I.L.; data curation, T.C., V.A. and S.I.L.; writing—original draft preparation, K.K. and T.C.; writing—review and editing, K.K., T.C., V.A., E.B. and S.I.L.; visualization, K.K.; supervision, C.G.A. and S.I.L.; project administration, S.I.L. All authors have read and agreed to the published version of the manuscript.

Funding: This research received no external funding.

Institutional Review Board Statement: Not applicable.

Informed Consent Statement: Not applicable.

Data Availability Statement: All the data are contained within the article.

Conflicts of Interest: The authors declare no conflict of interest.

Abbreviations

The following abbreviations are used in this manuscript:

COX	Calculated Oxidizability Value
HH	Hypocholesterolemic/Hypercholesterolemic Ratio
HPI	Health-Promoting Index
IA	Index of Atherogenicity
IT	Index of Thrombogenicity
MUFA	Monounsaturated Fatty Acids
PUFA	Polyunsaturated Fatty Acids
SCG	Spent Coffee Grounds
SFA	Saturated Fatty Acids
TM	*Tenebrio molitor*
UFA	Unsaturated Fatty Acids

References

1. Kaza, S.; Yao, L.; Bhada-Tata, P.; Van Woerden, F. *What a Waste 2.0 Introduction—"Snapshot of Solid Waste Management to 2050." Overview Booklet*; Urban Development Series; World Bank Group: Washington, DC, USA, 2018; pp. 1–38.
2. FAO 2011. *The Methodology of the FAO Study: "Global Food Losses and Food Waste—Extent, Causes and Prevention"*; Vol. SIK Report; FAO: Rome, Italy, 2013; ISBN 9789172903234.
3. Patwa, N.; Sivarajah, U.; Seetharaman, A.; Sarkar, S.; Maiti, K.; Hingorani, K. Towards a Circular Economy: An Emerging Economies Context. *J. Bus. Res.* **2021**, *122*, 725–735. [CrossRef]
4. Mussatto, S.I.; Carneiro, L.M.; Silva, J.P.A.; Roberto, I.C.; Teixeira, J.A. A Study on Chemical Constituents and Sugars Extraction from Spent Coffee Grounds. *Carbohydr. Polym.* **2011**, *83*, 368–374. [CrossRef]
5. Chatzimitakos, T.; Athanasiadis, V.; Kotsou, K.; Palaiogiannis, D.; Bozinou, E.; Lalas, S.I. Optimized Isolation Procedure for the Extraction of Bioactive Compounds from Spent Coffee Grounds. *Appl. Sci.* **2023**, *13*, 2819. [CrossRef]
6. Getachew, A.T.; Chun, B.S. Influence of Pretreatment and Modifiers on Subcritical Water Liquefaction of Spent Coffee Grounds: A Green Waste Valorization Approach. *J. Clean. Prod.* **2017**, *142*, 3719–3727. [CrossRef]
7. Seo, H.S.; Park, B.H. Phenolic Compound Extraction from Spent Coffee Grounds for Antioxidant Recovery. *Korean J. Chem. Eng.* **2019**, *36*, 186–190. [CrossRef]
8. Panusa, A.; Zuorro, A.; Lavecchia, R.; Marrosu, G.; Petrucci, R. Recovery of Natural Antioxidants from Spent Coffee Grounds. *J. Agric. Food Chem.* **2013**, *61*, 4162–4168. [CrossRef] [PubMed]
9. Smyrnakis, G.; Stamoulis, G.; Palaiogiannis, D.; Chatzimitakos, T.; Athanasiadis, V.; Lalas, S.I.; Makris, D.P. Recovery of Polyphenolic Antioxidants from Coffee Silverskin Using Acid-Catalyzed Ethanol Organosolv Treatment. *ChemEngineering* **2023**, *7*, 72. [CrossRef]
10. Santana-Gálvez, J.; Cisneros-Zevallos, L.; Jacobo-Velázquez, D.A. Chlorogenic Acid: Recent Advances on Its Dual Role as a Food Additive and a Nutraceutical against Metabolic Syndrome. *Molecules* **2017**, *22*, 358. [CrossRef]
11. Kang, N.J.; Lee, K.W.; Shin, B.J.; Jung, S.K.; Hwang, M.K.; Bode, A.M.; Heo, Y.S.; Lee, H.J.; Dong, Z. Caffeic Acid, a Phenolic Phytochemical in Coffee, Directly Inhibits Fyn Kinase Activity and UVB-Induced COX-2 Expression. *Carcinogenesis* **2009**, *30*, 321–330. [CrossRef]

12. Haider, K.; Haider, M.R.; Neha, K.; Yar, M.S. Free Radical Scavengers: An Overview on Heterocyclic Advances and Medicinal Prospects. *Eur. J. Med. Chem.* **2020**, *204*, 112607. [CrossRef]
13. Zarrinbakhsh, N.; Rodriguez, A.; Misra, M.; Mohanty, A. Characterization of Wastes and Coproducts from the Coffee Industry for Composite Material Production. *BioRes* **2016**, *11*, 7637–7653. [CrossRef]
14. Bordiean, A.; Krzyżaniak, M.; Stolarski, M.J. Bioconversion Potential of Agro-Industrial Byproducts by *Tenebrio molitor*—Long-Term Results. *Insects* **2022**, *13*, 810. [CrossRef] [PubMed]
15. Aidoo, O.F.; Osei-Owusu, J.; Asante, K.; Dofuor, A.K.; Boateng, B.O.; Debrah, S.K.; Ninsin, K.D.; Siddiqui, S.A.; Chia, S.Y. Insects as Food and Medicine: A Sustainable Solution for Global Health and Environmental Challenges. *Front. Nutr.* **2023**, *10*, 1113219. [CrossRef] [PubMed]
16. Thrastardottir, R.; Olafsdottir, H.T.; Thorarinsdottir, R.I. Yellow Mealworm and Black Soldier Fly Larvae for Feed and Food Production in Europe, with Emphasis on Iceland. *Foods* **2021**, *10*, 2744. [CrossRef]
17. Zielińska, E.; Baraniak, B.; Karaś, M.; Rybczyńska, K.; Jakubczyk, A. Selected Species of Edible Insects as a Source of Nutrient Composition. *Food Res. Int.* **2015**, *77*, 460–466. [CrossRef]
18. Jantzen da Silva Lucas, A.; Menegon de Oliveira, L.; da Rocha, M.; Prentice, C. Edible Insects: An Alternative of Nutritional, Functional and Bioactive Compounds. *Food Chem.* **2020**, *311*, 126022. [CrossRef]
19. Andreadis, S.S.; Panteli, N.; Mastoraki, M.; Rizou, E.; Stefanou, V.; Tzentilasvili, S.; Sarrou, E.; Chatzifotis, S.; Krigas, N.; Antonopoulou, E. Towards Functional Insect Feeds: Agri-Food by-Products Enriched with Post-Distillation Residues of Medicinal Aromatic Plants in *Tenebrio molitor* (Coleoptera: Tenebrionidae) Breeding. *Antioxidants* **2022**, *11*, 68. [CrossRef]
20. Bordiean, A.; Krzyżaniak, M.; Stolarski, M.J.; Peni, D. Growth Potential of Yellow Mealworm Reared on Industrial Residues. *Agric.* **2020**, *10*, 599. [CrossRef]
21. Bordiean, A.; Krzyżaniak, M.; Aljewicz, M.; Stolarski, M.J. Influence of Different Diets on Growth and Nutritional Composition of Yellow Mealworm. *Foods* **2022**, *11*, 3075. [CrossRef]
22. Biasato, I.; Gasco, L.; De Marco, M.; Renna, M.; Rotolo, L.; Dabbou, S.; Capucchio, M.T.; Biasibetti, E.; Tarantola, M.; Sterpone, L.; et al. Yellow Mealworm Larvae (*Tenebrio molitor*) Inclusion in Diets for Male Broiler Chickens: Effects on Growth Performance, Gut Morphology, and Histological Findings. *Poult. Sci.* **2018**, *97*, 540–548. [CrossRef]
23. Rumpold, B.A.; Schlüter, O.K. Nutritional Composition and Safety Aspects of Edible Insects. *Mol. Nutr. Food Res.* **2013**, *57*, 802–823. [CrossRef] [PubMed]
24. Truman, J.W. The Evolution of Insect Metamorphosis. *Curr. Biol.* **2019**, *29*, R1252–R1268. [CrossRef] [PubMed]
25. May, J.C.; Wheeler, R.M.; Grim, E. The Gravimetric Method for the Determination of Residual Moisture in Freeze-Dried Biological Products. *Cryobiology* **1989**, *26*, 277–284. [CrossRef] [PubMed]
26. Kotsou, K.; Chatzimitakos, T.; Athanasiadis, V.; Bozinou, E.; Rumbos, C.I.; Athanassiou, C.G.; Lalas, S.I. Enhancing the Nutritional Profile of *Tenebrio molitor* Using the Leaves of *Moringa oleifera*. *Foods* **2023**, *12*, 2612. [CrossRef] [PubMed]
27. Son, Y.; Hwang, I.; Nho, C.; Kim, S.; Kim, S. Determination of Carbohydrate Composition in Mealworm. *Foods* **2021**, *10*, 640. [CrossRef] [PubMed]
28. Al-Amiery, A.A.; Al-Majedy, Y.K.; Kadhum, A.A.H.; Mohamad, A.B. Hydrogen Peroxide Scavenging Activity of Novel Coumarins Synthesized Using Different Approaches. *PLoS ONE* **2015**, *10*, e0132175. [CrossRef]
29. Ocampo, E.T.M.; Libron, J.A.M.A.; Guevarra, M.L.D.; Mateo, J.M.C. Phytochemical Screening, Phenolic Acid Profiling and Antioxidant Activity Analysis of Peels from Selected Mango (*Mangifera* spp.) Genotypes in the Philippines. *Food Res.* **2020**, *4*, 1116–1124. [CrossRef]
30. Makris, D.P.; Kefalas, P. Characterization of Polyphenolic Phytochemicals in Red Grape Pomace. *Int. J. Waste Resour.* **2013**, *3*, 2–5. [CrossRef]
31. Cicco, N.; Lanorte, M.T.; Paraggio, M.; Viggiano, M.; Lattanzio, V. A Reproducible, Rapid and Inexpensive Folin-Ciocalteu Micro-Method in Determining Phenolics of Plant Methanol Extracts. *Microchem. J.* **2009**, *91*, 107–110. [CrossRef]
32. Pinotti, L.; Ottoboni, M. Substrate as Insect Feed for Bio-Mass Production. *J. Insects Food Feed* **2021**, *7*, 585–596. [CrossRef]
33. LeCato, G.; Flaherty, B. Description of Eggs of Selected Species of Stored-Product Insects (Coleoptera and Lepidoptera). *J. Kansas Entomol. Soc.* **1974**, *39*, 308–317.
34. Pakshir, K.; Dehghani, A.; Nouraei, H.; Zareshahrabadi, Z.; Zomorodian, K. Evaluation of Fungal Contamination and Ochratoxin A Detection in Different Types of Coffee by HPLC-Based Method. *J. Clin. Lab. Anal.* **2021**, *35*, e24001. [CrossRef] [PubMed]
35. Kröncke, N.; Benning, R. Self-Selection of Feeding Substrates by *Tenebrio molitor* Larvae of Different Ages to Determine Optimal Macronutrient Intake and the Influence on Larval Growth and Protein Content. *Insects* **2022**, *13*, 657. [CrossRef] [PubMed]
36. Hussein, H.; Abouamer, W.; Ali, H.; Elkhadragy, M.; Yehia, H.; Farouk, A. The Valorization of Spent Coffee Ground Extract as a Prospective Insecticidal Agent against Some Main Key Pests of Phaseolus Vulgaris in the Laboratory and Field. *Plants* **2022**, *11*, 1124. [CrossRef]
37. Thanasoponkul, W.; Changbunjong, T.; Sukkurd, R.; Saiwichai, T. Spent Coffee Grounds and Novaluron Are Toxic to Aedes Aegypti (Diptera: Culicidae) Larvae. *Insects* **2023**, *14*, 564. [CrossRef] [PubMed]
38. Kröncke, N.; Wittke, S.; Steinmann, N.; Benning, R. Analysis of the Composition of Different Instars of *Tenebrio molitor* Larvae Using Near-Infrared Reflectance Spectroscopy for Prediction of Amino and Fatty Acid Content. *Insects* **2023**, *14*, 310. [CrossRef]
39. Kröncke, N.; Benning, R. Determination of Moisture and Protein Content in Living Mealworm Larvae (*Tenebrio molitor* L.) Using Near-Infrared Reflectance Spectroscopy (NIRS). *Insects* **2022**, *13*, 560. [CrossRef]

40. Costa, S.; Pedro, S.; Lourenço, H.; Batista, I.; Teixeira, B.; Bandarra, N.M.; Murta, D.; Nunes, R.; Pires, C. Evaluation of *Tenebrio molitor* Larvae as an Alternative Food Source. *NFS J.* **2020**, *21*, 57–64. [CrossRef]

41. Nowak, V.; Persijn, D.; Rittenschober, D.; Charrondiere, U.R. Review of Food Composition Data for Edible Insects. *Food Chem.* **2016**, *193*, 39–46. [CrossRef]

42. Iriondo-Dehond, A.; Uranga, J.A.; Del Castillo, M.D.; Abalo, R. Effects of Coffee and Its Components on the Gastrointestinal Tract and the Brain–Gut Axis. *Nutrients* **2021**, *13*, 88. [CrossRef]

43. Siulapwa, N.; Mwambungu, A.; Lungu, E.; Sichilima, W. Nutritional Value of Four Common Edible Insects in Zambia. *Int. J. Sci. Res.* **2014**, *3*, 876–884.

44. Melo, V.; Garcia, M.; Sandoval, H.; Jiménez, H.D.; Calvo, C. Quality Proteins from Edible Indigenous Insect Food of Latin America and Asia. *Emirates J. Food Agric.* **2011**, *23*, 283–289.

45. Daniel, J.R.; Vidovic, N. *Carbohydrates, Role in Human Nutrition*; Reference Module in Food Science; Elsevier: Amsterdam, The Netherlands, 2018. [CrossRef]

46. Abidin, N.A.Z.; Kormin, F.; Abidin, N.A.Z.; Anuar, N.A.F.M.; Bakar, M.F.A. The Potential of Insects as Alternative Sources of Chitin: An Overview on the Chemical Method of Extraction from Various Sources. *Int. J. Mol. Sci.* **2020**, *21*, 4978. [CrossRef] [PubMed]

47. Dong, T.; Guo, M.; Zhang, P.; Sun, G.; Chen, B. The Effects of Low-Carbohydrate Diets on Cardiovascular Risk Factors: A Meta-Analysis. *PLoS ONE* **2020**, *15*, e0225348. [CrossRef] [PubMed]

48. Santos, F.L.; Esteves, S.S.; da Costa Pereira, A.; Yancy, W.S.J.; Nunes, J.P.L. Systematic Review and Meta-Analysis of Clinical Trials of the Effects of Low Carbohydrate Diets on Cardiovascular Risk Factors. *Obes. Rev. Off. J. Int. Assoc. Study Obes.* **2012**, *13*, 1048–1066. [CrossRef]

49. Ghaly, A.E.; Alkoaik, F.N. The Yellow Mealworm as a Novel Source of Protein. *Am. J. Agric. Biol. Sci.* **2009**, *4*, 319–331. [CrossRef]

50. Dreassi, E.; Cito, A.; Zanfini, A.; Materozzi, L.; Botta, M.; Francardi, V. Dietary Fatty Acids Influence the Growth and Fatty Acid Composition of the Yellow Mealworm *Tenebrio molitor* (Coleoptera: Tenebrionidae). *Lipids* **2017**, *52*, 285–294. [CrossRef]

51. Ghosh, S.; Lee, S.M.; Jung, C.; Meyer-Rochow, V.B. Nutritional Composition of Five Commercial Edible Insects in South Korea. *J. Asia. Pac. Entomol.* **2017**, *20*, 686–694. [CrossRef]

52. Jajic, I.; Popovic, A.; Urosevic, M.; Krstovic, S.; Petrovic, M.; Guljas, D.; Samardzic, M. Fatty and Amino Acid Profile of Mealworm Larvae (*Tenebrio molitor* L.). *Biotechnol. Anim. Husb.* **2020**, *36*, 167–180. [CrossRef]

53. Pipoyan, D.; Stepanyan, S.; Stepanyan, S.; Beglaryan, M.; Costantini, L.; Molinari, R.; Merendino, N. The Effect of Trans Fatty Acids on Human Health: Regulation and Consumption Patterns. *Foods* **2021**, *10*, 2452. [CrossRef]

54. Yun, J.M.; Surh, J. Fatty Acid Composition as a Predictor for the Oxidation Stability of Korean Vegetable Oils with or without Induced Oxidative Stress. *Prev. Nutr. Food Sci.* **2012**, *17*, 158–165. [CrossRef]

55. Baum, S.J.; Kris-Etherton, P.M.; Willett, W.C.; Lichtenstein, A.H.; Rudel, L.L.; Maki, K.C.; Whelan, J.; Ramsden, C.E.; Block, R.C. Fatty Acids in Cardiovascular Health and Disease: A Comprehensive Update. *J. Clin. Lipidol.* **2012**, *6*, 216–234. [CrossRef] [PubMed]

56. Paszczyk, B.; Polak-śliwińska, M.; Łuczyńska, J. Fatty Acids Profile, Trans Isomers, and Lipid Quality Indices in Smoked and Unsmoked Cheeses and Cheese-like Products. *Int. J. Environ. Res. Public Health* **2020**, *17*, 71. [CrossRef] [PubMed]

57. Atik, I.; Karasu, S.; Sevik, R. Physicochemical and Bioactive Properties of Cold Press Wild Plum (*Prunus spinosa*) and Sour Cherry (*Prunus cerasus*) Kernel Oils: Fatty Acid, Sterol and Phenolic Profile. *Riv. Ital. Sostanze Grasse* **2022**, *99*, 13–20.

58. Wali, J.A.; Jarzebska, N.; Raubenheimer, D.; Simpson, S.J.; Rodionov, R.N.; O'Sullivan, J.F. Cardio-Metabolic Effects of High-Fat Diets and Their Underlying Mechanisms-A Narrative Review. *Nutrients* **2020**, *12*, 1505. [CrossRef] [PubMed]

59. Al-Amrousi, E.F.; Badr, A.N.; Abdel-Razek, A.G.; Gromadzka, K.; Drzewiecka, K.; Hassanein, M.M.M. A Comprehensive Study of Lupin Seed Oils and the Roasting Effect on Their Chemical and Biological Activity. *Plants* **2022**, *11*, 2301. [CrossRef]

60. Nantapo, C.T.W.; Muchenje, V.; Hugo, A. Atherogenicity Index and Health-Related Fatty Acids in Different Stages of Lactation from Friesian, Jersey and Friesian × Jersey Cross Cow Milk under a Pasture-Based Dairy System. *Food Chem.* **2014**, *146*, 127–133. [CrossRef]

61. Kumar, M.; Kumari, P.; Trivedi, N.; Shukla, M.K.; Gupta, V.; Reddy, C.R.K.; Jha, B. Minerals, PUFAs and Antioxidant Properties of Some Tropical Seaweeds from Saurashtra Coast of India. *J. Appl. Phycol.* **2011**, *23*, 797–810. [CrossRef]

62. Chen, J.; Liu, H. Nutritional Indices for Assessing Fatty Acids: A Mini-Review. *Int. J. Mol. Sci.* **2020**, *21*, 5695. [CrossRef]

63. Makhutova, O.N.; Nokhsorov, V.V.; Stoyanov, K.N.; Dudareva, L.V.; Petrov, K.A. Preliminary Estimation of Nutritional Quality of the Meat, Liver, and Fat of the Indigenous Yakutian Cattle Based on Their Fatty Acid Profiles. *Foods* **2023**, *12*, 3226. [CrossRef]

64. Doley, J. *Vitamins and Minerals in Older Adults: Causes, Diagnosis, and Treatment of Deficiency*; Elsevier Inc.: Amsterdam, The Netherlands, 2017; ISBN 9780128092996.

65. Oonincx, D.G.A.B.; Finke, M.D. Insects as a Complete Nutritional Source. *J. Insects Food Feed* **2023**, *9*, 541–543. [CrossRef]

66. Wu, C.T.; Agrawal, D.C.; Huang, W.Y.; Hsu, H.C.; Yang, S.J.; Huang, S.L.; Lin, Y.S. Functionality Analysis of Spent Coffee Ground Extracts Obtained by the Hydrothermal Method. *J. Chem.* **2019**, *2019*, 4671438. [CrossRef]

67. Kapała, A.; Szlendak, M.; Motacka, E. The Anti-Cancer Activity of Lycopene: A Systematic Review of Human and Animal Studies. *Nutrients* **2022**, *14*, 5152. [CrossRef] [PubMed]

68. Anand, R.; Mohan, L.; Bharadvaja, N. Disease Prevention and Treatment Using β-Carotene: The Ultimate Provitamin A. *Rev. Bras. Farmacogn.* **2022**, *32*, 491–501. [CrossRef]

69. Payne, C.L.R.; Scarborough, P.; Rayner, M.; Nonaka, K. A Systematic Review of Nutrient Composition Data Available for Twelve Commercially Available Edible Insects, and Comparison with Reference Values. *Trends Food Sci. Technol.* **2016**, *47*, 69–77. [CrossRef]
70. Andrade, C.; Perestrelo, R.; Câmara, J.S. Bioactive Compounds and Antioxidant Activity from Spent Coffee Grounds as a Powerful Approach for Its Valorization. *Molecules* **2022**, *27*, 7504. [CrossRef]
71. Turck, D.; Bohn, T.; Castenmiller, J.; De Henauw, S.; Hirsch-Ernst, K.I.; Maciuk, A.; Mangelsdorf, I.; McArdle, H.J.; Naska, A.; Pelaez, C.; et al. Safety of Frozen and Dried Formulations from Whole Yellow Mealworm (*Tenebrio molitor* Larva) as a Novel Food Pursuant to Regulation (EU) 2015/2283. *EFSA J.* **2021**, *19*, e06778. [CrossRef]

Article

Sustainable Strategies for the Recovery and Valorization of Brewery By-Products—A Multidisciplinary Approach

Alina Soceanu [1], Simona Dobrinas [1,*], Viorica Popescu [1], Alina Buzatu [2] and Anca Sirbu [3]

1   Chemistry and Chemical Engineering Department, Faculty of Applied Chemistry and Engineering, "Ovidius" University from Constanta, 900470 Constanta, Romania; asoceanu@univ-ovidius.ro (A.S.); vpopescu@univ-ovidius.ro (V.P.)
2   Romanian Philology, Classical and Balkan Languages Department, Faculty of Letters, "Ovidius" University from Constanta, 900470 Constanta, Romania; alina.buzatu@univ-ovidius.ro
3   Department of Fundamental Sciences and Humanities, Constanta Maritime University, 900663 Constanta, Romania; anca.sirbu@cmu-edu.eu
*   Correspondence: sdobrinas@univ-ovidius.ro; Tel.: +40-241606434

Abstract: The prevention of environmental pollution is a current concern of the population, which is looking for ways to reduce the production of industrial waste. The brewing industry generates huge amounts of waste, with difficult management from an economic point of view. The waste obtained from the technological process of beer production is used in various branches, such as the food industry, mainly as feed, additives, or food ingredients; as animal feed; in biofuel production; and in building or packaging materials. The valuable by-products obtained from brewery waste can serve as raw materials for further processing or become functional ingredients for the production of new functional products. Reusing and recycling are essential strategies for transforming waste into new valuable resources, and such strategies enable circular solutions to maintain the value of products and resources for as long as possible. The chemical composition of the waste obtained from beer manufacturing can vary slightly depending on the type and quality of the ingredients used and the prevailing conditions during each stage of the manufacturing process. This paper focuses on sustainable strategies for the recovery and valorization of brewery by-products. Experimentally, the aim was to determine the chemical characteristics of different types of brewery waste, such as moisture content, ash, pH, total content of phenolic compounds, and total protein content. The experimental values obtained have shown that brewery waste is a valuable by-product.

Keywords: waste; brewery; antioxidants; phenolic compounds; sustainability; strategies

Citation: Soceanu, A.; Dobrinas, S.; Popescu, V.; Buzatu, A.; Sirbu, A. Sustainable Strategies for the Recovery and Valorization of Brewery By-Products—A Multidisciplinary Approach. *Sustainability* **2024**, *16*, 220. https://doi.org/10.3390/su16010220

Academic Editors: Dario Donno and Dimitris Skalkos

Received: 9 December 2023
Revised: 21 December 2023
Accepted: 24 December 2023
Published: 26 December 2023

1. Introduction

Beer is a special drink; it has been known since 6000 years ago, from the time of the Sumerians. At that time, it was a divine drink intended only for the Goddess of Fertility, and later, it was used by monks in the Middle Ages as a sweetener for meat, being considered a food rich in carbohydrates, proteins, vitamins B1, B6, B12, PP and E, folic acid, nicotinic acid, potassium, magnesium, etc. Moderate consumption of beer has a beneficial effect on human health mainly due to antioxidants, suppression of the rise of blood plasma lipoproteins [1], anti-inflammatory effects, and reduced risk of cardiovascular disease [2–4]. In beers, a considerable part of the antioxidant capacity is linked to the content of phenolic compounds [5].

Beer is one of the most popular alcoholic beverages in the world. It is undistilled and obtained by fermentation in the traditional way with only four basic ingredients: water, wort (made from malt), hops, and yeast; the malt can be partly replaced by unmalted grains (corn, barley, or rice) or possibly with some enzymes. Barley malt is usually used in brewing, which, when wort is obtained, provides both the enzyme equipment and the substrate necessary for enzyme action and, at the same time, gives finished beers a typical malt flavor.

The classification of the types of beer denotes the character, even the origin of this drink, the ingredients used, and also the manufacturing technique, giving clear clues in terms of taste, appearance, color, and smell of beer. The aroma, color, and texture of beer may differ depending on the type of fermentation, which may be low or high frequency. Based on the types of fermentation, two main types of beer are distinguished: Lager beer, obtained by a low fermentation process, and Ale beer, obtained by high fermentation. Apart from the main types of beer mentioned, there are other types, such as: wheat beer, spontaneous fermentation beer, non-alcoholic beer, or craft beer. Of course, there are other classifications of the types of beer established according to the ingredients used in the production or certain particular characteristics of the brewing process, and this is the case with the types of hybrid beer. Beer assortments also differ depending on the brewer's recipe.

However, there is a global trend in the production of new types of beer, known as "styles", that differ from traditional ones. Beer styles vary significantly due to wort changes in the production technology through different microorganisms used, the use of atypical malts, or the addition of various adjuvants such as salt, herbs, or spices. The use of fruits in beer production is one of the most prominent trends found globally in local craft breweries or industrial breweries [6,7].

Modern brewing technologies that take into account technical, economic, sanitary, and quality standards have emerged. The implementation of unconventional technologies in the technological process of beer manufacturing aims to simplify technological operations, reduce production costs, reduce manufacturing time, diversify beer assortments, as well as reduce the amount of waste resulting from the technological flow [8,9].

The use of synthetic antioxidants in the food industry is restricted by legislation [10]. For this reason, a sustainable and renewable alternative is to obtain these compounds from waste materials, such as agroindustrial residues, plants, vegetables, and fruits, which are efficient sources of phenolic compounds [11–14].

2. Product Identity and Sales Strategies

If the scientific discourse of chemistry addresses beer in terms of denoted, clear, transitive, non-ambiguous, measurable information, talking about ingredients, properties, technological processes, etc., it is interesting to see how, having become a commercial product, beer becomes the pretext for fictional constructions, stories with legitimizing value.

Product identity, chemical composition, and sales strategies are some factors that can influence beer consumers [15].

Its commercial product status is known to entail a hybrid identity, codified by both image and written word. A product's labels become a semiotic space in which, on the one hand, expert, mandatory, legally regulated information (type of beer, ingredients, proportions, etc.) and, on the other hand, advertising information and telling a story, place the commercial product in a fictional world, attaching experiences and values projected by the imagination. Beer brands devote considerable financial and creative resources to building these stories—because they sell alongside the product and also sell the product—in fierce competition. To be recognizable and easy to remember, stories have semantic-symbolic labels, which are both iconic (the brand image) and textual—short, concise, simple, sometimes striking slogans, often delivered in a certain manner because the intonation or rhythm of delivery increases the chances that the prospective buyer will remember the product and be persuaded to buy it. There is another essential feature of these stories that are attached to a product through advertising discourse—they always add positive qualifiers to a product because they are meant to legitimize it in the commercial markets by enhancing its value.

The best-known local beer brands build their identities through often comparable advertising strategies. Most of them advertise premium quality and the fact that they are produced in Romania. Existence for several years adds value; some brands boast a foundation dating back to 1718 or 1874, another claims existence since 1974, and craft beer brands show more recent entry into commercial markets.

Image emblems symbolically link beer brands to Romanian identity, building advertising fictions around animals or birds that are representative of the local fauna—the brown bear, the bearded vulture, etc.—which imply strength, robustness, superiority, and royalty. The texts on the label enhance the image emblems through identity emotion, giving value to the commercial story by over-emphasizing a characteristic that is always positive; one brand with an animal emblem defines itself as the king of beer in Romania, another says that the story goes further, accompanying the message with the image of the national flag.

Other image emblems seem to highlight the production process itself as a guarantee of quality; a representation of the brewery appears on the label, which is called—also in the official, registered name—"the good beer brewery". Another brand also adds to its label the image of a building without any other text specifications, leading us to believe that it may be the production site or another space (a house?) that conveys to the buyer the idea of community, conviviality, and shared well-being.

Regardless of the strategies of advertising persuasion and the stories through which they assert their identity, local beer brands insist on communicating that they are part of our lives and that they share our experiences by responsibly facilitating joviality, euphoria, friendship, communication, and humor.

Expert information chimes with advertising fiction: some aimed at communities with scientific expertise, others at the general public. As with humans, a product's identity is part hard, denoted, verifiable information, part imaginative projection, an illusion of reality or make-believe.

3. Brewery Technology

In recent years, the beer industry has recorded important progress worldwide in terms of technology and the provision of new types of equipment and installations [16,17]. The technological flow is fully automated, monitored, and controlled from the "command and control room", where there are computers containing special programs for monitoring the entire technological process of brewing beer.

The classic technological process of beer manufacturing consists of obtaining malt, which is realized through the following sequence of phases: cleaning, sorting, weighing, germination, drying, and root cleaning of barley grains; obtaining the beer wort is achieved by grinding, scalding, saccharification, filtering the scald, boiling the malt wort with hops; primary fermentation and secondary fermentation; and final operations include filtration, pasteurization, and bottling.

3.1. Raw Materials

The malt represents the main raw material for brewing beer and is obtained by germinating barley seeds in order to obtain the source of hydrolytic enzymes, which, through their action on the substrate, determine the formation of the extract. The quality of the beer is due to the quality of the malt, which is why it is also called the "soul of beer".

The hops are the indispensable raw material; they represent the "seasoning of beer", giving it the bitter taste, specific aroma, color, clarity, improvement of the foam, and preservation power of the beer (natural preservative), which it influences thanks to the compounds that it contains, especially bitter substances (resins ~12–21%) and essential oils 0.5–2.5%.

The water must be drinkable because, following the technological flow of obtaining beer, it has a contribution of approximately 88% and requires a certain content of salts not to influence the technological process. The hardness of the water depends on the type of beer, and the pH of the wort and beer is influenced by the composition of the water [18].

3.1.1. Unmalted Cereals

Unmalted cereals are used to obtain a lighter-colored beer with abundant and stable foam: maize is used in the form of flour in a proportion of 5–30% compared to the amount of malt; rice is used to obtain bottom-fermented blonde beer by separately processing barley and rice bran mixed at 65 °C; barley, oats, and wheat are used in special types of beer.

3.1.2. Yeast

The yeast used in the technological flow of beer production is part of the *Saccharomyces carlsbergensis* group, which always ferments alcoholically. It was derived either from pure laboratory cultures or from the recovery of cells grown from a previous batch of fermentation. The most important property of yeast is its ability to agglutinate. From a fermentation point of view, yeast can be: top fermentation yeast, which ferments at high temperatures (greater than 10 °C) and rises to the surface at the end of fermentation, or lower fermentation yeast, which ferments at low temperatures and settles to the bottom of the vessel at the end of fermentation.

3.2. Stages of the Technological Process of Brewing Beer

Following the qualitative and quantitative reception of the malted and unmalted cereals, they are sent for cleaning, weighing, and removal of impurities, which is done through a magnetic separator or with six-roll mills. After that, the malt is ground so that the enzymes can act in the fermentation and saccharification stages, transforming the insoluble macromolecular substances into soluble products that will enter the composition of the beer wort.

Grinding the malt is a relatively long process, determined not only by the size and quality of the grain but also by the temperature of the water used. In order for the milling process to be efficient, the malt must undergo the polishing process to slide easily into the mill, avoiding adhesion to the walls of the soaking bunker.

Malting and saccharification form a complex process of enzymatic hydrolysis and consist of mixing ground malt with unmalted cereals and water, where the malt enzymes will destroy flour components and nitrogen-assimilable substances will accumulate in the must, forming brazing. The amount of water used is variable and is added depending on the type of malt used. As a result of this process, the disaggregation of proteins and the transformation of starch into dextrin and maltose are obtained [19].

The next stage is the filtering of the lees, which aims to obtain the beer wort and takes place by separating the wort from the insoluble parts that will later form the malt wort. In the first phase of the process, the must will flow freely with the formation of the first must, and in the second phase, the wort will be washed with water (50–75 °C) in order to recover the retained extract.

Boiling the wort with hops is the stage obtained after filtering the lees with the addition of hops and aims to solubilize the bitter and aromatic substances of the hops, protein coagulation, enzyme inactivation, and must sterilization, concentration and coloring of the must, cooling and clarifying the must, and elimination of some sulfur substances.

After boiling the wort and separating it from the hops, it is necessary to cool and retain the wort so that the temperature is brought to the values at which the seeding with yeast cultures is done. It is a complex process that takes place in the presence of air; it causes chemical transformations as a result of the oxidation of maltose, glucose, protein substances, etc.

The cooling of the wort takes place in two stages: the first time, there is a pre-cooling from 100 °C to 65 °C, followed by a deep cooling that must reach the yeast seeding temperature of 6–18 °C.

The hot wort is formed in the boiling step of the wort with hops and separates in the pre-cooling step. It consists of coarser particles (30–80 μm) and is removed from the must by sedimentation, centrifugation, or settling cyclones.

The cold tube is formed during deep cooling after the hot tube has been separated. It consists of fine particles, and its separation is done at a maximum percentage of 80–85% by sedimentation, centrifugation, or flotation.

Sterilization of the must is done by inactivating the foreign microflora that occurs during the uncontrolled acidification of the must. It is sterilized by bringing the must to the boiling stage. In order to be seeded, the must is cooled to 6–7 °C, to which the yeast suspension is added, obtained through pure laboratory cultures. There is also bubbling with purified air to stimulate fermentation.

After all the primary wort phases of the technological flow and its fermentation, the young beer product is obtained. Fermentation can be of two kinds: primary and secondary.

The primary fermentation takes place in several stages, and the result is young beer (fermentable extract of 1.5%). Young beer is characterized by an unpleasant taste and aroma due to secondary fermentation products, contains insufficient CO_2, and is slightly cloudy due to the presence of yeast or other suspended particles.

Secondary fermentation helps primary fermentation obtain a finished product with pleasant organoleptic qualities for consumption. It takes place slowly, where the rest of the fermentable extract is transformed into alcohol and CO_2 and is called maturation or storage. The duration of the secondary fermentation can be reduced by stirring the must, fermentation under pressure, or fermentation in bioreactors. If, toward the end of the secondary fermentation, the beer remains cloudy, it means that the yeasts are wild or some bacteria are present. Cloudy beer resulting from secondary fermentation can be clarified with the help of substances such as tannins, bentonite, silica gel, activated carbon, enzyme preparations, reducing substances, polyamides, etc.

Beer, as a finished product, is distributed in glass bottles, tinplates, or aluminum and stainless-steel barrels. Regardless of the packaging method, the beer is packaged isobarometrically (equal pressures in the tank and in the packaging). The bottling technological process takes place with the help of machines that can be simple or complex, automated or semi-automated [20,21].

3.3. Sustainable Strategies for Recovery and Valorization of Brewery Waste

"Waste" from the beer industry represents any substance obtained as a result of the technological process that the holder eliminates, intends to eliminate, or has the obligation to eliminate.

Due to its complex chemical composition, the waste resulting from the processing of food products generally presents optimal conditions for the appearance and development of numerous microbiological and biochemical processes, which ultimately lead to their total degradation. For this reason, it is necessary to first take measures to ensure the conservation of waste until the moment of its subsequent use or processing.

Following the technological flow of obtaining beer, waste results can be used in the manufacturing process of other products in different industries, such as: the pharmaceutical industry, the chemical industry, or the animal feed industry.

The waste resulting from beer production is part of the category of non-hazardous production waste, being classified into specific waste and non-specific waste (packaging, containers for storing chemicals used in the cleaning and disinfection of facilities).

Spent grains or brewer's spent grains (BSGs) represent approximately 85% of all residues produced by the brewing industry. BSG is the solid residue left after filtration of the beer wort, a by-product obtained after the saccharification of the malted cereal grains [22,23]. These are formed at the beginning of the brewing process and are removed before the boiling step. This solid residue is composed of barley hulls, the husk pericarp, and seed coat. This is a heterogeneous material formed from lignocellulosic biomass and is rich in protein (20–30%), fiber (30–70%), lipids, vitamins, and minerals. It contains approximately 12–28% lignin, 12–25% cellulose, and 28% non-cellulosic polysaccharides, mainly arabinoxylans. Spent grains have been proven to contain vitamins, minerals, and many amino acids. This by-product is also rich in oligo- and polysaccharides and phenolic compounds.

Recycling strategies for BSG were identified in the literature as animal feed, human food, composting, biogas, substrate for mushroom production, absorbents, ceramic material, paper, bricks, bioethanol, and food packaging [24,25].

Wet, spent grain can serve as feed for ruminants; the high moisture content makes it easily digestible for animals [26]. Dried spent grains are also an alternative protein source for ruminants and have a positive influence on production efficiency in cattle without affecting fertility. It improves the production and composition of milk and increases the fat and protein content [27,28]. The addition of spent grain to a lamb's diet has been found to have a positive effect on its growth performance and meat quality [12,29].

BSG represents an innovative source for the food industry that can improve the value of food products, such as snacks, bread, pasta, cereals, sausages, or meat patties [30,31], mainly because of its health-related bioactive components: alkaloids, antibiotics, plant growth factors, food-grade pigments, and phenolic acids [12,32].

Composting is another interesting alternative to BSG because it is a rich source of nitrogen and organic materials for soil nutrition. A sustainable strategy that brings advantages to the problems involving climate change issues is biogas generation from the conversion of organic brewery waste. The biogas can be reused for energy generation in the brewery. The use of BSG as a by-product for the generation of substrate for mushroom production or to produce heavy metal absorbers or ceramic materials has been successfully tested. Bioethanol is also a product that can be obtained from malt bagasse [25].

There is a great demand for biodegradable materials for food packaging applications, and BSG is an ideal resource for sustainable food packaging because of its components, such as cellulose, lignin, AX, and protein [24].

Malt tusks are obtained following the germination stage of malt, and the correct name is malt radicele. The removal of the tusks takes place even after the drying stage because the very dry state of the malt tusks allows for easy detachment by rubbing the grains with each other and by hitting them against the walls of the machine. Malt tusks are used for fodder purposes and for the composition of culture media in the compressed yeast industry. For cattle feed, malt tusks are used mixed with coarse fodder (chaff, straw). This feed is not eaten with pleasure by animals because of its bitter taste. It is recommended to be used in the feeding of dairy cows, as it favorably influences milk production.

Having a high content of nutrient substances such as vitamins, provitamins, and other growth stimulants, they are used to prepare molasses for the manufacture of fodder yeast and to manufacture lactic acid by fermentation [33].

Lately, research has shown that the methanolic extract of malt tusks is an effective antioxidant and that, due to its low price, it could constitute a rational capitalization of this waste from breweries [14]. The use of malt tusk extract or flour as an antioxidant is of particular importance in the food industry, firstly due to its lack of harmfulness compared to other antioxidants, and secondly, due to its high content of protein substances, vitamins, a favorable flavor for some preparations, and the color and content of certain enzymes that may be valuable for some foods. Malt tusk flour is used in various pastries. It has also yielded good results in the preservation of peanut oil, butter, and margarine. The malt tusk has the form of thin grains of yellowish-brown color, containing vitamins of the B complex, vitamin E, provitamins A and D, as well as other growth stimulants.

Treating the wort or beer with this extract from malt tusks leads to an increase in the physico-chemical stability of the beer. The treated and pasteurized beer has a stability 2–3 times higher than the untreated one [14].

Malt wort is the one that results in the highest quantity from the technological process of brewing, being the result of the filtration stage of the wort at the boiling section. The resulting malt wort has a humidity of about 80%, with a sweet taste and the smell of malt.

Beer wort is a special fodder for feeding animals and, in particular, dairy cows. However, it can only be used on a large scale in its fresh state because, after a relatively short storage time, as such, it undergoes butyric fermentation, which produces an unpleasant smell, so that the animals refuse to consume it or do not consume it with pleasure.

Its silage in special cemented basins must be done carefully. If the fodder is made with wort that has undergone butyric fermentation, the taste of butyric acid is also transmitted to the milk, thus making it unfit for consumption.

To preserve the wort for a longer time, it can be dried, which requires additional fuel consumption but ensures the long-term preservation of its fodder qualities and facilitates long-distance transportation.

The wort contains an appreciable amount of protein and non-hydrogen extractive substances, which gives it a high nutritional value for the manufacture of animal feed. The wort contains 70–75% of the fat and protein of the malt used in the brewing process. It has important forage value.

Besides its use as fodder, the wort has also found use as an addition to pickling polished cast iron. The wort, after treatment with hydrochloric acid, is used in pickling solutions at a concentration of 10–12 g/L [20,21].

Hop wort results from boiling the wort with hops. After boiling the beer wort together with the hops, it is pumped into the hop separator. The hop pulp, according to its components and the degree of assimilation (50%), corresponds to the meadow hay. To date, it has not found use as fodder due to the fact that it has a strong bitter taste, and the cattle refuse to consume it as such.

Animal feeding experiences with mixed waste (hops and malt) proved not only that the waste is suitable for feeding cattle but also that it exerts a favorable effect on their weight gain as well as on the quantity and quality of milk obtained.

Due to its chemical composition, which is similar to that of manure, hop wort can be used as fertilizer in agriculture [20,21].

Protein sediment (Hot Trub) is a by-product obtained mainly in the separator after the brewing process. It consists of hops, inactive yeast, fat, and protein.

The protein sediment appears in the beer wort after boiling with hops, continuing in the cooling phase. Its separation begins with the removal of the hop cones, where a part is retained on them, but the greater part accompanies the wort until the cooling phase. Here, following the cooling process, another part of the protein precipitates and is also separated. The protein sediment is found as large flakes (coarse trub) or in the form of small particles, known as fine trub. The more or less complete separation of the dregs from the beer wort depends on the equipment and operations used in the cooling phase. The amount of wort obtained depends on a whole series of factors, including the quality of the malt, the grinding degree of the wort, the amount of hops used, the concentration of the wort when boiling with hops, the duration and intensity of cooling, etc.

The chemical composition of hot trub is very varied. It contains albuminoid substances, tannins, hop resins, and mineral substances. The highest percentage is represented by albuminoid substances. The ethereal extract includes not only fats but also bitter substances from the hops, which give the protein sediment its bitter taste, which prevents its direct use as feed, although the degree of assimilability, in general, is quite high.

The protein sediment, as such, can be used primarily as feed but only as an addition in small quantities to other feeds. In this way, it can be added to the malt wort from boiling. It is also used to feed fish in ponds, as a binder in road works, or as fertilizer [20,21].

The foam layer from fermentation is formed in the primary fermentation process of beer wort and is composed of hop resins and precipitated protein substances.

Before passing the young beer to secondary fermentation, the layer formed on the surface of the beer is carefully removed because the bitter resins that it contains would render the beer too bitter and unpleasant a taste. Although the foam layer, which accounts for 0.2% of the beer, is not yet used, it was concluded that, as with the protein sediment, extraction of the hop resins with organic solvents is possible. With the extract obtained, approximately 20% of the fresh hops used in beer production can be replaced. The extract obtained from the foam layer contains 14% more hard resins compared to the extract obtained from fresh hops when benzine is used as a solvent.

Almost all the bitter substances were removed from the foam layer by extraction. The high content of digestible proteins shows that, after the removal of bitter substances, the residue can be used as feed.

Residual spent yeast or brewer's spent yeast (BSY) results in the form of waste in beer manufacturing, both from the fermentation tanks (primary fermentation yeast) and from the storage tanks (secondary fermentation yeast). The main shortcoming in relation to baker's yeast or fodder yeast is the pronounced bitter taste due to the hop resins it contains. Due to its valuable principles, brewer's yeast is used, as such, or after a preliminary drying process, in animal husbandry as a component of concentrated fodder.

In its wet form, as it results from beer production, yeast is extremely degradable: at ambient temperature, it enters a degradation process after a few hours. If it is dehydrated so that it reaches a moisture content of less than 10%, it can be stored without deterioration for a very long time [20,21].

4. The Chemical Characteristics of Different Types of Beer Waste

BSG hot trub and residual spent yeast samples were collected from a brewery located in the Region of Dobrudja, Romania. The studied samples were collected immediately after their generation in the brewing process stages for the production of traditional lager-type beer—the brewery spent grain after lautering, the hot trub after wort boiling, and residual spent yeast after the yeast fermentation step (Figure 1) [34]. All used reagents were of high grade and purchased from Sigma Aldrich or Merck.

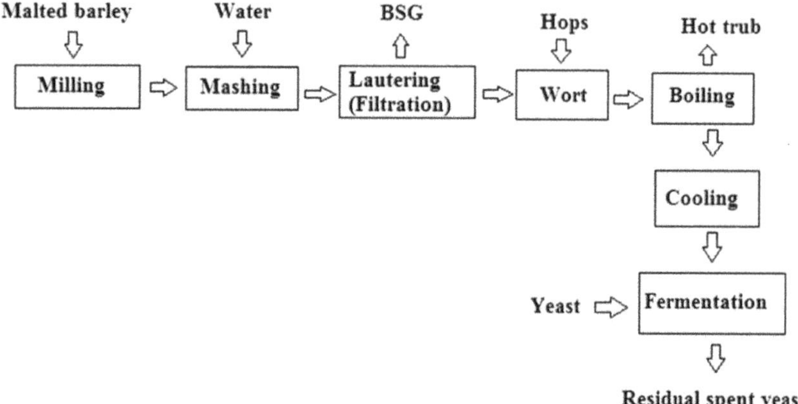

Figure 1. Steps of the brewing process to obtain the most important by-products.

The determination of moisture content, ash, pH, total content of phenolic compounds, and total protein content was aimed. Thermal drying was used for the determination of the moisture content of the studied samples. The moisture content was determined as the difference between the weight of studied samples and the weight recorded after 1 h and 30 min heating at 130 °C in an oven. Ash content was made by calcination in a muffle at 700 °C. The pH was measured with a pH meter. Total protein content was determined by the Kjeldahl method. The method consists of the mineralization of the sample with concentrated sulfuric acid when the nitrogen from the organic combinations is released and passed as ammonium sulfate. Reaction with a strong base releases ammonia, which is removed from the solution by distillation and captured in an excess solution of 0.1 N sulfuric acid. The total protein content was determined using the Kjeldahl system consisting of the Turbosog and Vapodens apparatus.

Extraction of phenolic compounds in studied samples was conducted in 80% methanol for 2 h at room temperature, then the extracts were centrifuged, and the supernatant was retained and stored at -20 °C until analysis. Total phenolic content was determined using the Folin–Ciocalteu method [35], and the results were expressed as mg gallic acid equivalents (GAE) per 100 g of studied sample. Aliquots of 0.5 mL extracts were mixed with 1 mL of 1:2 (*v/v*) Folin–Ciocalteu reagent in 50 mL calibrated flasks. Then, 1 mL ethanol and 1 mL sodium carbonate solution (20%) was added. The solution was vigorously mixed, allowed to react for 10 min at room temperature, and filled up to the mark with distilled water. The resulting mixtures were maintained at room temperature for 30 min, and the absorbance was measured at 675 nm against the corresponding blank with a UV-VIS spectrophotometer (Jasco 550). Quantification was performed by a calibration curve using gallic acid as a phenolic standard ($R^2 = 0.9997$, y = 0.0041x).

5. Results and Interpretations

Table 1 presents the experimental values obtained for the studied samples of the brewery waste.

Table 1. Values of the moisture content, ash content, pH, total protein content, and total content of phenolic compounds of the studied samples.

Samples	Moisture Content, %	Ash Content, %	pH	Total Protein Content, %	Total Content of Phenolic Compounds, mg GAE */100 g dw **
Brewer's spent grains (BSG)	5.4	4.1	5.1	17.46	354.26
Hot trub	5.8	5.3	4.5	12.37	285.35
Residual spent yeast	7.6	15.1	4.4	34.38	112.45

* GAE = gallic acid equivalents. ** dw = dry weight.

For the studied samples, pH values ranged between 4.4 and 5.1; the highest value was obtained for the brewer's spent grains. Ash content is related to mineral content, and for the studied samples, values were between 4.1 and 15.1%; the highest value was obtained for residual spent yeast. The ash content of residual spent yeast was in agreement with the value reported by Vieira et al. [36]. For the studied samples, the moisture content was between 5.4 and 7.6%. The highest protein content was obtained for residual spent yeast (34.38%), while the lowest value was obtained for hot trub (12.37%).

The phenolic compounds in beer originate mainly from beer ingredients used in the brewery process; phenolic acids and flavonoids are the most studied in beer [37]. Brewery wastes like spent grain, hot trub, or residual spent yeast are also valuable by-products thanks to their phenolic composition. Results showed that the total phenolic compound content in the brewer's spent grains was 354.26 mg GAE/100 g dw, in the hot trub was 285.35 mg GAE/100 g dw, while in the residual spent yeast, it was 112.45 mg GAE/100 g dw. A decrease was observed in the total phenolic compounds that can be associated with the precipitation of compounds during the maturation process or their adsorption to the hot trub during processing or to the yeast cells during the fermentation process [38]. These phenolic compounds are considered natural antioxidants with effects on human health, such as cardiovascular diseases, neurogenerative diseases, diabetes (types I and II), and some types of cancer [14,39].

An efficient way of stabilizing beer is to reduce the polyphenol content by adsorption with polyvinylpolypyrrolidone (PVPP), which forms hydrogen bonds between the hydroxyphenolic group and the amide of PVPP. After use, the PVPP resin is regenerated in the brewery, in a washing process that generates large quantities of a waste stream, which contains large amounts of polyphenols, being a promising alternative and economical source of natural antioxidants. The main phenolic compounds present in the crude extracts are simple phenolic compounds, mainly phenolic acids and flavan-3-ol catechin, which are the main causes of beer haze [14].

Sustainable strategies that valorize by-products from brewery processes are more convenient than expenses attributed to the disposal of industrial by-products or waste treatment costs [12,40,41]. Brewing waste is used mainly as animal feed, in biofuel production, building, or packaging materials [42]. The recovery of proteins and polyphenols from brewing wastes may represent an innovative source for the food industry and biotechnological processes. An adequate approach that emphasizes the value of brewing industry wastes is to provide knowledge about chemical composition, marketability, and research implications [34].

Brewery waste is a negative cost factor for the brewery industry; its rich content of complex carbohydrates, nitrogen, and minerals is very important to the production of biotechnological products such as biofuel, enzymes, organic acids, and bioactive compounds [43].

6. Conclusions

The use of green technologies, the exploitation of food industry by-products, and outside-the-box thinking are sustainable strategies to reduce the pollution arising from industrial activities. The concept of zero waste means implementing a circular economy, where companies should not consider residues as waste but as raw materials for their use in other processes or to valorize these by-products as functional ingredients to create a sustainable food system. This concept also means to transform society by changing resource management, influencing consumer behavior, and creating the appropriate legal framework and economic incentives.

Author Contributions: Conceptualization: A.S. (Alina Soceanu); data curation: A.S. (Alina Soceanu), S.D., A.B. and A.S. (Anca Sirbu); formal analysis: V.P.; investigation: A.S. (Alina Soceanu) and V.P.; methodology: A.S. (Alina Soceanu), S.D., A.S. (Anca Sirbu), A.B. and V.P.; project administration: S.D., V.P. and A.S. (Anca Sirbu); software: A.B.; supervision: V.P.; validation: S.D.; visualization: A.S. (Alina Soceanu) and V.P.; roles/writing—original draft: A.S. (Alina Soceanu), S.D. and A.S. (Anca Sirbu); writing—review and editing: A.S. (Alina Soceanu), S.D., A.B., A.S. (Anca Sirbu) and V.P. All authors have read and agreed to the published version of the manuscript.

Funding: This research received no external funding.

Data Availability Statement: No new data were created or analyzed in this study. Data sharing is not applicable to this article.

Conflicts of Interest: The authors declare no conflict of interest.

References

1. Rozsypal, J.; Sevcik, J.; Bartosova, Z.; Papouskova, B.; Jirovsky, D.; Hrbac, J. Automated electrochemical determination of beer total antioxidant capacity employing microdialysis online coupled with amperometry. *Microchem. J.* **2022**, *183*, 107955. [CrossRef]
2. Chiva-Blanch, G.; Magraner, E.; Condines, X.; Valderas-Martínez, P.; Roth, I.; Arranz, S.; Casas, R.; Navarro, M.; Hervas, A.; Sisó, A.; et al. Effects of alcohol and polyphenols from beer on atherosclerotic biomarkers in high cardiovascular risk men: A randomized feeding trial. *Nutr. Metab. Cardiovasc. Dis.* **2015**, *25*, 36–45. [CrossRef] [PubMed]
3. De Gaetano, G.; Costanzo, S.; Di Castelnuovo, A.; Badimon, L.; Bejko, D.; Alkerwi, A.; Chiva-Blanch, G.; Estruch, R.; La Vecchia, C.; Panico, S.; et al. Effects of moderate beer consumption on health and disease: A consensus document. *Nutr. Metab. Cardiovasc. Dis.* **2016**, *26*, 443–467. [CrossRef] [PubMed]
4. Yang, D.; Gao, X. Research progress on the antioxidant biological activity of beer and strategy for applications. *Trends Food Sci. Technol.* **2021**, *110*, 754–764. [CrossRef]
5. Piazzon, A.; Forte, M.; Nardini, M. Characterization of phenolics content and antioxidant activity of different beer types. *J. Agric. Food. Chem.* **2010**, *58*, 10677–10683. [CrossRef] [PubMed]
6. Nardinia, M.; Garaguso, I. Characterization of bioactive compounds and antioxidant activity of fruit beers. *Food Chem.* **2020**, *305*, 125437. [CrossRef]
7. Zapata, P.J.; Martínez-Esplá, A.; Gironés-Vilaplana, A.; Santos-Lax, D.; Noguera-Artiaga, L.; Carbonell-Barrachina, A.A. Phenolic, volatile, and sensory profiles of beer enriched by macerating quince fruits. *LWT Food Sci. Technol.* **2019**, *103*, 139–146. [CrossRef]

8. Kerpes, R.; Fischer, S.; Becker, T. The production of gluten-free beer: Degradation of hordeins during malting and brewing and the application of modern process technology focusing on endogenous malt peptidases. *Trends Food Sci. Technol.* **2017**, *67*, 129–138. [CrossRef]

9. Dragone, G.; Mussatto, S.I.; Almeida e Silva, J.B. High gravity brewing by continuous process using immobilised yeast: Effect of wort original gravity on fermentation performance. *J. Inst. Brew.* **2007**, *113*, 391–398. [CrossRef]

10. European Union. Regulation (EC) No 1333/2008 of the European Parliament and of the Council of 16 December 2008 on food additives. *Off. J. Eur. Union* **2008**, *354*, 16–33.

11. Yadav, R.; Yadav, N.; Saini, P.; Kaur, D.; Kumar, R. Potential Value Addition from Cereal and Pulse Processed By-Products: A Review. *Sustain. Food Waste Manag.* **2021**, *9*, 155–176. [CrossRef]

12. Rachwał, K.; Waśko, A.; Gustaw, K.; Polak-Berecka, M. Utilization of brewery wastes in food industry. *PeerJ* **2020**, *8*, e9427. [CrossRef] [PubMed]

13. Hajji, T.; Mansouri, S.; Vecino-Bello, X.; Cruz-Freire, J.M.; Rezgui, S.; Ferchichi, A. Identification and characterization of phenolic compounds extracted from barley husks by LC-MS and antioxidant activity in vitro. *J. Cereal Sci.* **2018**, *81*, 83–90. [CrossRef]

14. Barbosa-Pereira, L.; Bilbao, A.; Vilches, P.; Angulo, I.; LLuis, J.; Fité, B.; Paseiro-Losada, P.; Cruz, J.M. Brewery waste as a potential source of phenolic compounds: Optimisation of the extraction process and evaluation of antioxidant and antimicrobial activities. *Food Chem.* **2014**, *145*, 191–197. [CrossRef] [PubMed]

15. Betancur, M.I.; Motoki, K.; Spence, C.; Velasco, C. Factors influencing the choice of beer: A review. *Food Res. Int.* **2020**, *137*, 109367. [CrossRef]

16. Bezerril, F.; Pimentel, C.T.; Peixoto de Aquino, K.; Schabo, D.C.; Heli Paiva Rodrigues, M.; dos Santos Lima, M.; Schaffner, D.W.; Furlong, E.B.; Magnani, M. Wheat craft beer made from AFB1-contaminated wheat malt contains detectable mycotoxins, retains quality attributes, but differs in some fermentation metabolites. *Food Res. Int.* **2023**, *172*, 112774. [CrossRef]

17. Panda, R.; Zoerb, H.F.; Cho, C.Y.; Jackson, L.S.; Garber, E.A.E. Detection and Quantification of Gluten during the Brewing and Fermentation of Beer Using Antibody-Based Technologies. *J. Food Prot.* **2015**, *78*, 1167–1177. [CrossRef]

18. Dabija, A. *Biotechnologies in the Fermentative Food Industry*; Oim Publishing House: Iasi, Romania, 2010.

19. Ezati, M.; Ghavamipour, F.; Adibi, H.; Pouraghajan, K.; Arab, S.S.; Sajedi, R.H.; Khodarahmi, R. Design, synthesis, spectroscopic characterizations, antidiabetic, *in silico* and kinetic evaluation of novel curcumin-fused aldohexoses. *Spectrochim. Acta Part A Mol. Biomol. Spectrosc.* **2023**, *285*, 121806. [CrossRef]

20. Banu, C. *Malt and Beer Technology*; Technical Publishing House: Bucharest, Romania, 2003.

21. Berzescu, P. *Beer and Malt Technology*; Ceres Publishing House: Bucharest, Romania, 1981.

22. Pires, E.J.; Ruiz, H.A.; Teixeira, J.A.; Vicente, A.A. A new approach on brewer's spent grains treatment and potential use as lignocellulosic yeast cells carriers. *J. Agric. Food Chem.* **2012**, *60*, 5994–5999. [CrossRef]

23. Outeiriño, D.; Costa-Trigo, I.; Paz, A.; Deive, F.J.; Rodríguez, A.; Domínguez, J.M. Biorefining brewery spent grain polysaccharides through biotuning of ionic liquids. *Carbohydr. Polym.* **2019**, *203*, 265–274. [CrossRef]

24. Qazanfarzadeh, Z.; Ganesan, A.R.; Mariniello, L.; Conterno, L.; Kumaravel, V. Valorization of brewer's spent grain for sustainable food packaging. *J. Clean. Prod.* **2023**, *385*, 135726. [CrossRef]

25. Bonato, S.V.; De Jesus Pacheco, D.A.; Schwengber ten Caten, C.; Caro, D. The missing link of circularity in small breweries' value chains: Unveiling strategies for waste management and biomass valorization. *J. Clean. Prod.* **2022**, *336*, 130275. [CrossRef]

26. Kerby, C.; Vriesekoop, F. An overview of the utilisation of brewery by-products as generated by British craft breweries. *Beverages* **2017**, *3*, 24. [CrossRef]

27. De Souza, L.; Zambom, M.; Alcalde, C.R.; Fernandes, T.; Castagnara, D.D.; Radis, A.C.; Santos, S.M.; Possamai, A.P.; Pasqualoto, M. Feed intake, nutrient digestibility, milk production and composition in dairy cows fed silage of wet brewers' grain. *Semin. Ciênc. Agrár.* **2016**, *37*, 1069–1080. [CrossRef]

28. Faccenda, A.; Zambom, M.; Castagnara, D.; de Avila, A.S.; Fernandes, T.; Eckstein, E.I.; Anschau, F.A.; Schneider, C.R. Use of dried brewers' grains instead of soybean meal to feed lactating cows. *Rev. Bras. Zootec.* **2017**, *46*, 39–46. [CrossRef]

29. Radzik-Rant, A.; Rant, W.; Niznikowski, R.; Swiatek, M.; Szymanska, Z.; Slezak, M.; Niemiec, T. The effect of the addition of wet brewers' grain to the diet of lambs on body weight gain, slaughter value and meat quality. *Arch. Anim. Breed.* **2018**, *61*, 245–251. [CrossRef]

30. Kim, H.; Hwang, K.; Song, D.; Lee, S.; Choi, M.; Lim, Y.; Choi, J.; Choi, Y.; Kim, H.; Kim, C. Effects of dietary fiber extracts from brewer's spent grain on quality characteristics of chicken patties cooked in convective oven. *J. Am. Soc. Brew. Chem.* **2013**, *33*, 45–52. [CrossRef]

31. Choi, M.S.; Choi, Y.S.; Kim, H.W.; Hwang, K.E.; Song, D.H.; Lee, S.Y.; Kim, C.J.; Hwang, K.; Song, D.; Lee, S.; et al. Effects of replacing pork back fat with brewer's spent grain dietary fiber on quality characteristics of reduced-fat chicken sausages. *Food Sci. Anim. Resour.* **2014**, *34*, 158–165. [CrossRef]

32. Waters, D.M.; Jacob, F.; Titze, J.; Arendt, E.K.; Zannini, E. Fibre, protein and mineral fortification of wheat bread through milledand fermented brewer's spent grain enrichment. *Eur. Food Res. Technol.* **2012**, *235*, 767–778. [CrossRef]

33. Xie, Y.; Bao, J.; Li, W.; Sun, Z.; Gao, R.; Wu, Z.; Yu, Z. Effects of Applying Lactic Acid Bacteria and Molasses on the Fermentation Quality, Protein Fractions and In Vitro Digestibility of Baled Alfalfa Silage. *Agronomy* **2021**, *11*, 91. [CrossRef]

34. Rodriguez, L.M.; Camina, J.L.; Borroni, V.; Perez, E.E. Protein recovery from brewery solid wastes. *Food Chem.* **2023**, *407*, 134810. [CrossRef] [PubMed]

35. Singleton, V.L.; Rossi, J.A. Colorimetry of Total Phenolics with Phosphomolybdic-Phosphotungstic Acid Reagents. *Am. J. Enol. Vitic.* **1965**, *16*, 144–158. [CrossRef]
36. Vieira, E.F.; Carvalho, J.; Pinto, E.; Cunha, S.; Almeida, A.A.; Ferreira, I.M.P.L.V.O. Nutritive value, antioxidant activity and phenolic compounds profile of brewer's spent yeast extract. *J. Food Compos. Anal.* **2016**, *52*, 44–51. [CrossRef]
37. Viana, A.C.; Colombo Pimentel, T.; Borges do Vale, R.; Santos Clementino, L.; Ferreira, E.T.J.; Magnani, M.; dos Santos Lima, M. American pale Ale craft beer: Influence of brewer's yeast strains on the chemical composition and antioxidant capacity. *LWT Food Sci. Technol.* **2021**, *152*, 112317. [CrossRef]
38. Leitao, C.; Marchioni, E.; Bergaentzl, B.; Zhao, M.; Didierjean, L.; Taidi, B.; Ennahar, S. Effects of processing steps on the phenolic content and antioxidant activity of beer. *J. Agric. Food Chem.* **2011**, *59*, 1249–1255. [CrossRef] [PubMed]
39. Humia, B.V.; Santosa, K.S.; Kleveston Schneider, J.; Leal, I.L.; de Abreu Barreto, G.; Batista, T.; Souza Machado, B.A.; Druzian, J.I.; Krause, L.C.; da Costa Mendonça, M.; et al. Physicochemical and sensory profile of Beauregard sweet potato beer. *Food Chem.* **2020**, *312*, 126087. [CrossRef]
40. Puligundla, P.; Mok, C.; Park, S. Advances in the valorization of spent brewer's yeast. *Innov. Food Sci. Emerg. Technol.* **2020**, *62*, 102350. [CrossRef]
41. Guan, Y.; Xu, X.; Liu, C.; Wang, J.; Niu, C.; Zheng, F.; Li, Q. Evaluating the physiology and fermentation performance of the lager yeast during very high gravity brewing with increased temperature. *LWT Food Sci. Technol.* **2023**, *173*, 114312. [CrossRef]
42. Bachmann, A.A.L.; Calvete, T.; Feris, L.A. Potential applications of brewery spent grain: Critical overview. *J. Environ. Chem. Eng.* **2022**, *10*, 106951. [CrossRef]
43. Atalay, P.; Perendeci, N.A.; Goksungur, M.Y. Valorization of brewery waste. *J. Eng. Sci.* **2020**, *26*, 1257–1266. [CrossRef]

 *sustainability*

Article

How Do Existing Organizational Theories Help in Understanding the Responses of Food Companies for Reducing Food Waste?

Ramakrishnan Ramanathan [1,*], Usha Ramanathan [1], Katarzyna Pelc [2] and Imke Hermens [3]

1   Department of Management, College of Business Administration, University of Sharjah, Sharjah P.O. Box 27272, United Arab Emirates; uramanathan@sharjah.ac.ae
2   Business and Management Research Institute, Business School, University of Bedfordshire, Luton LU1 3JU, UK; katarzyna.pelc@beds.ac.uk
3   Whysor BV, Brandemolen 65, 5944 ND Arcen, The Netherlands; imke@whysor.com
*   Correspondence: rramanathan@sharjah.ac.ae; Tel.: +971-6-5050579

Abstract: Food waste is a serious global problem. Efforts to reduce food waste are closely linked to the concepts of circular economy and sustainability. Though food organizations across the world are making efforts to reduce waste in their supply chains, there is currently no theoretical explanation that would underpin the responses of food companies in reducing food waste. Based on interactions with food companies over a nearly 5-year period, we explore the applicability of some well-known and not so well-known organizational theories in the operations management literature to underpin the observed responses of companies in reducing food waste. This paper is one of the first attempts to study food waste from an operations and supply chains point of view, especially from the lens of existing theories in the operations management literature and newer sustainability theories borrowed from other disciplines. Our research findings not only show that existing organizational theories and societal theories can help explain the motivations of firms engaging in food waste reduction, but also call for more research that could help explain some interesting observations that are not apparent when existing theories are used. This paper contributes to the UN's Sustainable Development Goals 1, 2 and 12.

Keywords: food supply chains; food waste; circular economy; organizational theories

Citation: Ramanathan, R.; Ramanathan, U.; Pelc, K.; Hermens, I. How Do Existing Organizational Theories Help in Understanding the Responses of Food Companies for Reducing Food Waste? *Sustainability* **2024**, *16*, 1534. https://doi.org/ 10.3390/su16041534

Academic Editor: Dimitris Skalkos

Received: 2 January 2024
Revised: 26 January 2024
Accepted: 8 February 2024
Published: 11 February 2024

1. Introduction

Food waste is a serious global problem. It has close links with the concepts of circular economy (CE) and sustainability. From a CE point of view, food waste is a kind of waste that needs attention in terms of the 4Rs, namely reduce, reuse, recycle, and recover [1]. Waste prevention is an integral part of CE approaches [2,3]. In terms of sustainability, food waste has economic, environmental, and social implications. In this sense, saving food waste contributes to several of the UN's Sustainable Development Goals.

From an operations management (OM) point of view, food waste can be reduced or eliminated via productivity improvement and lean mechanisms. The food waste sector faces huge challenges in their supply chains [4]. Any food waste that is unavoidable can then be reused or recycled in a suitable way to complete the CE cycle before the food quality deteriorates. Several initiatives have been considered to reduce and avoid food waste. The first objective of this paper is to understand the nature of food waste and its impact across food supply chains (FSCs).

While food waste is generated at various levels and at different stages, in this study, we focus on food waste occurring in FSCs. Due to the focus on reducing food waste and using circular economy principles, we use the term circular food supply chains (CFSCs) in this paper and look at opportunities for applying the 4Rs to food waste in CFSCs. A detailed review of the food waste literature shows that most studies are practice-oriented and efforts to understand food waste practices from a theoretical point of view is missing. This is the primary research gap that this paper aims to address. Based on an extensive interaction with food companies as part of a large European research project [5], we examine whether popular existing theories in the OM discipline (e.g., [6,7]) can help us to understand the behavior of food companies and their supply chain partners. Accordingly, the remaining objectives of this research are (ii) to identify suitable theories (that are popular in the OM discipline and other theories borrowed from other disciplines) that are relevant for reducing food waste in CFSCs, (iii) to examine how these theories can help understand the observed behavior of food companies, and (iv) to identify the scope for future research in explaining certain interesting observations that are not apparent when existing theories are used. The main research question addressed in this research to fill the above research gap is as follows: how can existing organizational theories be used to explain the observed behavior of food companies in reducing food waste?

Our research will contribute to the literature in several ways. For the first time, the issue of food waste will be studied from an OM point of view, focusing on the nature of food waste and its impact across food supply chains. A review of existing OM theories and newer theories borrowed from other disciplines that are relevant for food waste and sustainability issues in CFSCs is another novel contribution of this paper. Another novel contribution is the examination of how these theories can be used to explain certain observed behaviors of food companies. Finally, we also support future researchers by highlighting the scope for future research; specifically, we focus on some of the behaviors of food companies that need to be explored in further research and beyond existing theories. Thus, our research contributes to the OM literature, CE literature, and sustainability literature.

The paper is organized as follows. The next section will provide a review of the relevant literature. Section 3 will elaborate on the interactions with agribusinesses and opportunities for theory building. The European Union has been leading efforts on CE and hence this section will provide an overview of the CE implications of food waste in the EU context. Section 4 will review theories that are relevant to understand food waste issues in CFSCs. A number of organizational theories commonly used in the OM literature will be presented in this section. In addition, we will try to borrow interesting theories from other disciplines and study how they can be used to understand issues related to food waste in CFSCs and sustainability. Section 5 will provide more discussions and the last section will provide our conclusions.

2. Literature Review

The term circular economy (CE) is generally defined as the practices aimed at maximizing resource efficiency in organizations [8]. Geissdoerfer et al. [9] defines it as "a regenerative system in which resource input and waste, emission, and energy leakage are minimised by slowing, closing, and narrowing material and energy loops. This can be achieved through long-lasting design, maintenance, repair, reuse, remanufacturing, refurbishing, and recycling." (Page 759). By focusing on resource circularity and optimization, CE practices contribute to increasing productivity [10,11].

The European Commission has pioneered the ideas of CE in its action plan [12], where it highlights three areas for a sustainable policy framework—designing sustainable products, empowering consumers, and implementing circularity principles in production processes. In their action plan, among other targets, they have committed to targeted food waste reduction. A CE perspective will identify opportunities that extend a product's own life cycle (e.g., via product repair), the life of its constituent parts (e.g., refurbishing or remanufacturing), or find use for the materials in the product at the end of its life cycle (e.g., recycling). From a CE point of view, waste prevention is an integral part of CE approaches [2] and needs attention in terms of the 4Rs, namely reduce, reuse, recycle, and recover [1].

The literature on CE usually focuses on business models for achieving the desired CE outcomes (e.g., Ref. [13]). Using a multiple case study approach, Vermunt et al. [1] link the 4R framework (reduce, reuse, recycle, and recover) of the CE with important CE business models—the product-as-a-service model, product life extension model, resource recovery model, circular supplies model, and hybrid models. They observe that supply chain-related barriers are not prevalent in product-as-a-service business models. In a similar study, De Angelis and Feola [14] have used a single case study approach to underline the salient characteristics of circular economy based on the ReSOLVE (regenerate, share, optimize, loop, virtualize, and exchange) framework of the Ellen MacArthur Foundation.

There is a debate in the literature about whether the concepts of CE and sustainability lead to the same level of sustainable development [15,16]. These authors highlight that CE prioritizes economic systems from an environmental point of view [12], with not much emphasis on social sustainability, while the Triple-Bottom-Line approach to sustainability will yield equal importance to all the three—economic, environmental, and social—pillars of sustainability. Accordingly, they and other similar authors [9,17] feel that CE business models (which are defined as the way in which CE principles are embedded in the value propositions in value chains) may not contribute much to social sustainability.

2.1. Circular Economy and Food Waste

As per WRAP [18], nearly one-third of produced food is lost or wasted. This provides an adequate background for applying CE principles in the food industry. When CE ideas are applied to the food industry, the effort is to reduce, recycle, or reuse food waste, or recover value from food waste that cannot be either recycled or reused. Reducing food waste improves the financial bottom-line for food companies and increases food availability with societal benefits. Food waste that ends up in landfills emits significant greenhouse gases and hence reducing food waste has significant environmental benefits. Thus, a circular economy business model aimed at zero food waste in circular food supply chains will be able to reach all the three pillars of sustainability [5]. Thus, saving food waste contributes to the UN's Sustainable Development Goal 1 (No Poverty), 2 (Zero Hunger), and 12 (Responsible Consumption and Production).

Thus, tackling food waste will help improve circularity and sustainability significantly. Food waste is further linked to various CE aspects such as reverse logistics, remanufacturing, servitization (or product–service systems), and sustainable supply chain management. Food waste reduction, like the focus of CE-based studies, can help organizations improve their environmental performance (e.g., waste reduction, pollution reduction, and improved ecological/carbon footprint), financial performance (e.g., profitability and economic efficiency), operational performance (e.g., productivity, product quality, and attractiveness), and social performance (health, employee morale, increased employment, and improved food security) [8].

Food waste can occur in multiple ways—at the upstream level by the producer at the production site, at the downstream level by consumers, and in between when food is moved along supply chains. There are huge consumer-behavior studies focusing on how to change the behavior and lifestyle of consumers to facilitate the reduction and complete elimination of food waste at the consumer level. Productivity studies at farms and food manufacturing plants are focusing on the upstream level. However, food waste in supply chains is a relatively unexplored area. While waste minimization in general has been a hot topic in sustainability research, understanding the mechanisms by which food companies reduce food waste in their supply chains is a relatively less explored topic. This finding has been confirmed by Kalmykova et al. [2], who, based on a literature review, observe that manufacturing and distributions are not widely studied in the context of CE. One reason for the relative under-exploration of food waste in supply chains could be because food waste that occurs in supply chains is generally considered as an unavoidable food loss [5].

2.2. Food Waste in Circular Food Supply Chains

Food waste is a global problem and has significant economic, environmental, social, and ethical implications. Nearly one-third of produced food ends up as waste [18]. It has been estimated that the EU produces around 88 million tons of food waste annually, equivalent to EUR 143 billion, highlighting the economic impacts of food waste. Food waste in other parts of the world paint an equally, if not more, bleak picture. Using a Life Cycle Analysis (LCA), it has been estimated that food waste alone is responsible for 8–10% of global GHG emissions.

The EU has committed to halving food waste by 2030. Target 12.3 of the UN's Sustainable Development Goals has called for halving global food waste by 2030. Several research studies have been carried out with a view to achieving these ambitious targets. For example, research studies are being conducted about when, where, and how much food waste occurs (e.g., Ref. [18]).

Based on the work from a project named FUSIONS, Parry et al. [19] have stressed the importance of preventing food waste in the first place. As per their calculations, the redistribution of food to people before it becomes waste will save 3090 kg of CO_2 equivalent per ton of food waste. This prevention strategy is the best strategy to fight food waste and associated greenhouse gas emissions. The calculations from their report provide very valuable information about options for treating food waste and can be linked to the 4R principles of CE. Thus, redistributing food to people before the food becomes waste is the best option, as it has the potential for saving a very high level of carbon emissions. Converting the food to animal feed is the next best option, saving 220 kg CO_2 equivalent per ton of food waste. Sending food waste to landfills is the least preferred option, as this will generate additional GHG emissions in landfills (about 536 kg per ton of food waste).

About 20–30 percent of food waste in food businesses occur in their supply chains—when the food is being transported or stored from the production to the final consumers [19]. A part of this loss is due to improper food storage conditions—temperature, humidity, etc.—when food produce is on the move (e.g., in a truck) or in an intermediate warehouse [20–23]. The appropriate treatment (i.e., reduce, reuse, recycle, or recover) of food waste will help food supply chains move from being linear to circular, and enable them to create circular food supply chains (CFSCs). For this study, we document our interactions with firms in European CFSCs about the behavior of food supply chain companies and their motivations to reduce food waste. We briefly discuss the food companies in the next section.

3. Interactions with Food Businesses and Opportunities for Theory Building

Thanks to generous funding for a large European project [5], we had close interactions with several food companies in the UK and EU for over nearly five years on our quest to support reductions in food waste. The details are given in Table 1 below.

Table 1. The companies involved in reducing food waste. Source: Ref. [5].

	Description	Country
1.	Food processing in an abattoir	UK
2.	Food processing in an abattoir	Republic of Ireland
3.	Food storage in a frozen food company	UK
4.	Milk transportation	UK
5.	Food transport	UK
6.	Food transport	The Netherlands
7.	Food storage and transport in multiple stages of the supply chain	Luxembourg
8.	Food storage and transport in multiple stages of the supply chain	Germany
9.	Food storage	The Netherlands
10.	Food processing—wine manufacturing	UK
11.	Food production—raising cattle	UK
12.	Food transportation	UK
13.	Food production, storage, and transport	Germany

While we were involved in interacting with several of these companies, some companies had only limited interactions for several reasons such as the COVID-19 pandemic. A detailed description of the nature of our involvement with several companies has already been published [22,23]. It is not our intention to describe our experiences working with these companies again in this paper, but rather to use some of our experiences to create links with theories in OM, CE, and sustainability.

4. Methodology

We have followed a rigorous methodology (Figure 1) for answering the research question and achieving the research objectives specified in Section 1. This includes a review of prominent organizational theories, identifying major themes advocated by these theories, linking our observations when working with food companies to reduce food waste with these major themes, and identifying the behaviors of firms that are not directly apparent from these theories.

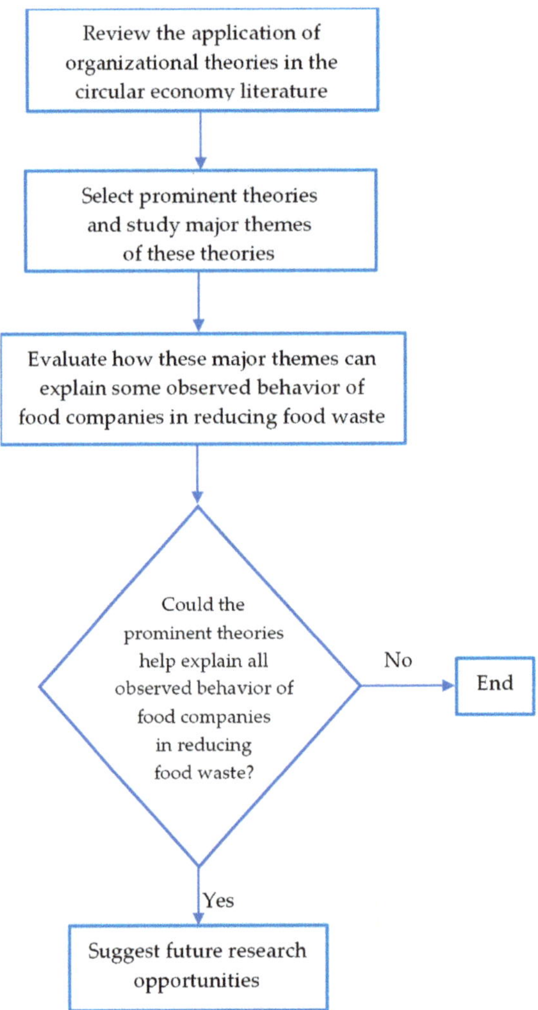

Figure 1. Research methodology steps.

As per these steps, we first provide a brief overview of some important theories in the next section.

5. Theoretical Underpinnings for Food Waste

We first provide an overview of some important theories in this section and use these theories to explain the behavior of the food companies listed in Table 1. These theories are reviewed from the CE and sustainability literature, where we highlight those theories that are used by a majority of OM researchers and those that are not well employed in OM research. As highlighted earlier, our emphasis is not to provide a detailed disposition of these theories, but rather to link important tenets of these theories to our observations about the behavior of food companies. Readers interested in understanding the theories in detail may refer to other suitable references [6,7,24]. Here, we highlight that there has been no study that has applied these theories directly to food waste practices, and hence we draw on the theories in the field of circular economy.

Various organizational theories have been employed by CE researchers to understand the behavior of firms, though it has been observed that these theoretical approaches and business models are under-researched topics when applied to circular economy [8,25]. Ref. [26] has listed a total of 15 management theories in their review of the literature on CE business models. The theories include some theories that are familiar to OM researchers (e.g., the business model theory, dynamic capabilities perspective, industrial network theory, marketing systems theory, principal-agency theory, systems theory, transaction cost theory, stakeholder theory, and institutional theory), and some that are not so familiar (e.g., eco-efficient value creation, ecological economy and social/solidarity theory, industrial ecology theory, life-cycle thinking theory, prospect theory, and the ecological modernization theory). Similarly, Sehnem et al. [8] have reviewed the literature and found a limited set of theories (including the institutional theory, stakeholder theory, resource-based theory (RBT), industrial ecology, transaction cost economics, and social network analysis) that have been applied to the CE literature. Based on a review of the literature on GSCM and CE, Liu et al. [6] have identified that twelve theories have been applied to both the disciplines, while seven additional theories have been applied only to GSCM, and eight more have been applied only to CE. Lahti et al. [27] describe in detail how six commonly used organization theories—contingency theory, TCE, RBT, path-dependence theory, and agency theory—can be used to understand the behavior of firms engaged in CE.

The major organizational theories in the CE literature include the institutional theory [28], the stakeholder theory [15], and the resource-based theory [25]. Based on a literature review, Gusmerotti et al. [29] have identified the institutional theory and the RBT theory as the most appropriate organizational theories that can explain behavior of firms engaged in circular economy practices. However, there seems to be a confusion in the development of theories in the CE literature, as many authors (e.g., Ref. [29]) seem to borrow theories from the sustainability literature directly without providing adequate attention to the distinction between sustainability and CE.

Ranta et al. [28] highlight that the institutional theory can be used to explain the drivers and barriers of CE. They focused on institutional drivers and barriers for the 3R (reduce, reuse, and recycle) principles of the CE. Drawing from Scott's institutional theory framework [30], they have used the three pillars—regulative, normative, and cultural–cognitive—to highlight the differing pressures on institutions engaging with circular economy ideas.

The stakeholder theory [31] is commonly employed in sustainability research. As CE is closely linked to sustainability, the need for considering requirements of all stakeholders has been discussed in several articles that discuss enhancing the value propositions of CE-based business models [15,32,33].

Lopez et al. [25] identify five main categories of barriers (institutional, market, organizational, behavioral, and technological) for resource efficiency in firms. Vermunt et al. [1] have prepared a slightly larger set of barriers, adding supply chain barriers, financial barriers, and knowledge barriers to the above list. For example, the lack of appropriate partners, low availability of raw materials, higher dependence on external parties, and lack of information exchange/conflict of interests between supply chain actors are crucial supply chain-related barriers that could affect circular economy business models. These barriers support the application of the RBT to CE.

In an interesting recent development, De Angelis [14] has linked several tenets of the paradox theory with CE business models. Paradoxes exist when contradictory, yet interrelated elements occur simultaneously in a system. Several categories of paradoxical tensions (viz. learning, organizing, belonging, and performing paradoxes) for CE principles and CE value loops have been identified [14].

There has been more interest in the literature about the use of technologies in reducing food waste. For example, Li et al. [34] have reviewed the literature on the use of blockchain in food supply chains. Li et al. [35] have focused on the specific application of internet of things sensors for perishable food management. Stefanini and Vignali [36] have further highlighted how new technologies can help food companies achieve the three pillars of sustainability. Traditional innovation theories such as the innovation diffusion theory [37], the technology–organization–environment (TOE) framework [38], and the technology acceptance model (TAM) [39], including its latest versions (e.g., the unified theory of acceptance and use of technology or UTAUT [40]), are some common theories that can be associated with the use of technology for reducing food waste.

There is a general awareness in the CE literature about the importance of technologies. For example, Bressanelli et al. [41] explain how new digital technologies such as IoT and big data analytics can help support circular economy practices using a single case study. Tseng et al. [42] highlight the paucity of research articles on circular economy that exploit the power of newer digital technologies (including IoT technology, big data analytics, cyber–physical systems, cloud computing, artificial intelligence, and more) and explain that these technologies have a huge scope for the transition from the current linear economy to the more sustainable circular economy. Others have suggested OM-linked CE business models using the ReSOLVE framework of CE [26], where they stressed the value of new digital technologies for CE business models for operations, logistics, and supply chain activities.

Given the importance of technology in CE, efforts were focused on how we can best utilize technologies to achieve desired outcomes. Accordingly, theories developed in the context of technology have been employed. Using a literature survey and based on the ReSOLVE framework, Ref. [26] brings out multiple propositions linking CE with big data principles. The authors have suggested key stakeholders for each stage of the ReSOLVE framework and go on to explain how the stakeholder theory and institutional theory can form bases for understanding the links in greater detail in future studies. In fact, speculating on the directions of CE and big data in the future, they feel that multiple theories can help in future research in this direction. The theories they suggest include the resource-based theory and the dynamic capabilities theory that are more commonly used in operations and supply chain management research. In addition, they suggest borrowing newer theoretical ideas from other domains, including those in social and policy studies (e.g., ecological modernization theory) and information technology fields (e.g., technology acceptance models). In fact, there is huge scope for a rich understanding of CE from multiple theoretical frameworks. Amidst the development of OM theories, there is also a great concern with the evolution of novel theories. The dynamic nature of multiple theoretical frameworks creates avenues for the applicability and invention of new theories or the adaptation of existing theories. For example, some theories can be disproved by looking at events that are contrary to estimations [43,44].

Grover and Dresner [45] presented a theoretical model explaining how political resources could be aligned with supply chain strategies. This can be compared to the resource dependent theory or resource-based view. On a different note, another research paper [46] discussed publicness theory and supply chain integration. Sarkis et al. [7] presented a detailed literature review on green supply chain management theories to understand green concepts from several fields. They considered complexity theory, ecological modernization theory, and information theory alongside resource theories. Since 2010, technological intervention in agribusinesses has taken a great role in business performance.

Though not yet studied in detail in the context of food waste and circular economy, we feel that the institutional entrepreneurship theory could play a significant role in explaining entrepreneurial motivations in CFSCs. One of the most influential papers on institutional entrepreneurship to date has been developed by Battilana et al. [47]. Institutional entrepreneurs are change agents, who (i) initiate divergent changes and (ii) actively participate in the implementation of these changes. Only actors who initiate divergent changes, that is, changes that break with the institutionalized template for organizing within a given institutional context, can be regarded as institutional entrepreneurs. Battilana at al. [47] list two conditions enabling institutional entrepreneurship—field-level conditions and an actor's social position. In the context of CE, institutional entrepreneurship theory can be applied to understand how some actors in agri-food supply chains break with the dominant logic/template/way of doing things and introduce a new way of doing things.

The discussion above brings out some prominent theories used in OM and CE—the stakeholder theory, institutional theory, resource-based theory, paradox theory, resource dependence theory, and institutional entrepreneurship theory. This study will focus on these theories in the next section. Another important observation is that all these theories have been discussed in the context of CE, while there has been no effort to link these theories to the motivation of CFSCs in reducing food waste. This is a significant research gap. The next section contributes to the literature by filling this important research gap.

5.1. Applying Theoretical Underpinnings to Study Motivations for Food Waste Reduction in CFSCs

We believe that several of the organizational theories mentioned so far in this section can be used to explain the motivation of organizations engaging in food waste in CFSCs. While all the Rs of the CE [28] are relevant for food waste, reusing and recycling are more important in terms of social, economic, and environment impacts. By reusing food before it becomes unfit for human consumption, significant carbon emissions can be reduced [19].

In the paragraphs below, we utilize our interactions with food companies to explain the applicability of common organizational theories for reducing food waste.

5.1.1. The Stakeholder Theory

Some principles on the application of the stakeholder theory to food waste can be borrowed from the sustainability literature. This theory can be used to explain some interesting behavior of food businesses engaging in the reduction of food waste in their supply chains. Several stakeholders are influencing food companies to reduce food waste. Based on our interactions with food companies, we identified that the government, via legislation, is one of the most influential stakeholders. Governmental regulations play a strong role here. For example, regulatory systems can force companies to comply with the regulations and hence help them reduce food waste. This is apparent in the agri-food industry, with European directives such as the Hazard Analysis & Critical Control Point (HACCP) EU Directive. This regulation, introduced in the EU in the 1990s and modified in subsequent years, expects EU food business operators to put in place, implement, and maintain a permanent procedure or procedures based on the HACCP principles. Without an appropriate plan to avoid hazards such as the contamination of food with bacteria, fungi, viruses, and parasites, food items may cause several food-borne illnesses in consumers. The plan could include, for example, maintaining the correct atmospheric conditions (temperature, humidity, etc.) that would keep the shelf life of food long enough, which in turn would avoid them becoming waste quickly.

These regulatory pressures can act as barriers if companies perceive that these regulations are not effectively enforced by local/regional governments. Other stakeholders are also important for reducing food waste. Top management commitment and commitment from employees play a strong role in reducing food waste. Other downstream supply chain partners, by virtue of their position as customers, also exert pressures on reducing food waste.

5.1.2. The Institutional Theory

Some tenets of the institutional theory can be used to explain the behavior of the food companies engaged in technology demonstrations. The three pillars of the institutional theory [28,30] can be applied to understand why and how firms in agri-food supply chains can engage in actions to reduce food waste. The three pillars—regulative, normative, and cultural–cognitive—of this theory can provide motivations and inhibitors for reducing food waste. Reducing food waste can be internalized by food companies using the normative pillar of the institutional theory. This can be implemented if all stakeholders of a firm believe that disposing food waste in landfills is less preferable to, for example, donating to charities. This internalization is important to motivate companies when they perceive that the costs of technology investments in reducing food waste are larger than the market value of avoiding food waste. Explicit associations with established food charities can be a good motivator for the normative pillar. The normative pillar was manifested in the companies listed in Table 1 when they prioritized their own survival and were hesitant to engage in the innovative activities of the project, even though they knew that working on the project would benefit them in due course. We experienced another manifestation of the normative pillar when some of the food companies joined the project for the green image it generated. The cultural–cognitive pillar of the institutional theory represents practices that involve mostly unconsciously adopted decisions. For example, there is in general a higher emphasis on reducing food waste and adopting sustainable food practices in modern days compared to a few decades earlier, which can explain why all food companies are putting more and more efforts, even when some of them are not required by law, to reduce food waste in their supply chains.

The institutional theory has also been used to explain the barriers to the implementation of CE in organizations. Lopez et al. [25] identify five main categories of barriers (institutional, market, organizational, behavioral, and technological) for resource efficiency in firms. Of these, we had opportunities to observe three—organizational, behavioral, and technological—categories. Specifically, we observed that a willingness of firms in terms of favorable changes in behavioral and organizational efforts are critical for the right application of technology and the maximum reduction of food waste. Vermunt et al. [1] have prepared a slightly larger set of barriers, adding supply chain barriers, financial barriers, and knowledge barriers to the above list. For example, the lack of appropriate partners, low availability of raw materials, higher dependence on external parties, and lack of information exchange/conflict of interests between supply chain actors are crucial supply chain-related barriers that could affect circular economy business models. This is equally true for reducing food waste in food supply chains. In fact, during our interactions with food companies, we experienced that generally negative perceptions of IT projects were a significant barrier to overcome.

5.1.3. The Resource-Based Theory (RBT) and NRBT

We believe that the resource-based theory (RBT) [48] and the natural resource-based theory (NRBT) [49] provide very good opportunities to explain the behavior of food companies engaging in food waste reduction efforts in their supply chains. The NRBT is required if we view food waste in the context of the imputed natural resources (energy, labor, soil, fertilizers, water, and more) needed to produce the food. The issue of valorizing food waste can be used from the lens of the RBT. Although the RBT has not yet been applied to the case of food waste, a related concept, called the resource-based paradigm, has been shown to help view waste in terms of resources [50]. This idea can be extrapolated further to bring the RBT into the food waste context, if we emphasize that the unique, inimitable knowledge generated in firms that have started to view waste as another resource, enables firms to gain competitive advantages. The knowledge of the composition of what is currently termed as waste and understanding the potential utility of waste can become accumulated into an inimitable knowledge in the long run. Continuously looking for technological innovations in-house and elsewhere to valorize what is currently termed as waste in a firm can not only generate more revenues via extra sales but also can lead to a reduction in costs via reduced raw-material consumption and reduced waste disposal costs. This observation has been somewhat echoed by Ref. [2] in their discussion of the Circular Economy Strategies Database that captures 45 CE strategies (e.g., material substitution, green procurement, product labelling, eco design, re-use, recycling, extended producer responsibility, and more) in various stages (e.g., material sourcing, design, manufacturing, distribution, consumption, collection, recycling, and more) of the economy. Implementing these strategies efficiently will result in valuable and inimitable knowledge for improving the resource efficiencies of operations in businesses. Thus, understanding opportunities for CE in a business will provide the scope for a firm to gain a competitive advantage from the point of view of the RBT.

Klassen and Whybark [51] and similar researchers working on pollution prevention have used the RBT to differentiate between pollution-prevention technologies (i.e., technologies that prevent pollution from occurring) and pollution-control technologies (i.e., technologies that try to reduce the impact of pollution once it has occurred). In a similar way, the RBT can also help us to understand the implications of waste prevention vs. waste control in CE organizations. In the context of food waste, waste prevention would mean that efforts are made to avoid food waste from occurring. There are a number of strategies available to organizations to achieve this. For example, effective information sharing from supply chain partners can help produce food in the right quantity for a given purpose with little waste. Effective scheduling in food-processing industries or in the farming industry can help reduce waste. Using appropriate technologies can improve productivity and reduce waste. The main aim of our interactions with food companies is to avoid waste from occurring in the supply chain by ensuring that produce is kept in the right conditions. These strategies will help to prevent food waste from occurring. The RBT can help successful firms to mobilize their available resources to prevent waste and gain competitive advantages. By preventing waste, firms are able to reduce the cost of raw materials consumed, improve quality, reduce their waste disposal costs, and thus gain competitive advantages. Waste-control strategies are useful to limit the impact once waste has occurred. They do not result in as much of a competitive advantage compared to waste-prevention strategies. Thus, the RBT helps firms to look for opportunities to prevent waste first.

The economic advantage derived by reduced food waste translates into competitive advantages to firms. This economic angle provides another way that the RBT can be used to explain the behavior of firms in reducing food waste. One rather more interesting way that the RBT can be used to explain the motivations of agribusiness companies in reducing food waste is the quality angle. The food companies we worked with mentioned that they do not incur food waste anymore after engaging with us about the use of technologies. On their view, digital technologies primarily help them in improving the quality of their produce. Using the appropriate monitoring of quality-related variables, these food companies are more confident that their food produce will have high quality in the market. The literature on quality management [52] explains that investments in improving quality help firms in gaining competitive advantages via improved market prices and reduced waste.

5.1.4. The Paradox Theory

The paradox theory [53] can also help understand the motivations of firms in the food waste context. Given that food waste can be reduced via soft means (e.g., behavioral changes) or relatively hard means (e.g., using technological support), there is a learning paradox in deciding on the relative importance of these two means. To reduce food waste in food supply chains, there is a need for supply chain collaboration, but at the same time, different supply chain partners need to maintain their identities, giving raise to both organizational and belonging paradoxes. A paradox is apparent when one needs to verify whether technological investments in reducing food waste consume more resources compared to the resources imputed to the food that is likely to be saved.

5.1.5. The Resource Dependence Theory (RDT) and NRDT

The resource dependence theory [54] and the natural resource dependence theory (NRDT) [55] will be relevant in the context of food waste in FSCs. The RDT emphasizes external dependence on scarce resources and the uncertainty that it creates for organizations to survive [54], while the NRDT explicitly emphasizes additional dependence on natural resources [55]. Members of a supply chain are traditionally dependent on each other for the continued success of the supply chains and hence their own survival, which explains the applicability of the RDT for CFSCs. For example, firms depend on data from other supply chain partners for making prompt business decisions and consider data sharing and security (including threats from hackers) as crucial limiting issues for their growth. The use of modern technologies exacerbates these dependencies, for example, for global connectivity. In some cases, a food producer may not be able to install gateways to transmit sensor signals from a truck if lorry drivers (which are another crucial resource in CFSCs) object to having too many transmitting devices in their cabin. As highlighted earlier, since food waste involves significant but scarce natural resources for producing food, there is a crucial dependence on natural resources too.

5.1.6. The Institutional Entrepreneurship Theory

As mentioned earlier, due to its ability to explain entrepreneurial motivations, the institutional entrepreneurship theory [47] could play a significant role in helping us to understand the motivations of firms in reducing food waste. Several EU policy makers have highlighted that the European Commission has only recently started to coordinate national policies about food waste in member states, and in some cases (where such policies did not exist) it has started to push member states to develop such policies. The European Commission has used its 'social position' in the field to push for a major institutional change—at both the member state and EU level—to develop policies for food waste at the national and EU levels. Investment firms have highlighted that profit was not the key decisive criterion for these firms to invest in sustainability. There is growing evidence that firms are breaking away from the dominant stereotypical ways of behaving. For example, firms are explicit in stating that profit generation is not enough anymore; the food business needs to also promote animal welfare and respect the environment and natural resources. Some of the previous discussions did highlight many additional reasons for companies to engage in the reduction of food waste in their CFSCs, including legal requirements and quality enhancement. These considerations, we think, are examples of institutional entrepreneurship—breaking with the dominant template and way of doing things.

As a summary of this section, Table 2 presents major tenets of important organizational theories and how they can be interpreted in the context of food waste reduction in circular FSCs.

Table 2. Elements of various organizational theories for managing food waste in circular food supply chains.

Theory	Element	Links to Food Waste
The stakeholder theory	Multiple stakeholders	• Regulatory stakeholder (HACCP directive) • Top management commitment • Employees • Supply chain partners

Table 2. *Cont.*

Theory	Element	Links to Food Waste
The institutional theory	Regulative, normative, and cultural–cognitive pillars	• Firm belief in reducing food waste • Preference for donating to food charities than sending to landfill • Explicit association with established food charities • Prioritizing survival to innovation during COVID-19 lockdowns • Associating with the green image • Voluntary initiatives on reducing food waste and sustainable food practices • General lack of trust in IT projects
The resource-based theory and the natural resource-based theory	VRIN (valuable, rare, inimitable, and non-substitutable) resources and competitive advantage	• Wasted food is a waste of valuable imputed natural resources (energy, labor, soil, fertilizers, water, and more) • Inimitable knowledge on reducing, recycling, and reusing waste, and valorizing food waste is a source of competitive advantage • Efficient operations management for reducing raw material consumption for competitive advantage • Waste prevention vs. waste control in CE organizations • Efficient quality control for competitive advantage
The paradox theory	Learning, organizational, and belonging paradoxes	• Soft means (e.g., behavioral changes) vs. relatively hard means (e.g., using technological support) for reducing food waste • Supply chain collaboration vs. maintaining individual identities of partners • Comparing costs of technological investments in reducing food waste with the resources imputed to the saved food
The resource dependence theory and the natural resource dependence theory	Supply chain dependency	• Dependence on data from other supply chain partners for making prompt business decisions • Data sharing and security issues • Use of modern technologies exacerbates these dependencies
Institutional entrepreneurship	Breaking away from dominant ways of doing things	• Entrepreneurs and enterprises do not consider profit as their single motive anymore. Other considerations including social and environmental impacts are increasingly being employed in entrepreneurial decision-making.

6. Discussion

As highlighted in Table 2, the six theories we have selected in the previous section can help us to understand the several observed behaviors of food companies. For example, the stakeholder theory helps to visualize the government as the regulatory stakeholder and explain how food companies approach the HACCP directive. It also helps to see why food companies should consider the views of other stakeholders, such as the top management, employees, or supply chain partners, while making decisions on how to reduce food waste. The institutional theory has helped us to explain why food companies prioritized their own survival rather than engaging in innovations for reducing food waste during the pandemic. It also explains the general lack of trust in IT projects for reducing food waste. The resource-based theory has helped us to explain the trade-offs in food companies between investing in waste-prevention vs. waste-control options. While the waste in food supply chains can be reduced by sharing information, the issues of privacy and security has been explained using the resource dependence theory. The fact that some companies attempt to lead in the development of strategies for food waste reduction more than others has been explained using the institutional entrepreneurship theory.

In summary, we believe that there is no single theory for explaining all the behaviors of food companies. Interestingly, each theory is able to explain only a part of these behaviors. We further believe that all the six theories listed in Table 2, together, are able to explain a majority of the behaviors of food companies in reducing food waste. This might call for a proposal of a *unified theory of food waste* by bringing the features of all six theories together. This calls for interesting research by future researchers.

Many of the theories discussed in this article have been developed in the past several decades using organizational, social, economic, ethical, and sustainable viewpoints. Among their diverse approaches, all of them point towards the enhancement of the subject being considered. Two main themes, namely sustainability and circular economy, are brought together in the 21st century to visualize our green future with financial and social prosperity. Our paper has discussed the theories and their contribution to CFSCs in detail; this could lead to many future research projects in the area of supply chains and to collaborations aimed at implementing sustainability and circular economy practices with societal involvement.

In spite of the support of these theories in explaining some observed behaviors of food companies, some other observations could not be readily explained using any of the six theories above. For example, we observed a strange behavior when we attempted to install some new technologies to control and monitor storage conditions of food in trucks. Since any new piece of technology (e.g., IoT sensors) installed in trucks could also track the location of vehicles, drivers were reluctant to engage in the installations. This negates, for example, the resource dependency theory. Similarly, when we could not support a specific request of a company due to technological limitations (for example, due to a lack of the availability of a suitable low-cost technology to continuously facilitate international shipments in multiple continents), the company did not wish to engage with us further for other parts of their operations. Similarly, another company decided not to work with us after their internal restructuring. We may need newer theories or borrow from several other disciplines to explain this behavior.

We believe that the discussions in the preceding sections could provide a basis for classifying theories in the context of food waste. In addition to the prominent theories discussed in this paper, other theories can be potentially applied to understand the behavior of firms regarding food waste. They include systems theory, systems thinking theory, systems dynamics theory [56,57], the intermediary actors theory [58], social practice theory [59], perceived behavioral control/the theory of planned behavior [60], information systems theory [61], the network design theory [62], creating shared value theory [63], and more. A comprehensive view of several theories discussed above could be integrated around the theme of circular economy and circular food supply chains. Figure 2 is representative of the achievable pathways for CFSCs and global sustainability. Each of the theories from various perspectives can be brought together to visualize sustainability in the long run. We have positioned the theories that are enabling and enriching the sustainability of CFSCs under six main perspectives (refer to Figure 2). These theories can be considered as imperative to understand the concepts of sustainability in agribusiness supply chains and are positioned in the inner circle of Figure 2. These perspectives are named as dynamic and innovative perspectives, institutional and business perspectives, economic perspectives, social networking perspectives, resources perspectives, and ecology and environment perspectives. The theories in the outer circle in Figure 2 are either based on technology usage or borrowed from other disciplines (e.g.,predator–prey theory [64], chaos theory [65,66], complexity theory [7], and the institutional entrepreneurship theory [47]). We call these theories supporting theories, as we have yet to see a large-scale application of these theories to food waste issues; however these theories could provide additional support for the cause of food waste reduction.

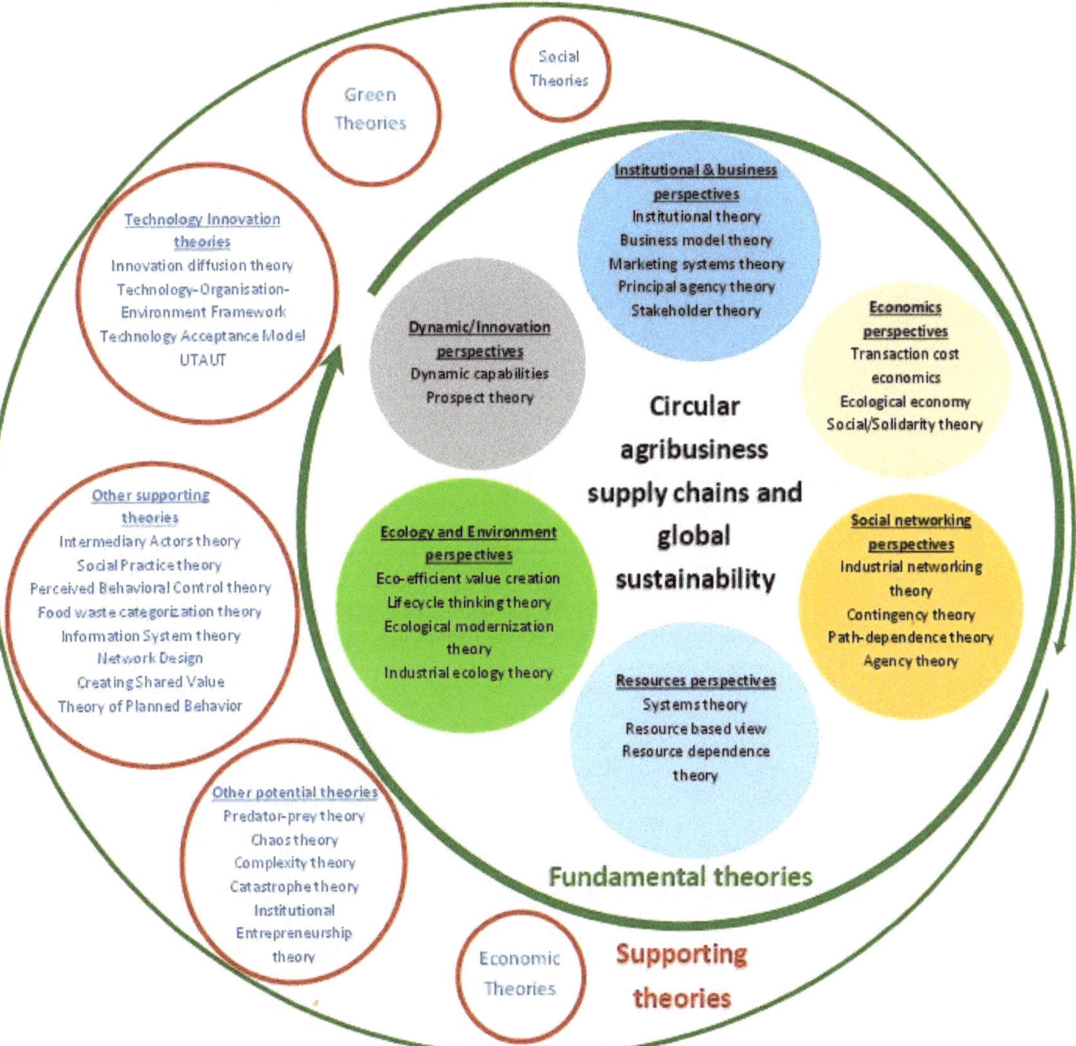

Figure 2. A pictorial representation of various theoretical perspectives woven around CFSCs and global sustainability principles.

Some theories have been criticized for their applications to practical matters. For example, although the RBT has been used widely in OM area, there is a series of conversations about making improvements to the theory and its applications. While Singh et al. [67]

used the theoretical lens of resource dependence theory to understand ISO 9000 for organizational green management, Hitt et al. [68] suggested RBT enhancement through the adoption of various applications. These on-going conversations clearly specify the richness of the theory and its impact in practice. We can say that OM theories, in combination with theories from other disciplines, are the backbone for the development of applications, new practices, and new methods. The concept of circular economy in agri-business is a relatively new concept that needs enrichment and support from theories across disciplines to show its real potential for economic, social, and environmental practices.

In summary, we can classify theories into two main categories: (i) fundamental theories (six perspectives given in Figure 2) of food waste that strengthen our understanding and the importance of identifying areas of potential food waste reduction, and (ii) food waste supporting theories that support actual food waste-reduction intiatives (in the outer circle of Figure 2). Using our empirical research related to the European food sector, we classify the paradox theory, the resource-based theory, institutional theory, and stakeholder theory as the fundamental theories, and we classify the institutional entrepreneurship theory and technology organization theory as the supporting theories. In simple terms, the fundamental theories will act as inputs and the supporting theories will serve as outputs for any research that considers the socio-economic and environmental aspects of waste reduction and sustainability.

7. Conclusions

Given the growing importance of food waste in meeting several of the UN's Sustainable Development Goals, we studied the issue of food waste from an operations management point of view in this paper. Our main research question was to understand how existing organizational theories could help to understand the observed behavior of food companies in reducing food waste. We have addressed this research question via Table 2. The research objectives have also been achieved as we have identified suitable theories and used them to understand food waste practices. With the help of the literature on circular economy, we studied circular food supply chains (CFSCs) using multiple theories. We specifically looked the stakeholder theory, institutional theory, the resource-based theory, the paradox theory, the resource dependence theory, and the institutional entrepreneurship theory in greater detail in the context of food waste.

This paper contributes to the literature on operations, supply chains, and circular economy in multiple ways. Though the importance of food waste was recognized long ago, this is the first time the issue of food waste has been studied from an OM and supply chains point of view. We reviewed existing theories related to circular economy and discussed some newer theories that may hold promise to support a better understanding of circular economy. We used at least six of these theories, for the first time, in the context of food waste.

In spite of these contributions, there are limitations to our approach in this paper. The data from our qualitative study come from a small sample of 13 companies listed in Table 1. We think that our continuous association with these 13 companies over a period of more than 4 years has yielded rich insights in explaining organizational behavior. However, data could be gathered from more companies to yield more generalizable results. Similarly, our sample focused only on companies based in North-West Europe. For better generalizations, more companies from other parts of the world could be studied. Finally, we only focused on six important theories and applied them to understand the food-waste context. However, several more theories (e.g., the agency theory, the contingency theory) can also be applied; we could not focus on them due to the limited time and space. Some exciting new theories, borrowed from the engineering literature, can also help to understand the circular economy principles, but we did not discuss them due to lack of time and space. For example, the catastrophe theory [69] has the capability to explain why, how, and when public perceptions about specific features of circular economy will change. Akin to the sudden change in the public perception of the use of plastics, public perceptions of waste streams can also quickly change, which can be studied using the catastrophe theory. Future papers can consider this theory to support circular economy practices in greater detail. Another area of interest that we could not examine in more detail in this paper is the distinction between food-waste prevention and food-waste control. Borrowing from the pollution-prevention/control literature, we explained how the resource-based theory can help understand the relative merits of food-waste prevention and control. This issue can also be studied in more depth in future studies.

We are confident that the analysis of food waste for CFSCs will be useful to researchers engaged in theoretical studies of food waste and to policy makers engaged in food policy and circular economy.

Author Contributions: Conceptualization, R.R. and U.R.; formal analysis, R.R. and U.R.; funding acquisition, R.R., U.R., K.P. and I.H.; investigation, R.R., K.P. and I.H.; methodology, R.R. and U.R.; project administration, R.R. and K.P.; resources, R.R. and K.P.; writing—original draft, R.R.; writing—review and editing, U.R., K.P. and I.H. All authors have read and agreed to the published version of the manuscript.

Funding: This research was funded by Interreg North-West Europe (NWE831).

Institutional Review Board Statement: The study was conducted after gaining ethical approval (ref BMRI/Ethics/Staff/2018-19/005) from the University of Bedfordshire, UK.

Informed Consent Statement: Informed consent was obtained from all subjects involved in the study.

Data Availability Statement: The original contributions presented in the study are included in the article, further inquiries can be directed to the corresponding author.

Acknowledgments: This research has benefited from the insights about food waste reduction in agribusiness supply chains from the REAMIT project. We acknowledge the contribution by members of the project though we could not include all their names as authors in this article.

Conflicts of Interest: Author Imke Hermens was employed by the company Whysor BV. The remaining authors declare that the research was conducted in the absence of any commercial or financial relationships that could be construed as a potential conflict of interest.

References

1. Vermunt, D.A.; Negro, S.O.; Verweij, P.A.; Kuppens, D.V.; Hekkert, M.P. Exploring barriers to implementing different circular business models. *J. Clean. Prod.* **2019**, *222*, 891–902. [CrossRef]
2. Kalmykova, Y.; Sadagopan, M.; Rosado, L. Circular economy—From review of theories and practices to development of implementation tools. *Resour. Conserv. Recycl.* **2018**, *135*, 190–201. [CrossRef]
3. WRAP. WRAP and the Circular Economy, Waste & Resources Action Programme. 2021. Available online: https://wrap.org.uk/taking-action/climate-change/circular-economy (accessed on 30 December 2021).
4. Akkerman, R.; Buisman, M.; Cruijssen, F.; de Leeuw, S.; Haijema, R. Dealing with donations: Supply chain management challenges for food banks. *Int. J. Prod. Econ.* **2023**, *262*, 108926. [CrossRef]
5. Ramanathan, R.; Duan, Y.; Ajmal, T.; Pelc, K.; Gillespie, J.; Ahmadzadeh, S.; Condell, J.; Hermens, I.; Ramanathan, U. Motivations and challenges for food companies in using IoT sensors for reducing food waste: Some insights and a road map for the future. *Sustainability* **2023**, *15*, 1665. [CrossRef]
6. Liu, J.; Feng, Y.; Zhu, Q.; Sarkis, J. Green supply chain management and the circular economy: Reviewing theory for advancement of both fields. *Int. J. Phys. Distrib. Logist. Manag.* **2018**, *48*, 794–817. [CrossRef]
7. Sarkis, J.; Zhu, Q.; Lai, K.-H. An organizational theoretic review of green supply chain management literature. *Int. J. Prod. Econ.* **2011**, *130*, 1–15. [CrossRef]
8. Sehnem, S.; Vazquez-Brust, D.; Pereira, S.; Campos, L. Circular economy: Benefits, impacts and overlapping. *Supply Chain. Manag. Int. J.* **2019**, *24*, 784–804. [CrossRef]
9. Geissdoerfer, M.; Savaget, P.; Bocken, N.M.; Hultink, E.J. The Circular Economy—A new sustainability paradigm? *J. Clean. Prod.* **2017**, *143*, 757–768. [CrossRef]
10. Missemer, A. Natural Capital as an economic concept, history and contemporary issues. *Ecol. Econ.* **2018**, *143*, 90–96. [CrossRef]
11. Linder, M.; Williander, M. Circular Business Model Innovation: Inherent Uncertainties. *Bus. Strat. Environ.* **2017**, *26*, 182–196. [CrossRef]
12. EEA. *Circular Economy in Europe. Developing the Knowledge Base*; EEA Report No 2/2016; European Environment Agency: Copenhagen, Denmark, 2016.

13. Rosa, P.; Sassanelli, C.; Terzi, S. Circular business models versus circular benefits: An assessment in the waste from electrical and electronic equipments sector. *J. Clean. Prod.* **2019**, *231*, 940–952. [CrossRef]
14. De Angelis, R. Circular economy and paradox theory: A business model perspective. *J. Clean. Prod.* **2021**, *285*, 124823. [CrossRef]
15. Manninen, K.; Koskela, S.; Antikainen, R.; Bocken, N.M.P.; Dahlbo, H.; Aminoff, A. Do circular economy business models capture intended environmental value propositions? *J. Clean. Prod.* **2018**, *171*, 413–422. [CrossRef]
16. Suárez-Eiroa, B.; Fernández, E.; Méndez-Martínez, G.; Soto-Oñate, D. Operational principles of circular economy for sustainable development: Linking theory and practice. *J. Clean. Prod.* **2019**, *214*, 952–961. [CrossRef]
17. Tukker, A. Product services for a resource-efficient and circular economy—A review. *J. Clean. Prod.* **2015**, *97*, 76–91. [CrossRef]
18. WRAP. Food Surplus and Waste in the UK—Key Facts, Waste & Resources Action Programme. 2020. Available online: https://wrap.org.uk/sites/default/files/2020-11/Food-surplus-and-waste-in-the-UK-key-facts-Jan-2020.pdf (accessed on 1 November 2021).
19. Parry, A.; James, K.; LeRoux, S. Strategies to Achieve Economic and Environmental Gains by Reducing Food Waste, Waste & Resources Action Programme (WRAP). 2015. ISBN 978-1-84405-473-2. Available online: https://wrap.org.uk/sites/default/files/2020-12/Strategies-to-achieve-economic-and-environmental-gains-by-reducing-food-waste.pdf (accessed on 4 January 2022).
20. Gillespie, J.; da Costa, T.P.; Cama-Moncunill, X.; Cadden, T.; Condell, J.; Cowderoy, T.; Ramsey, E.; Murphy, F.; Kull, M.; Gallagher, R.; et al. Real-time anomaly detection in cold chain transportation using IoT technology. *Sustainability* **2023**, *15*, 2255. [CrossRef]
21. Maiyar, L.M.; Ramanathan, R.; Roy, I.; Ramanathan, U. A decision support model for cost-effective choice of temperature-controlled transport of fresh food. *Sustainability* **2023**, *15*, 6821. [CrossRef]
22. Ramanathan, U.; Ramanathan, R.; Adefisan, A.; Da Costa, T.; Cama-Moncunill, X.; Samriya, G. Adapting digital technologies to reduce food waste and improve operational efficiency of a frozen food company—The case of Yumchop Foods in the UK. *Sustainability* **2022**, *14*, 16614. [CrossRef]
23. Ramanathan, U.; Pelc, K.; da Costa, T.P.; Ramanathan, R.; Shenker, N. A case study of human milk banking with focus on the role of IoT sensor technology. *Sustainability* **2022**, *15*, 243. [CrossRef]
24. Tsoukas, H.; Knudsen, C. (Eds.) *The Oxford Handbook of Organization Theory*; Oxford Handbooks Online; Oxford University Press: Oxford, UK, 2003.
25. Diaz Lopez, F.J.; Bastein, T.; Tukker, A. Business model innovation for resource-efficiency, circularity and cleaner production: What 143 cases tell us. *Ecol. Econ.* **2019**, *155*, 20–35. [CrossRef]
26. Lopes de Sousa Jabbour, A.B.; Rojas Luiz, J.V.; Rojas Luiz, O.; Jabbour, C.J.C.; Ndubisi, N.O.; Caldeira de Oliveira, J.H.; Junior, F.H. Circular economy business models and operations management. *J. Clean. Prod.* **2019**, *235*, 1525–1539. [CrossRef]
27. Lahti, T.; Wincent, J.; Parida, V. A definition and theoretical review of the circular economy, value creation, and sustainable business models: Where are we now and where should research move in the future? *Sustainability* **2018**, *10*, 2799. [CrossRef]
28. Ranta, V.; Aarikka-Stenroos, L.; Ritala, P.; Mäkinen, S.J. Exploring institutional drivers and barriers of the circular economy: A cross-regional comparison of China, the US, and Europe. *Resour. Conserv. Recycl.* **2018**, *135*, 70–82. [CrossRef]
29. Gusmerotti, N.M.; Testa, F.; Corsini, F.; Pretner, G.; Iraldo, F. Drivers and approaches to the circular economy in manufacturing firms. *J. Clean. Prod.* **2019**, *230*, 314–327. [CrossRef]
30. Scott, W.R. *Institutions and Organizations: Ideas and Interests*, 3rd ed.; Sage Publications: Thousand Oaks, CA, USA, 2008.
31. Freeman, R. *Strategic Management: A Stakeholder Approach*; Pitman: Boston, MA, USA, 1984.
32. Schaltegger, S.; Hansen, E.G.; Lüdeke-Freund, F. Business models for sustainability: Origins, present research, and future avenues. *Organ. Environ.* **2016**, *29*, 3–10. [CrossRef]
33. Stubbs, W.; Cocklin, C. Conceptualizing a "sustainability business model". *Organ. Environ.* **2008**, *21*, 103–127. [CrossRef]
34. Li, K.; Lee, J.-Y.; Gharehgozli, A. Blockchain in food supply chains: A literature review and synthesis analysis of platforms, benefits and challenges. *Int. J. Prod. Res.* **2023**, *61*, 3527–3546. [CrossRef]
35. Li, L.; Tang, O.; Zhou, W.; Fan, T. Backroom effect on perishable inventory management with IoT information. *Int. J. Prod. Res.* **2023**, *61*, 4157–4179. [CrossRef]
36. Stefanini, R.; Vignali, G. The influence of Industry 4.0 enabling technologies on social, economic and environmental sustainability of the food sector. *Int. J. Prod. Res.* **2023**, 1–18. [CrossRef]
37. Rogers, E.M. *The Diffusion of Innovations*, 3rd ed.; Free Press: New York, NY, USA, 1983.
38. Tornatzky, L.G.; Fleischer, M. *The Processes of Technological Innovation*; Lexington Books: Lexington, MA, USA, 1990.
39. Davis, F.D. Perceived usefulness, perceived ease of use, and user acceptance of information technology. *MIS Q.* **1989**, *13*, 319–340. [CrossRef]
40. Wang, J.; Li, X.; Wang, P.; Liu, Q.; Deng, Z.; Wang, J. Research trend of the unified theory of acceptance and use of technology theory: A bibliometric analysis. *Sustainability* **2022**, *14*, 10. [CrossRef]
41. Bressanelli, G.; Adrodegari, F.; Perona, M.; Saccani, N. The role of digital technologies to overcome circular economy challenges in PSS business models: An exploratory case study. *Procedia CIRP* **2018**, *73*, 216–221. [CrossRef]
42. Tseng, M.-L.; Tan, R.R.; Chiu, A.S.; Chien, C.-F.; Kuo, T.C. Circular economy meets industry 4.0: Can big data drive industrial symbiosis? *Resour. Conserv. Recycl.* **2018**, *131*, 146–147. [CrossRef]
43. Schmenner, R.W.; Seink, M.L. On theory in operations management. *J. Oper. Manag.* **1998**, *17*, 97–113. [CrossRef]
44. Schmenner, R.W.; Van Wassenhove, L.; Ketokivi, M.; Heyl, J.; Lusch, R.F. Too much theory, not enough understanding. *J. Oper. Manag.* **2009**, *27*, 339–343. [CrossRef]

45. Grover, A.K.; Dresner, M. A theoretical model on how firms can leverage political resources to align with supply chain strategy for competitive advantage. *J. Supply Chain Manag.* **2022**, *58*, 48–65. [CrossRef]
46. Seepma APVan Donk, D.P.; De Blok, C. On Pulicness theory and its implications for supply chain integration: The case of criminal justice supply chains. *J. Supply Chain. Manag.* **2021**, *57*, 72–103. [CrossRef]
47. Battilana, J.; Leca, B.; Boxenbaum, E. How actors change institutions: Towards a theory of institutional entrepreneurship. *Acad. Manag. Ann.* **2009**, *3*, 65–107. [CrossRef]
48. Barney, J.B. Firm resources and sustained competitive advantage. *J. Manag.* **1991**, *17*, 99–120. [CrossRef]
49. Hart, S.L. A natural-resource-based view of the firm. *Acad. Manag. Rev.* **1995**, *20*, 986–1014. [CrossRef]
50. Park, J.Y.; Chertow, M.R. Establishing and testing the "reuse potential" indicator for managing wastes as resources. *J. Environ. Manag.* **2014**, *137*, 45–53. [CrossRef]
51. Klassen, R.D.; Whybark, D.C. The impact of environmental technologies on manufacturing performance. *Acad. Manag. J.* **1999**, *42*, 599–615. [CrossRef]
52. Powell, T.C. Total quality management as competitive advantage: A review and empirical study. *Strat. Manag. J.* **1995**, *16*, 15–37. [CrossRef]
53. Smith, W.K.; Lewis, M.W. Toward a theory of paradox: A dynamic equilibrium model of organizing. *Acad. Manag. Rev.* **2011**, *36*, 381–403.
54. Pfeffer, J.; Salancik, G.R. *The External Control of Organizations: A Resource Dependence Perspective*; Harper & Row: New York, NY, USA, 1978.
55. Tashman, P. A natural resource dependence perspective of the firm: How and why firms manage natural resource scarcity. *Bus. Soc.* **2021**, *60*, 1279–1311. [CrossRef]
56. Forrester, J.W. System dynamics systems thinking soft O.R. *Syst. Dyn. Rev.* **1994**, *10*, 245–256. [CrossRef]
57. Richmond, B.; Peterson, S. *An Introduction to Systems Thinking*; High Performance Systems, Inc.: Lebanon, NH, USA, 2001.
58. Frandsen, F.; Johansen, W. Organizations, stakeholders, and intermediaries: Towards a general theory. *Int. J. Strateg. Commun.* **2015**, *9*, 253–271. [CrossRef]
59. Hargreaves, T. Practiceing behaviour change: Applying social practice theory to pro-environmental behaviour change. *J. Consum. Cult.* **2011**, *11*, 79–99. [CrossRef]
60. Ajzen, I. Perceived behavioral control, self-efficacy, locus of control, and the theory of planned behavior. *J. Appl. Soc. Psychol.* **2002**, *32*, 665–683. [CrossRef]
61. Al-Qaysi, N.; Mohamad-Nordin, N.; Al-Emran, M. A systematic review of social media acceptance from the perspective of educational and information systems theories and models. *J. Educ. Comput. Res.* **2020**, *57*, 2085–2109. [CrossRef]
62. Jouzdani, J.; Govindan, K. On the sustainable perishable food supply chain network design: A dairy products case to achieve sustainable development goals. *J. Clean. Prod.* **2021**, *278*, 123060. [CrossRef]
63. Menghwar, P.S.; Daood, A. Creating shared value: A systematic review, synthesis and integrative perspective. *Int. J. Manag. Rev.* **2021**, *23*, 466–485. [CrossRef]
64. Berryman, A.A. The origins and evolution of predator-prey theory. *Ecology* **1992**, *73*, 1530–1535. [CrossRef]
65. Levy, D. Chaos theory and strategy: Theory, application, and managerial implications. *Strat. Manag. J.* **1994**, *15*, 167–178. [CrossRef]
66. Stapleton, D.; Hanna, J.B.; Ross, J.R. Enhancing supply chain solutions with the application of chaos theory. *Supply Chain Manag. Int. J.* **2006**, *11*, 108–114. [CrossRef]
67. Singh, P.J.; Power, D.; Chuong, S.C. A resource dependence theory perspective of ISO 9000 in managing organizational environment. *J. Oper. Manag.* **2011**, *29*, 49–64. [CrossRef]
68. Hitt, M.A.; Xu, K.; Carnes, C.M. Resource based theory in operations management research. *J. Oper. Manag.* **2016**, *41*, 77–94. [CrossRef]
69. Golubitsky, M. An introduction to catastrophe theory and its applications. *SIAM Rev.* **1978**, *20*, 352–387. [CrossRef]

sustainability

Article

Insect Production for Animal Feed: A Multiple Case Study in Brazil

Jaqueline Geisa Cunha Gomes [1,*], Marcelo Tsuguio Okano [1,2], Edson Luiz Ursini [2] and Henry de Castro Lobo dos Santos [2]

[1] UNIP—Graduate Program in Production Engineering, Universidade Paulista, São Paulo 04026-002, Brazil; marcelo.okano@unip.br

[2] UNICAMP—College of Technology, State University of Campinas, Limeira 13484-332, Brazil; ursini2@unicamp.br (E.L.U.); h190839@dac.unicamp.br (H.d.C.L.d.S.)

[*] Correspondence: jaqueline.gomes34@aluno.unip.br

Abstract: The production of insects as a sustainable protein source represents an innovation for animal feed. The objective of this research is to analyze the value chain of the use of edible insects in animal feed in Brazil through the framework of SWOT, the business model sustainable canvas, and a multiple case study, highlighting the sustainability characteristics. A qualitative approach of the descriptive exploratory type was used, and the multiple case study identified the actors in the chain and how value is generated. The young age of the sector explains the characteristics observed in the Brazilian chain, such as a large development deficit in terms of financing, technology and the qualification of human resources; a disorganized supply chain and supplier structure; and efforts undertaken by regulatory agencies to promote the development of regulations relating to the production and use of insects in animal feed, which, in turn, will lead those wishing to participate in this innovative venture into research and development in the area. Brazil's edible insect supply chain can become a more significant aspect of sustainable agriculture by closing nutrient and energy loops, promoting food security and minimizing climate change and biodiversity losses, all of which are associated with the achievement of the Sustainable Development Goals.

Keywords: value chain; edible insects; insect cultivation; animal nutrition

Citation: Gomes, J.G.C.; Okano, M.T.; Ursini, E.L.; Santos, H.d.C.L.d. Insect Production for Animal Feed: A Multiple Case Study in Brazil. *Sustainability* **2023**, *15*, 11419. https://doi.org/10.3390/su151411419

Academic Editor: Dimitris Skalkos

Received: 28 June 2023
Revised: 19 July 2023
Accepted: 21 July 2023
Published: 23 July 2023

1. Introduction

The global demand for feed, as well as the competition for protein, grows annually. Research is carried out worldwide to mitigate possible protein production shortages, which will need to be increased by sixty percent by 2050 to meet future world demand [1]. Another aspect affecting the animal protein production system is the high inputs required to produce feed, the global demand for which will reach more than 1 billion tons by 2050 (an increase of 60 to 70% compared to around 800 million tons in 2018) [2].

One of the biggest challenges related to this future demand is increasing the availability and simultaneously reducing the use of natural resources, such as land for soybean planting and water, as well as reducing greenhouse gas emissions. The use of insects as a partial or total ingredient in animal feed has been shown to represent an alternative source of protein and a substitute for traditional feed.

Some research has demonstrated that the consumption of red meat is associated with an increased probability of stroke, diabetes, colon cancer and lung cancer. Insects seem to have a more nutritious and healthier composition than meat-based foods, and they are also diverse in terms of nutritional value. As a result, they can be used as meat substitutes [3,4]. These problems encourage a decrease in meat consumption and its replacement with insect consumption. This is valid not only because they offer replacement protein, but because of insects being a more sustainable, healthy and economical product.

Studies such as those performed by Chia et al. [5] show that animal protein producers are aware of the potential related to using insects as a feed ingredient, and that producers' knowledge is directly proportional to their willingness to pay for this new type of feed.

In the study "Insects as a sustainable feed ingredient in pig and poultry diets—a feasibility study" [6], the actors involved in the production chain of using insects as a sustainable ingredient of pork and poultry feeds are demonstrated, along with a discussion of how their integration can influence the implementation of insects as an alternative source.

This article aims to analyze the value chain of using edible insects in animal feed in Brazil through the framework of SWOT and a sustainable business model canvas and multiple case study, highlighting the sustainable characteristics. As this value chain is still in its nascent stages in Brazil, the following actors were not considered: the poultry/pork sector and retail/consumers.

2. State of the Art

The most promising insects for use in industrial feed production are the black soldier fly (BSF) (*Hermetica Illucens*), the housefly (*Musca domestica*) and the yellow mealworm (*Tenebrio molitor*) [6].

Zhou et al. [7] summarized the nutritional value of different classifications of edible insects. The composition of the nutritional value of edible insects is illustrated in Figure 1.

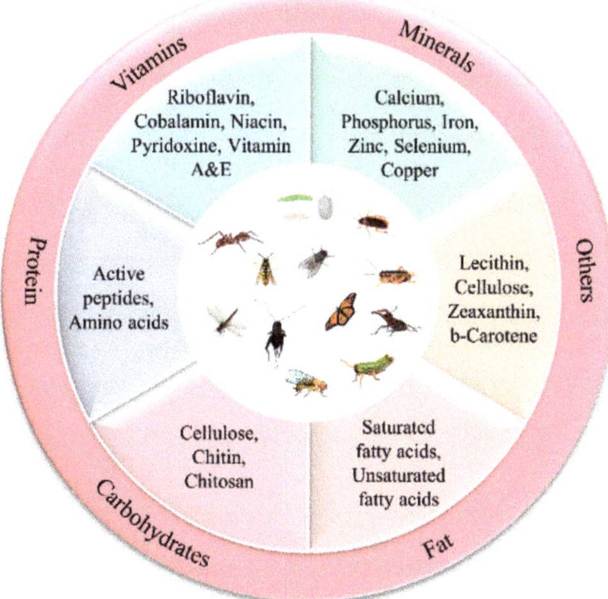

Figure 1. Nutrient composition of edible insects. Based on Zhou et al. [7].

The general composition of edible insects is listed in Figure 1. Different units are employed to represent the data due to the different sources from which they are derived, and the lack of their completeness is a result of the conversion units used. The material composition of insects is markedly different between species. In the dry matter, protein and fat are the most common substances [7].

Insects have tremendous potential at all life cycle stages as sources of nutritional value; they are a significant source of animal protein, contain essential amino acids and minerals (K, Na, Ca, Cu, Fe and Zn), and their fatty acids are unsaturated. The assimilation rate of insect proteins is 76–82%. Insect carbohydrates are primarily composed of chitin, which is present at a concentration range of 2.7 mg to 49.8 mg/kg of dry mass [8].

Table 1 presents Shah et al.'s [9] assessment of the types of economical insects and their chemical compositions and nutritive values.

Table 1. Types of economical insect and their chemical composition and nutritive value.

Insects Species	Percentage (%)									Milligram per Kilogram (mg/kg)				
	DM	CP	CF	Ash	Ca	P	Mg	K	Na	S	Zn	Cu	Mn	Fe
Black soldier fly larvae	27.40	56.10	23.20	9.85	2.14	1.15	0.39	1.35	0.13	27.04	13.10	11.20	23.20	20.40
Housefly larvae	83.47	33.29	6.20	6.25	0.49	1.09	0.23	1.27	0.54	ND	10.39	32.40	42.50	47.50
Mealworm larvae	94.60	55.83	25.19	4.84	0.21	1.06	0.30	1.12	0.21	ND	138.2	29.40	05.70	71.50

DM, dry matter; CP, crude protein; CF, crude fiber; Ash; Ca, calcium; P, phosphorous; Mg, magnesium; K, potassium; Na, sodium; S, sulfur; Zn, zinc; Cu, copper; Mn, manganese; Fe, iron. Source: Shah et al. [9].

The DM content in fresh BSFL (Table 1) is greater than that of other products (34.9 to 44.9%), which results in BSFL being more affordable and easier to make. Typically, BSFL has a composition of 41.1 to 43.6% CP, 15.0 to 34.8% EE, 7.0 to 10% CF, 14.6 to 28.4% ash, and 5278.49 kcal/kg GE, based on DM [10,11]. BSFL larvae are high in Ca (5 to 8%) and P (0.6 to 1.5%). Additionally, their mineral composition contains Cu (6.0 mg/kg), Fe (0.14–14%), Mn (246 mg/kg), Mg (0.39), Na (0.13), K (0.69%) and Zn (108 mg/kg) [9,12,13].

On average, housefly larvae contain 6.25% ash, 83% DM, 33.29% CP, 6.2% CF, 0.49% Ca, 1.09% P, 0.23% 1.27% Mg, 10.39 mg/kg Zn, 32.40 mg/kg Cu, 42.50 mg/kg Mn and 47.50 mg/kg Fe, based on DM (Table 1) [9]. Housefly larvae contain a lot of energy, protein and micronutrients (e.g., Cu, Fe and Zn), as well as EAA and FA. They are also inexpensive, have high nutritional value and are easier to access than other sources of animal protein [9].

According to Shah et al. [9], the average total DM composition of mealworms is 94.6%, comprising CP 55.83%, CF 25.19%, ash 4.84%, calcium 0.21%, phosphorous 1.06%, Mg 0.3%, K 1.12%, Na 0.21%, Zn 138.2 mg/kg, Cu 19.4 mg/kg, Mn 5.7 mg/kg and Fe 71.50 mg/kg (Table 1).

The edible insect sector has attracted global attention, and this has led to a re-assessment of the practice of entomophagy, or the consumption of insects, in countries that typically show reservations, as well as in countries that are entering the developmental stages of the practice [14].

In Europe, the aquatic insect feed market constitutes approximately half of the entire animal-based insect feed market. Its growth is expected to reach 75% in the next 6 years, and European breeders of insects currently carry around 1000 tons of protein-based insect feeds [15]. Since 2021, in the European Union and member states, processed animal proteins (PAPs) have been permitted for use in poultry and swine feed. This approval represents one of the first steps in the general authorization process, and will ultimately guarantee long-term resource utilization in animal protein production, thus addressing environmental concerns [15,16].

According to Lähteenmäki-Uutela et al. [16], in many parts of North America, insects have historically been incorporated into the food culture. The farming of insects intended for food and feed production began to increase following 2012. The modern insect industry comprises companies that already cultivate crickets and mealworms for animal food. To avoid the high costs of labor, many American and Canadian insect farms have invested in robots, automation, sensors and data aggregation.

Many countries in Africa have insufficient or no insect-specific laws, regulations, standards, or labeling in place to regulate the production and distribution of insects in food or feed chains. The absence of a solid foundation is a significant obstacle to the establishment of markets for insects and their related products [17]. There is a need for more enhanced technology in the rearing of insects to address the increasing pressures of population growth, as we can no longer rely on the catching of wild insects [17].

In several Asian countries, insects have historically been regarded as a form of food, and been used as a significant source of protein. In China, there are no specific laws that regulate their production. Other insects can also be utilized as food additives, and in this context, producers must follow the regulations established in the Administrative Measures for Food and Feed Additives [18]. In South Korea, insects have been a part of the human diet for centuries; they are also included in animal feeds, and there are no specific rules restricting the food and feed industry in relation to insects due to a legislative liberalization that took place in 2015 [19].

In Brazil, the food insect chain is still in its early stages of development, and there is still a significant lack of development in terms of funding, technology and the qualification of human resources. There have also been efforts made by regulatory agencies to promote the development of regulations pertaining to the production and utilization of insects in animal feeds, which will, in turn, encourage those who want to participate in this innovative endeavor [20].

Currently, in Europe, there is a supply–demand gap in the edible insect chain [15]. This scenario also applies to Brazil, as agrofactories still cannot access the necessary volumes (of consistent quality) required for processing insects, and the feeds produced are not competitive in terms of their costs compared to conventional protein sources. In the Brazilian scenario, logistical issues make the process more expensive, and hinder the growth of the chain. Furthermore, there is no knowledge on the part of protein producers about the effects of insects as a feed ingredient.

Another major challenge is that induvial companies are involved from the beginning to the end of the process. As such, they represent a significant part of the chain, from insect breeding to processing, thus raising production costs. As has been seen in Europe [15], it is expected that in the future, Brazilian companies will specialize as the value chain matures.

Shah et al. [9] have shown that a hypothetical expansion of commercial farms (that spend EUR 1000 per month on SBM-derived protein) would involve completely replacing SBM with BSF, HF or MW. The extra costs associated with these species are EUR 88,230, 3980 and 13,010, respectively. If we consider that the farmers farm every season, it would be extremely expensive for them to turn over the production of a whole field. In order for insects to be considered a viable alternative to SBM and FM, both cost and nutritional value must be maintained. For this to be successful, expenditure on insects must be reduced to EUR 0.4 per kilogram of direct weight based on 35% DM substances [9]. The economic values of different insects compared to other sources of protein are shown in Table 2.

A very important element of the end of the chain is the acceptance of insect-based feeds. The study carried out by Ankamah-Yeboah et al. [21] showed that 77% of the people interviewed are indifferent regarding the use of insects in animal feed for fish, which is a promising result for the aquaculture industry.

Not all insects can only be used as ingredients in animal feed. According to Čičková et al. [22], some insects can play dual roles within the chain, such as via the recycling of their organic by-products in compost fertilizers, and the use of their protein as feed. Another challenge encountered at the beginning of the chain relates to the food that insects can consume. These include residues, such as perishable organic by-products. These by-products are a valuable energy source for insects, but require producers to employ preservation techniques that both maintain the nutritional quality of the biomass and are economically viable [23].

In 2022, each person in Brazil generated an average of 1.043 kg of waste per day; that is, almost 1 kg per person. In general, 81.8 million tons of domestic waste were produced, which corresponds to 224 thousand tons per day (2022 edition of the *Panorama of Solid Waste in Brazil*, by ABRELPE) [24].

Table 2. The economic value of insects compared to other protein sources.

Potential Source	Housefly Maggot	Black Soldier Fly	Mealworm	Fishmeal	Soybean Meal
CP (%)	50.4	42.1	52.8	75.4	52.00
Lysine (%)	6.1	6.6	5.4	7.5	6.3
Methionine (%)	2.2	2.1	1.5	2.8	1.3
PPR (EUR/kg)	1.08	20	3.7	1.24	0.2
PP (EUR/kg)	2.14	47.51	7.01	1.64	0.54
PL (EUR/kg)	0.13	3.14	0.38	0.12	0.03
PM (EUR/kg)	0.05	1.00	0.11	0.05	0.01
PP TO PP SBM [1]	3.98	88.23	13.01	3.05	1.00
PL TO PL SBM [1]	3.85	92.23	11.15	3.64	1.00
PM TO PM SBM [1]	6.73	142.52	15.02	6.58	1.00

CP, crude protein; PPR, product price; PP, protein price; PL, price of lysine; PM, price of methionine; PB, protix biosystems; AP, agriprotein. [1] PP to PP SBM, price of replacing 1 kg of protein derived from SBM with other protein sources; PL to PL SBM, cost of replacing 1 kg of lysine derived from SBM with lysine from other protein sources; PM to PM SBM, cost of replacing 1 kg of methionine from SBM with methionine derived from other protein sources. Source: Shah et al. [9].

In Brazil, organic waste represents approximately 50% of all solid waste generated. Dry recyclables (28%) and waste (22%) are the next most prolific elements. The size of the organic fraction affirms the importance of using these residues in different ways, and avoiding their unnecessary disposal [25]. In addition, less than 2% of organic waste is currently composted in Brazil [25], and the storage of large batches of waste is a problem, since it cannot be immediately offered to insects [23].

Black soldier flies (BSFs) have received increased attention in recent years because of their capacity to contribute to sustainable waste management and renewable energy. In Brazil, the species that currently receives the most attention is the BSF. BSF larvae are voracious feeders and have the capacity to convert food waste into protein and lipid products [26]. Within the field of sustainability, BSFs are considered an interesting solution to reduce the ecological impacts of food waste. In transforming waste into valuable products, BSFs contribute to reducing the emission of environmental pollutants and greenhouse gases [26]. We highlight below some experiments carried out on BSFs and reported in published scientific articles and Master's theses.

Silva et al. [27] aimed to create a demonstration unit for black soldier fly larvae (*Hermetia illucens*) using organic waste from a restaurant located at the Federal University of South and Southeast Pará. It was known from previous experiments that the production of larvae of the black soldier fly can be used to efficiently decompose organic residues, transforming them into liquid composts and soils ready to be used in the cultivation of vegetables and the preparation of seedlings.

In Santos' Master's thesis [28], the objectives involved valuing the by-products of the food industry, obtaining a potential ingredient for animal feed, and also verifying the influence of diet on the weight, length and nutritional composition of BSF larvae. The results show that the diet significantly influenced the weight and average length of the larvae. Regarding the nutritional composition of the larvae, the protein content was not altered by the diet supplied, but the fat and ash contents of the diet directly influenced the composition of the larvae.

Teixeira Filho [29] proposed a possible solution to mitigate two issues: the disposal of solid organic waste and the pressure on the current supply of food protein. This solution is based on the mass production of larvae of *Hermetia illucens* (L., 1758) (Diptera: Stratiomyidae), also known as the black soldier fly, to degrade organic solid waste and also as an alternative source of animal protein. They achieved an 83.75% reduction in organic solid waste by employing a solid-waste-to-protein biomass conversion rate of 23.2%. This work provides a good indication that the mass production of the black soldier fly to enable the destruction of organic waste and for subsequent use as a source of animal protein is an excellent and sustainable alternative solution to problems related to solid waste and protein supply.

The treatment of organic waste is related to several SDGs, but specifically to numbers 2 (Zero Hunger and Sustainable Agriculture), 12 (Responsible Consumption and Production) and 13 (Action Against Global Climate Change).

Agricultural and food or agro-food supply chain (AFSC) encompasses all stages, from cultivation to harvesting, packaging, processing, transporting, marketing and distribution and final consumption. It not only entails general risks, including social, political, cultural and economic; due to the perishable nature of the products, seasonality, weather effects and quality and safety requirements, the chains are even more vulnerable [30–32]. The productive chain of edible insects operates like an agro-food chain, and to verify its characteristics, we will employ two frameworks.

All parts of the value chain of producing edible insects as an alternative protein source for animal feed, and the associated challenges, can be analyzed through the sustainable business model canvas, as shown in Figure 2.

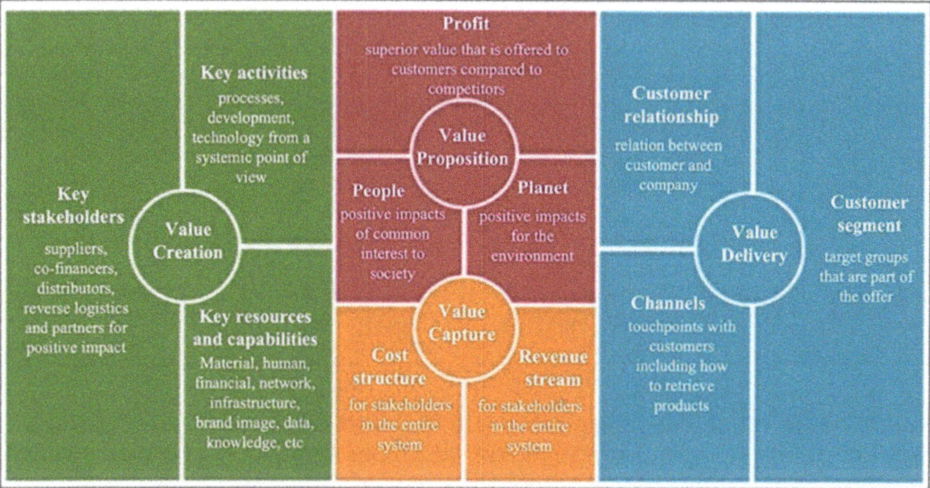

Figure 2. Screenshot of the sustainable business model canvas. Source: adapted with permission from Bocken et al. [33], Osterwalder and Pigneur [34] and Richardson [35].

Figure 2 depicts the business model canvas and introduces three new archetypes into the value proposition aspect: profit, people and planet. These make it a sustainable model with a holistic approach, the main (and most challenging) objective of which is to act sustainably in the future, with a simultaneous focus on environmental, economic and social changes [33].

The supply chain of edible insects will be evaluated using the SWOT matrix. SWOT is a helpful tool that can be used during the evaluation phase in order to yield a first interpretation of the possible future consequences. SWOT analysis is a simple method that provides a factual interpretation of the benefits and drawbacks of a business [36].

According to Benzaghta et al. [36], the SWOT matrix (Figure 3) can be summarized as follows:

- SO strategies—taking advantage of opportunities;
- ST strategies—avoiding threats;
- WO strategies—introducing new opportunities by reducing weaknesses;
- WT strategies—avoid threats by minimizing weaknesses.

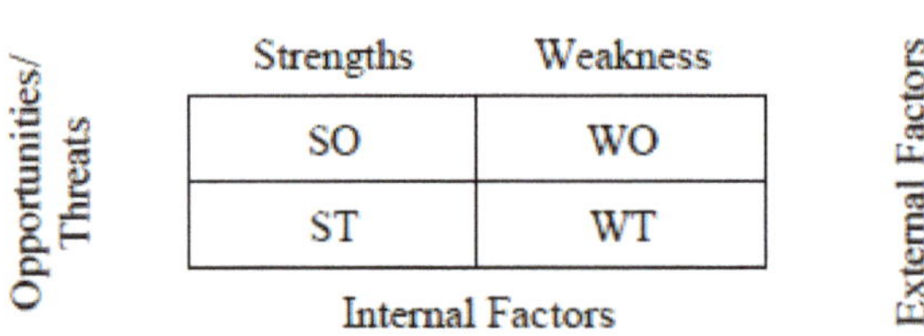

Figure 3. Screenshot of the sustainable business model canvas. Source: Based on Benzaghta et al. [36].

3. Methodology

This qualitative research approach considers the current state of the art of using insects for animal feed, organic waste and other sustainable actives. To enact this evaluation, bibliographical research was carried out, using the search strings "value chain" and "edible insects" in the databases Web of Science and Scopus.

To meet the research objective, articles with information on the use of edible insects for animal feed and organic waste were selected. The intention was to present the current state of the art in this field of research and to broadly contextualize it, highlighting the value chain of this agri-food chain in Brazil by focusing on two companies that operate in the area. A descriptive exploratory case study was carried out, and through interviews it was possible to analyze the business models of two Brazilian companies active in the edible insect sector.

For confidentiality reasons, the companies are named Alpha and Beta. Their main characteristics are described below.

Alpha company was created in 2015 with the aim of contributing to the development of sustainable food, focusing on insects with high nutritional quality as a protein source. The company's mission is to work towards resolving the food issue by 2050. The company seeks to have a social and environmental impact, in addition to generating profit. The owner of the company first encountered edible insects in 2013, according to a report by the FAO and a later interview undertaken with the researcher, professor doctor Van Huis.

Beta company was created in 2022 with the aim of ensuring food safety and quality in producing animal feed. The company produces protein and oil through a circular process. The company began operations about three and a half years ago with a specific project, and it has positioned itself in the market as a company that combats hunger, is sustainable and combines financial prosperity with the protection of the environment.

The methodological procedure employed consists of the following steps:

1. Sources of evidence for the case study—As a source of internal evidence, interviews were conducted with the owners of Alpha and Beta companies. External evidence sources were not used, as the companies do not have websites containing relevant data for this research, and it was not possible to access internal documentation, such as meeting minutes, process reports or quantitative data.

2. Research instruments—A pilot interview was carried out with one of the companies to assess whether the theoretical concepts reflect the day-to-day activities of the company. The pilot interview aimed to delineate the actors in the chain, including those related to biowaste, the rearing of insects, the processing of insects, the feed sector, the protein sector and retail/consumer. Subsequently, the final interview script was prepared based on the sustainable business model (Figure 1) [33–35];

3. Data collection—Interviews were conducted using a semi-structured script and the business owners were interviewed. These were scheduled and performed remotely, and were recorded and transcribed. Each interview was approached as a detailed case study of the parts of the respective business;

4. Data results—The collected data were compared with the theoretical assumptions. The interviews were carried out to understand the effectiveness of the actors in the chain, as well as the maturity of the companies and the difficulties encountered by entrepreneurs in the edible insect sector in Brazil;

5. Discussion and conclusion—The case studies yielded a practical understanding of the social, economic and environmental conditions of these companies. They elucidated the market transition that one company is currently undertaking, as well as the market positioning that both hope to achieve.

4. Results

4.1. Brazilian Edible Insects Supply Chain

The SWOT matrix was applied to analyze the Brazilian edible insects supply chain. This approach highlights the following:

Strengths:

- Insect farming generates less greenhouse gas than traditionally farmed cattle;
- The reproduction of insects using food waste also facilitates the dissipation of large amounts of organic waste;
- Reduced requirement for fresh fruit and grain flour (as currently used in insect breeding);
- The production of insect protein as a replacement for animal protein will reduce the consumption of red meat, and is more sustainable, healthy and economical;
- Insect farming requires less water than producing the same amount of animal protein.

Weaknesses:

- Legislative bans on insects and insect-based products that are intended for commercialization as food;
- Artificial diets based on food byproducts should be specifically studied for each species of edible insect;
- Only five companies currently work with edible insects.

Opportunities:

- Insects have a promising history of being used to produce proteins and fat, and they serve as an effective source of these two substances, thus helping to combat protein energy deficiency while minimizing the environmental impact of food production.

Threats:

- The presence of substances derived from organic materials (e.g., herbs) that are potentially harmful to insects;
- Cultural impediments to the introduction of edible insects into animal feed;
- The potential for heavy metals and mycotoxins to be bioaccumulated;
- The mechanisms of pesticide, drug and hormonal uptake are uncertain.

4.2. Characteristics of the Companies Studied

The sustainable business model canvas was used as a model for the analysis of companies. In addition to the nine blocks of the traditional business model canvas (customer segment, customer relationship, channels, cost structure, revenue stream, key activities, key resources and capabilities, key stakeholders and value proposition), this model incorporates the 3Ps of value proposition related to sustainability:

- Profit—superior value that is offered to customers compared to competitors;
- People—positive impacts of common interest to society;
- Planet—positive impacts for the environment.

Via the two case studies, it was possible to identify the actors in the chain, their roles, difficulties, and how value is generated.

4.2.1. Value Proposition

Alpha Company

- Profit: The company's value proposition relates to the production of sustainable feed, aiming at preserving the environment, as well as aiding the economic development of several Brazilian communities. The company's ambition is to create mechanisms that can prolong life and restore Brazilian ecosystems. At present, the development in this company of insect processing for animal feed relies on a single pilot plant, but one with high capacity and the potential to become the largest insect processing plant in Brazil and Latin America. The plant will begin serving the aquaculture, poultry and pet sectors of the animal protein market. Wirth a vision of future markets, the company invests in research into the use of insects in the pharmaceutical sector and has an expansive growth strategy extending up to 2027.

- People: The company's strategy is to open up the production process and the technology used in production, contributing to the expansion of the production chain, and consequently to increase employment. To this end, the company will supply insect eggs through partnerships, and will share their cultivation methods with new breeders who will become part of the chain. In this way, it will be possible to impart better living conditions upon small producers, offering these families a better income and a better quality of life.
- Planet: The insect used to produce feed is BSF, which is fed on waste. Organic waste, for example, that was previously sent to landfills, is now reintroduced into a new chain as food. However, the company faces challenges in accessing this waste. Some of these challenges include the cost of transporting the waste to the destination, the collection of large volumes in different locations and the duration for which the waste can be deposited while awaiting collection. Other types of waste can be used, such as brewery waste. However, this waste is about seventy percent water, which makes freight and transport extremely expensive, increasing operating costs. Depending on the type of waste, the logistical strategy changes, as the requirements of waste removal are inconsistent. In addition, the requirements when storing waste in its place of origin also differ from those of its place of origin. To overcome the challenges of waste collection, the company visits market actors involved in composting and landfills, and it has discussions with managers in the urban and organic waste chain. Once these challenges are overcome, the use of waste can contribute to mitigating environmental problems.

Beta Company

- Profit: The company's value proposal is to extinguish hunger and accelerate the ecological regeneration of Brazil, via the production of high-quality protein and oil, thus combining financial profit and sustainability. The company is now exiting the laboratory stage and looking to build a pilot plant for growing and processing BSF, which will be registered with the Ministry of Agriculture. The company's ambition is to become the largest animal feed production company in Latin America. The company enacts the entire process, from the collection of organic waste to the production of flour and insect-based oil used for animal consumption. In addition, the company markets the eggs it produces. As such, the company is involved in the collection of urban waste, in the transformation of this organic residue into substrate, in the fattening of the larvae and in the processing of the larvae for product formulation.
- People: Via the decompression of the protein chain, the company can work towards the fight against hunger by reducing the competition between protein sources for animal feed. When insects are introduced as an integral component of animal feed, it will be possible to guarantee a better supply of protein for people. Furthermore, greater understanding will lead to behavioral changes that will contribute to responsible consumption.
- Planet: At the company's center of operations, approximately three hundred and sixty-six tons of urban organic waste are generated per hour. The work of transforming this waste into substrate for the BSFs further significantly impacts daily waste production, even if to a smaller degree. In this way, organic waste is transformed into yeast for the development of BSF larvae, which are the raw materials for dog, cat and ornamental fish feed. As these animals' diets compete with human diets over the same protein sources, the proposal is to reduce the pressure on the protein production chain, enabling a reduction in the use of natural resources.

4.2.2. Value Creation

Alpha Company

- Key stakeholders: The company is partnered with universities for biological and engineering research. This partnership is based on a win–win theory. As a result of this collaboration, the company incurs no costs in laboratory research. In addition, the company's interns use the universities' laboratories for research. In the production of BSFs, the company is partnered with another company that specializes in organic protein production. In addition, the company is also partnered with the Brazilian Micro and Small Business Support Service (SEBRAE), which facilitates national and international fairs, increasing the company's visibility.
- Key activities: As the BSF processing methodology is not yet perfected, the company offers insects to other companies that produce biological agents. Further, they encourage the use of insects in different stages of their life cycle as food—including pulps, young insects and adults. The company undertakes the entire rearing process, separates, weighs and packs them, issues invoices and dispatches the insects to the end customer.
- Key resources and capabilities: The company's greatest resource is the formula with which they feed the insects, which comprises wheat bran, cornmeal and corn, in addition to hydrated vegetables. Another important resource is knowledge of the parameters of the breeding process, such as temperature and humidity. This knowledge was originally acquired experimentally, and continues to be improved since the ideal production model remains unknown.

Beta Company

- Key stakeholders: The main business partner is in the organic food composting sector, and this company is located in the same place as Beta company. Thus, the logistical costs are very low. This partner is also a co-founder of Beta company. The company also has partnerships with rural federal universities.
- Key activities: The company currently produces fresh insects to meet animal production needs.
- Key resources and capabilities: The company's main resources are the equipment and physical space that comprise the factory, in addition to labor.

4.2.3. Delivery of Value

Alpha Company

- Customer segment: The company has three customer segments. The first segment is customers who buy live insects and resell them for animal feed. The second segment is the final customers, who buy the insects to feed their pets. The third segment is customers who use the insects as biological agents.
- Relationship with customers: As the market is small, the relationship with customers is close. Due to its origins as a family business and its few employees, customer needs are addressed promptly.
- Channels: The most commonly used communication channels are social networks and WhatsApp. The company has a website that is currently being restructured, as the company's business model is undergoing development.

Beta Company

- Customer segment: When the company's plant is up and running, the main customer segment will be feed producers. Subsequently, the company will have its own feed line, mainly for domestic animals.
- Relationship with customers: The relationship with customers will be developed as the company matures. However, the company is already positioned in the market, and sells live insects.

- Channels: The company has a structured website, through which it is possible to contact them and ask questions about the products offered. It is also possible to contact the company by email and LinkedIn.

4.2.4. Value Capture

Alpha Company

- Revenue stream: All three market segments offer similar revenues. However, the revenue source offering the highest margin is the segment of final customers who buy insects to feed their non-conventional pets. This market is seasonal, depending on the type of animal; for example, frogs are fed live insects only in parts of their life cycle.
- Cost structure: The company's biggest costs are raw materials and labor. Under the new model of producing insect-based feeds, the biggest cost will be the processing of the feed.

Beta Company

- Revenue stream: The company's revenue is negligible, as it is still under development. However, the company predicts that 80% of its revenue will come from insect flour.
- Cost structure: The highest costs are related to the construction of the factory and the purchasing of equipment for BSF production, as well as electricity and general maintenance.

5. Discussion

Brazil's edible insect supply chain will play a significant role in circular sustainable agriculture via its capacity for closing nutrient and energy loops, promoting food security and minimizing climate change and biodiversity loss. These aspects strengthen the chain and are associated with the achievement of the Sustainable Development Goals.

The network is still young and disorganized, as only five companies are in operation and there are some factors that contribute to slowing development, such as legal obstacles against and a lack of legislation for insect-based products, and the lack of cultural acceptance of the consumption of insects in Brazil, as verified in the weaknesses section and pointed out by others [20]. Another characteristic of this chain is that the companies work independently and autonomously; that is, each performs all parts of the production process, and there is no network of collaboration among them.

The case studies presented elucidate certain as-yet-unmet opportunities to contribute to the sustainable development of the chain, such as the use of land for soybean planting; reducing water consumption and greenhouse gas emissions; replacing animal proteins, and reducing organic material, among others.

The sustainable business model canvas enabled us to analyze how both companies are structured. Despite only focusing on two case studies, the conditions of these companies reflect the contemporary condition of the value chain associated with the production of edible insects for animal feed in Brazil. As such, we can conclude that the chain is still young and that its development will be complex, since the production process is still under development.

As regards the business models, we can observe adhesions between practice and theory. The following constitute the pillars of the business model.

Value proposition: The companies' value propositions reflect concerns about the planet, and especially about the future demand for protein and the competition that will arise between food and feed, as well as the demand for fishmeal for fisheries [5]. The desire to preserve the environment has encouraged businessmen to seek new solutions in structuring the animal feed market. However, international research on the protein capacity of insects, as well as millionaire-funded projects, have been decisive in encouraging investment in a new market, which features sustainability as its main motivation.

Similarly to Europe [15], in Brazil, insect-based protein products still cannot compete with established protein sources, in terms of costs. Further, agrofactories still do not have the capacities or consistency in quality required for processing.

As identified by [22], some insects can play dual roles within the chain, such as via the recycling of organic by-products into compost fertilizers, and this is also the case for BSFs. BSFs consume waste as a source of energy, and this offers numerous benefits to a developing country that is facing difficulties in the collection of urban waste and has little infrastructure for selective collection.

Furthermore, as described in [6], the chain of using insects as animal feed in Brazil begins with bio-waste, but currently ends with insect processing, given that both the active companies still do not produce processed feed for poultry, pork, or fish. The next steps will be to develop the poultry, fish, pork and pet (i.e., dogs and cats) markets via feeding with processed feed.

Despite the entrepreneurs' willingness to reveal their secrets related to the cultivation of insects, certain aspects related to processing remain concealed, as well as the technology used. Both companies are currently undertaking test phases, and their industrial plants are still in the pilot phase. As such, it is not possible to guarantee, despite the energy expended on research, that their models will work. To arrive at the most optimal model, more tests and improvements will be required until the whole process is matured.

The creation of value in companies is based on partnerships. The chain is developed through strategic partnerships. However, there are differences in the types of partnerships. To produce BSFs, Alpha company is partnered with another that specializes in organic protein, and this company invests in technology and capital; on the other hand, Beta company's main partner is an organic food composting company. In addition, Beta company has gone public, and thus receives investments from private individuals.

Despite BSFs showing promise and representing a good investment for Brazilian businessmen, there is a logistical bottleneck associated with the cost of feeding them, and this represents one of the biggest challenges for the chain. Beta company encounters fewer difficulties due to its partnership, while Alpha company must search for waste depository locations and logistical strategies, even though there are several sanitary landfills in Brazil, as shown in Figure 4.

Figure 4. Guatapará landfill in São José do Rio Preto, São Paulo. Source: personal field research archive.

6. Conclusions

Our analysis of Brazilian companies active in the production of insects for animal feed in Brazil has allowed us to understand the current organizational condition of these companies. Just as international research on the subject is underdeveloped, the companies involved are also young, with Alpha being the most experienced in the sector, but having only been on the market for seven years.

The young age of the sector explains the characteristics observed in the Brazilian chain, such as the high degree of underdevelopment in terms of financing, technology and the qualification of human resources; the disorganized supply chain and supplier structure; the efforts being made by regulatory agencies to develop regulations relating to the production and use of insects in animal feed, which, in turn, should encourage those interested in this innovative venture into further research and development.

As shown in previous surveys undertaken in other countries and in Brazil, for BSFs to be integrated as an animal feed that can reduce organic waste in a sustainable way, companies must invest time and financial resources into research focusing mainly on BSFs, according to the characteristics of this species. However, despite these investments, in Brazil, as well as in Europe, insect-protein-based products are still not competitive compared to established protein sources in terms of cost. The non-competitiveness of the Brazilian chain makes it difficult for it to develop a market that is regulated in terms of prices, products, supply and demand.

Because the productive chain is still in its nascent stages, several aspects of the business model are still being developed, as well as the teams, processes, labor, technology and space required. Further, more time will be required to allow the maturation of the blocks that make up value delivery and capture, further characterizing the difficulties encountered in the Brazilian chain in establishing itself and becoming competitive.

The companies' strategies for dealing with this process were not revealed. However, Alpha company seems to be employing a "homemade" strategy, founded on the time it has spent in the market and its acquired experience. Beta company, on the other hand, seems to believe that investing in advanced technology is the safest path. It will soon be clear which of these strategies is best.

The concepts of sustainability and environmental regeneration increase the value of the benefits offered by companies. However, both these companies desire a market that will grow annually, which is why they are currently racing to be the first to establish processed BSF for use in animal feed. The conditions in Brazil are optimal for the growth and production of BSF, as a large volume of organic waste is generated daily and the species is native to this country.

Previous research and our study of the Brazilian edible insect production chain together indicate that the value chain holds great promise and presents several opportunities. These mainly relate to sustainability, since the development and advancement of the chain will contribute to a reduction in the use of natural resources, being an alternative source of protein and a substitute for traditional feed and allowing for the recycling of organic products into compound fertilizers, thus contributing to the Planet, Profit and People model.

The use of insects as animal feed is related to several SDGs, but specifically to SDGs 2 (Zero Hunger and Sustainable Agriculture), 12 (Responsible Consumption and Production) and 13 (Action Against Global Climate Change), in addition to 6 (Drinking Water and Sanitation) and 14 (Life in Water).

Author Contributions: Conceptualization, J.G.C.G. and M.T.O.; methodology, J.G.C.G. and E.L.U.; investigation, J.G.C.G.; original draft preparation, J.G.C.G.; writing—review and editing, M.T.O. and H.d.C.L.d.S. All authors have read and agreed to the published version of the manuscript.

Funding: This research received in part external funding by the Coordination for the Improvement of Higher Education Personnel—Brazil (CAPES)—Finance Code 001.

Institutional Review Board Statement: Not applicable.

Informed Consent Statement: Not applicable.

Data Availability Statement: Not applicable.

Acknowledgments: Special thanks to the companies that participated in the interviews, allowing us to obtain the data used for this study.

Conflicts of Interest: The authors declare no conflict of interest.

References

1. United Nation. Available online: https://news.un.org/en/story/2015/01/488592#.VL6UBEfF8kR (accessed on 1 June 2023).
2. Makkar, H.P.S. Review: Feed demand landscape and implications of food-not feed strategy for food security and climate change. *Animal* **2018**, *12*, 1744–1754. [CrossRef] [PubMed]
3. Dobermann, D.; Swift, J.A.; Field, L.M. Opportunities and hurdles of edible insects for food and feed. *Nutr. Bull.* **2017**, *42*, 293–308. [CrossRef]
4. Kromhout, D.; Spaaij, C.J.K.; de Goede, J.; Weggemans, R.M. The 2015 Dutch food-based dietary guidelines. *Eur. J. Clin. Nutr.* **2016**, *70*, 869–878. [CrossRef] [PubMed]
5. Chia, S.Y.; Macharia, J.; Diiro, G.M.; Kassie, M.; Ekesi, S.; Van Loon, J.J.A.; Dicke, M.; Tanga, M.C. Smallholder farmers' knowledge and willingness to pay for insect-based feeds in Kenya. *PLoS ONE* **2020**, *15*, e0230552. [CrossRef] [PubMed]
6. Veldkamp, T.; Van Duinkerken, G.; Van Huis, A.; Lakemond, C.M.M.; Ottevanger, E.; Bosch, G.; Van Boekel, M.A.J.S. *Insects as a Sustainable Feed Ingredient in Pig and Poultry Diets—A Feasibility Study*; Wageningen UR Livestock Research: Lelystad, Holland, 2012.
7. Zhou, Y.; Wang, D.; Zhou, S.; Duan, H.; Guo, J.; Yan, W. Nutritional Composition, Health Benefits, and Application Value of Edible Insects: A Review. *Foods* **2022**, *11*, 3961. [CrossRef]
8. Gorbunova, N.A.; Zakharov, A.N. Edible insects as a source of alternative protein. A review. Теория Практика Переработки Мяса **2021**, *6*, 23–32. [CrossRef]
9. Shah, A.A.; Totakul, P.; Matra, M.; Cherdthong, A.; Hanboonsong, Y.; Wanapat, M. Nutritional composition of various insects and potential uses as alternative protein sources in animal diets. *Anim. Biosci.* **2022**, *35*, 317. [CrossRef]
10. De Marco, M.; Martínez, S.; Hernandez, F.; Madrid, J.; Gai, F.; Rotolo, L.; Belforti, M.; Bergero, D.; Katz, H.; Dabbou, S.; et al. Nutritional value of two insect larval meals (*Tenebrio molitor* and *Hermetia illucens*) for broiler chickens: Apparent nutrient digestibility, apparent ileal amino acid digestibility and apparent metabolizable energy. *Anim. Feed. Sci. Technol.* **2015**, *209*, 211–218. [CrossRef]
11. Jayanegara, A.; Yantina, N.; Novandri, B.; Laconi, E.B.; Ridla, M. Evaluation of some insects as potential feed ingredients for ruminants: Chemical composition, in vitro rumen fermentation and methane emissions. *J. Indones. Trop. Anim. Agric.* **2017**, *42*, 247. [CrossRef]
12. St-Hilaire, S.; Sheppard, C.; Tomberlin, J.K.; Irving, S.; Newton, L.; McGuire, M.A.; Mosley, E.E.; Hardy, R.W.; Sealey, W. Fly prepupae as a feedstuff for rainbow trout, Oncorhynchus mykiss. *J. World Aquac. Soc.* **2007**, *38*, 59–67. [CrossRef]
13. Cullere, M.; Tasoniero, G.; Giaccone, V.; Miotti-Scapin, R.; Claeys, E.; De Smet, S.; Dalle Zotte, A. Black soldier fly as dietary protein source for broiler quails: Apparent digestibility, excreta microbial load, feed choice, performance, carcass and meat traits. *Animal* **2016**, *10*, 1923–1930. [CrossRef]
14. Vantomme, P. Way forward to bring insects in the human food chain. *J. Insects Food Feed.* **2015**, *13*, 121–129. [CrossRef]
15. Veldkamp, T.; Nathan, M.; Frank, A.; David, D.; Van Campenhout, L.; Gasco, L.; Roos, N.; Smetana, S.; Fernandes, A.; Van der Fels-Klerx, H.J. Overcoming Technical and Market Barriers to Enable Sustainable Large-Scale Production and Consumption of Insect Proteins in Europe: A susichain perspective. *Insects* **2022**, *13*, 281. [CrossRef]
16. Lähteenmäki-Uutela, A.; Marimuthu, S.B.; Meijer, N. Regulations on insects as food and feed: A global comparison. *J. Insects Food Feed.* **2021**, *7*, 849–856. [CrossRef]
17. Kipkoech, C.; Jaster-Keller, J.; Gottschalk, C.; Wesonga, J.M.; Maul, R. African traditional use of edible insects and challenges towards the future trends of food and feed. *J. Insects Food Feed.* **2023**, *9*, 965–988. [CrossRef]
18. Sogari, G.; Amato, M.; Biasato, I.; Chiesa, S.; Gasco, L. The potential role of insects as feed: A multi-perspective review. *Animals* **2019**, *9*, 119. [CrossRef]
19. Han, R.; Shin, J.T.; Kim, J.; Choi, Y.S.; Kim, Y.W. An overview of the South Korean edible insect food industry: Challenges and future pricing/promotion strategies. *Entomol. Res.* **2017**, *47*, 141–151. [CrossRef]
20. do Nascimento, A.F.; Natel, A.S.; dos Santos Corsini, F.; Madureira, E.R.; da Costa, D.V. *Insetos: Alimento Sustentável para Nutrição Animal. Sustentabilidade no Agronegócio*; Universidade Federal de Minas Gerais: Belo Horizonte, Brazil, 2020.
21. Ankamah-Yeboah, I.; Jacobsen, B.J.; Olsen, B.S. Innovating out of the fishmeal trap: The role of insect-based fish feed in consumers' preferences for fish tributes. *BFJ* **2018**, *120*, 2395–2410. [CrossRef]
22. Čičková, H.; Pastor, B.; Kozánek, M.; Martínez-Sánchez, A.; Rojo, S.; Takáč, P. Biodegradation of Pig Manure by the Housefly, Musca domestica: A Viable Ecological Strategy for Pig Manure Management. *PLoS ONE* **2012**, *7*, e32798. [CrossRef]
23. Van Campenhout, L. Fermentation technology applied in the insect value chain: Making a win-win between microbes and insects. *J. Insects Food Feed.* **2021**, *7*, 377–381. [CrossRef]
24. Associação Brasileira de Empresas de Limpeza Pública e Resíduos Especiais–ABRELPE. Panorama dos Resíduos Sólidos No Brasil 2022. 2022. Available online: https://abrelpe.org.br/panorama/ (accessed on 10 July 2023).
25. Empresa Brasileira de Pesquisa Agropecuária (Embrapa). Resíduos Sólidos. 2023. Available online: https://www.embrapa.br/hortalica-nao-e-so-salada/secoes/residuos-organicos (accessed on 10 July 2023).
26. Oliveira, K.J.P.; Viana, M.L.; da Silva Rodrigueiro, M.M.; Neto, M.M.; Oliveira, K.S.M.; de Oliveira Morais, F.J.; Florindo, D.N.F.; Filho, L.R.A.G.; Junior, S.S.B.; dos Santos, P.S.B. Utilização De Resíduos Na Alimentação Das Larvas De Moscas Soldado Negro (*Hermetia Illucens*): Revisão Sistemática Da Literatura. *RECIMA21-Rev. Cient. Multidiscip.* **2023**, *4*, e452989.

27. Silva, K.R.D.; Costa, C.D.A.; Oliveira, I.V.D.; Cruz, W.P.D. Criação de larvas da mosca soldado negra (*Hermetia illucens*) e compostagem de resíduos orgânicos. *Cad. Agroecol.* **2018**, *13*, 1–5.
28. Santos, J.I.P. Bioconversão de Resíduos Orgânicos pela Larva da Mosca Soldado Negro e as suas Aplicações na Alimentação Animal e Humana. Master's Thesis, Universidade Lusófona de Humanidades e Tecnologias, Lisbon, Portugal, 2022. Available online: https://recil.ensinolusofona.pt/handle/10437/12873 (accessed on 10 July 2023).
29. Teixeira Filho, N.P. Devoradores de Lixo: Aspectos Biológicos, Produtivos e Nutricionais da Mosca Soldado *Hermetia illucens* (L., 1758) (Díptera; Stratiomyidae) em Resíduos Sólidos Orgânicos em Manaus, AM. Master's Thesis, Universidade Federal do Amazonas (UFAM), Manaus, Brazil, 2018. Available online: https://tede.ufam.edu.br/handle/tede/6377 (accessed on 10 July 2023).
30. Esposito, B.; Sessa, M.R.; Sica, D.; Malandrino, O. Towards circular economy in the agri-food sector. A systematic literature review. *Sustainability* **2020**, *12*, 7401. [CrossRef]
31. Yanes-Estévez, V.; Oreja-Rodríguez, J.R.; García-Pérez, A.M. Perceived environmental uncertainty in the agrifood supply chain. *Br. Food J.* **2010**, *112*, 688–709. [CrossRef]
32. Mehmood, A.; Ahmed, S.; Viza, E.; Bogush, A.; Ayyub, R.M. Drivers and barriers towards circular economy in agri-food supply chain: A review. *Bus. Strategy Dev.* **2021**, *4*, 465–481. [CrossRef]
33. Bocken, N.; Schuit, C.; Kraaijenhagen, K. Experimenting with a circular business model: Lessons from eight cases. *Environ. Innov. Soc. Transit.* **2018**, *28*, 79–95. [CrossRef]
34. Osterwalder, A.; Pigneur, Y. *Business Model Generation: A Handbook for Visionaries, Game Changers, and Challengers*; John Wiley & Sons: Hoboken, NJ, USA, 2010; Volume 1.
35. Richardson, J. The business model: An integrative framework for strategy execution. *Strateg. Change* **2008**, *17*, 133–144. [CrossRef]
36. Benzaghta, M.A.; Elwalda, A.; Mousa, M.M.; Erkan, I.; Rahman, M. SWOT analysis applications: An integrative literature review. *J. Glob. Bus. Insights* **2021**, *6*, 55–73. [CrossRef]

Article

A Pilot Study on Industry Stakeholders' Views towards Revalorization of Surplus Material from the Fruit and Vegetable Sector as a Way to Reduce Food Waste

Shelley Fox [1], Owen Kenny [1], Francesco Noci [2] and Maria Dermiki [1,*]

[1] Department of Health and Nutritional Sciences, Faculty of Science, Atlantic Technological University, Ash Lane, F91YW50 Sligo, Ireland; shelley.fox@research.atu.ie (S.F.); owen.kenny@atu.ie (O.K.)

[2] Department of Sports Exercise and Nutrition, School of Science and Computing, Atlantic Technological University, Dublin Road, H91T8NW Galway, Ireland; francesco.noci@atu.ie

* Correspondence: maria.dermiki@atu.ie

Abstract: Food waste is a global issue, with the fruit and vegetable sector accounting for higher losses compared with other sectors. The aim of this study was to gain an understanding into how industry stakeholders in Ireland manage surplus fruit and vegetable material remaining after their main processing. An explanatory sequential mixed methods approach was employed to collect data in the form of online surveys (n = 55) and one-to-one interviews (n = 7). The findings outlined several barriers to revalorization. Most respondents were measuring food waste and actively trying to minimize it, although this was for economic rather than environmental sustainability reasons. Environmental sustainability measures were an important factor for larger companies, although all respondents agreed it was important to manage this material from an environmental perspective. This material was mostly classified as "food waste" and usually composted or used for animal feed. Many stakeholders had identified opportunities for revalorization; however, for smaller businesses, this cannot become a reality without considerable investment. Joined-up thinking is required among all stakeholders, including consumers and policy makers, to create positive sustainable changes. Education and greater awareness about the extent of the food waste crisis may assist in achieving reduction targets and encourage revalorization in the industry.

Keywords: food loss; sustainability; surplus material; waste management; mixed methods

Citation: Fox, S.; Kenny, O.; Noci, F.; Dermiki, M. A Pilot Study on Industry Stakeholders' Views towards Revalorization of Surplus Material from the Fruit and Vegetable Sector as a Way to Reduce Food Waste. *Sustainability* **2023**, *15*, 16147. https://doi.org/10.3390/su152316147

Academic Editor: Dimitris Skalkos

Received: 23 October 2023
Revised: 9 November 2023
Accepted: 19 November 2023
Published: 21 November 2023

1. Introduction

Food waste is a global issue, impacting not only economic factors but also social and environmental [1]. Globally, a quarter of all food produced do not reach our tables [2,3]. To address this and other global issues, the United Nations has developed the sustainable development goals (SDGs), with SDG 12.3 specifically aimed towards a 50% reduction in overall food waste by the year 2030 [4]. In Europe, in order to achieve this target, a number of proposals have been adopted by European Union (EU) member states as part of the EU's Farm to Fork strategy, which at its center focuses on achieving a "fair, healthy and environmentally friendly food system" [5]. This process will not be without difficulties and will require the collaboration of all stakeholders across the food supply chain.

The first step in reducing food waste is accurate measurement. However, this has been challenging due to the many definitions of what constitutes food loss and waste, respectively [1,6]. According to the Food and Agriculture Organization (FAO), food loss is mostly associated with primary production and early stages in the food supply chain, while food waste is considered as happening at the latter stages of the supply chain, particularly at the retail and consumer stages [1]. The definitions varied considerably in the EU Fusions project [6], which included edible and inedible waste material and categorized all material as food waste. It also did not classify material that goes to biorefineries or for animal feed as waste or loss [6]. These subtle differences in definitions from study to study have been recognized as creating a barrier to accurate measurement and implementation of food waste reduction practices [7,8].

Other factors to consider in the classification of food waste and losses are food surplus and by-products. Food surplus refers to food that is produced in excess to meet potential orders that may never be realized. This phenomenon is threatening food security and generating food waste that could be avoidable [9,10]. By-products are specific materials that are generated during normal production practices but not required in the finished product and potentially discarded [11], like the peels and trimmings of vegetables discarded when preparing a soup mix, for example. With all these factors to consider, it is understandable how measurement can be difficult. Further classification could be required to identify what element of food surplus is potentially avoidable or categorically unavoidable.

In Ireland, the Food Waste Charter is a voluntary agreement led by the Environmental Protection Agency (EPA) that has measurement protocols to guide businesses in (i) identifying where waste is occurring and (ii) providing suggestions for how to avoid this waste [12]. The most recent Irish figures show that approximately 770,316 tons of food waste were generated across the food supply chain, with 31% coming from households, 29% from processing and manufacturing, 23% from hospitality and 9% from primary processing [13]. While fruit and vegetables may only account for a small quantity of food overall, they generate a considerably larger amount of food waste: 76% (fruit) and 41% (vegetables) across the supply chain [14]. Irish figures also confirmed that the largest contributor among primary processing industries was horticulture [14].

The food waste hierarchy [15] is used as a tool to understand the different ways in which food waste can be managed, with prevention the primary objective and disposal to landfill the least preferred option. This has been adapted in many studies to provide a clear direction for utilizing this material in beneficial ways to reduce food waste, such as re-using it for animal or human consumption and recovery of nutrients, as well as energy or revalorization of by-products [8,9,16].

Naturally, food producers have always considered using by-products or side-streams and creating new routes to market to improve profitability [17]. To date, the most popular method for using the waste material from the horticulture industry is as animal feed [9]. Recycling food waste into animal feed is positioned in the center of the food waste hierarchy after all measures have been exhausted to keep food in the human food supply chain [15]. Many horticulture farmers also rear animals; therefore, using this material as animal feed is a cost-effective way of managing surplus material [18]. It is also important to consider that if food waste is to be brought back into the human food supply chain, it may require further treatment or storage space to allow waste material to be held and used safely [19], which could be outside the smaller growers/farmers' capability in terms of infrastructure. Biomass conversion is a method that has been explored for managing excess loss or waste, not only from the horticulture sector but other sectors as well, and constitutes the recovery stage of the food waste hierarchy which treats unavoidable food waste [9,15]. This is a relatively new process, but research indicates it could be a lucrative one with the generation of value-added biobased food and feed ingredients [20,21]. Studies have also shown that waste material, such as the shells of nuts [22] or onion peels [23], can be used in the production of bio composite films with improved mechanical properties. Numerous studies have been conducted focusing on the extraction of bioactive compounds from fruit and vegetable waste streams, mostly on a small scale but with clearly demonstrated applications in both the food industry and cosmetic industry [19,24–35]. However, due consideration should be given to the cost of setting up biorefineries or purchasing new equipment for the extraction of functional ingredients from side-streams remaining after normal processing, as costs may be prohibitive [19,31,36]. The availability and the quality of the waste material present should also be considered, as factors like seasonality could impact this material negatively. It may also require treatment to avoid spoilage before going through revalorization, and this could cause the cost to spiral [26]. Time may also be required to find new markets for new functional ingredients, and new biobased foods may need to go through novel food legislation before being eligible for sale on the open market [36]. However, this effort could be valuable, particularly due to the quantity of waste generated from fruits and vegetables which results in nutrients being lost from the supply chain [25]. Bringing nutrients such as water-soluble vitamins and dietary fibers, along with phenolic compounds, back into the supply chain through waste valorization could add value and improve sustainability [25,27,37]. This method of waste valorization is often referred to as "circular eating", as it creates a circular process, using all of the surplus material where possible and avoiding food waste [38–41].

The literature identifies the exploration of fruit and vegetable waste valorization as a sustainable means of managing food waste but also demonstrates several barriers, mostly linked to economic factors such as scaling up processing. A review of the literature so far indicates information on volumes of food losses in primary production [14] and food waste across the supply chain [13], but there is a lack of information on current management practices of industry stakeholders in Ireland.

Therefore, the aim of this research was to identify stakeholders' current management of surplus fruit and vegetable material remaining after processing and what factors influence their decision to revalorize this surplus material. To satisfy this aim, the following research questions were employed:

Q1. How do stakeholders currently manage their fruit and vegetable surplus material or by-products?
Q1(a) What factors influence management of surplus material by the companies?
Q1(b) What types of product(s) if any, do the stakeholders have surpluses of?
Q2. What are the barriers or drivers impacting stakeholders' decision to add value to surplus material generated from the fruit and vegetable sector?

2. Materials and Methods

2.1. Methodology

The term stakeholders in this study refers to companies along the food supply chain working with fruit and vegetables: for example, farmers, wholesalers, food processing companies, restaurants, cafes and retail supermarkets.

An explanatory sequential mixed methods approach was employed to collect data from stakeholders on their current management of surplus fruit and vegetable material remaining after their main processing [42]. This was in the form of an online questionnaire, followed by semi-structured one-to-one online interviews that helped further explain the findings from the surveys. Quantitative and qualitative data provided reliable, valid and comprehensive information resulting in a more detailed understanding of the issues and the potential for circular eating through waste valorization in Ireland, where currently there is a lack of information. This study was approved by the Research Ethics Committee of Institute of Technology Sligo, Ireland in December 2021 (Reference No: 2020048, part 2).

2.2. Survey Design

The survey was developed using Qualtrics XM (first release 2005, copyright year 2021, US, available at https://www.qualtrics.com, accessed on 22 February 2022), and it was structured in a way to answer the research questions (Q1 and Q2). The survey was divided into four sections. The first section collected background information about the stakeholder, for example, the type and size of each stakeholder. The second section focused on the types of surplus material present, whether measurements are usually recorded, and if so, what the estimated annual quantities are. Surplus material was defined as any fruit or vegetable material remaining, after normal manufacturing operations are concluded. The survey respondents were then asked to read statements in relation to the management of surplus material in their business and rate how much they agreed or disagreed with each statement.

After these sections, the focus was narrowed to the fruit and vegetable surplus material present, how the respondents classified this material, if they had considered adding value to this material, and if not, how they currently deal with this surplus material. The question grid that presented each question and justified the reason for inclusion in the survey based on the literature can be viewed in the Supplementary Material (Table S1). A link to the survey can be found here: https://itsligo.fra1.qualtrics.com/jfe/form/SV_25 nlpyPUoFKSBim, accessed on 28 March 2022.

Those who completed the survey were asked if they wanted to progress to a one-to-one semi-structured interview, which was conducted and recorded through online videoconferencing platforms (MS Teams or Zoom). The interviews followed a format similar to that of the survey, firstly collecting some introductory information about the stakeholder and then asking questions related to the measurement of surplus material. This was followed by questions on the types of surplus material present and questions related to valorization of the surplus material. Then, the respondents were asked about the perceived benefits or barriers to valorization. Finally, the stakeholders were informed about the SDG 12.3 target goal of reducing food waste by 50% by the year 2030 and asked about their knowledge of this target and their belief that it was achievable. During both the survey and the interviews, if a stakeholder answered no to certain questions, for example, "no I have not identified a side-stream", they skipped to the next relevant question for them. The question grid including the questions used for the interviews, along with the justification, can be seen in in the Supplementary Material (Table S2). Both the interviews and survey were piloted before going live with stakeholders.

Stakeholders were purposefully recruited for the survey and interviews, to achieve sample triangulation (Figure 1) by advertising the study via organizations like the restaurants association, Enterprise Ireland (EI) and the Local Enterprise Offices (LEO), Irish Business Employers Confederation (IBEC), the Farmers Journal and social media platforms (LinkedIn, Twitter, Facebook). Snowball sampling was also employed for recruitment purposes to increase awareness of the study [18,43].

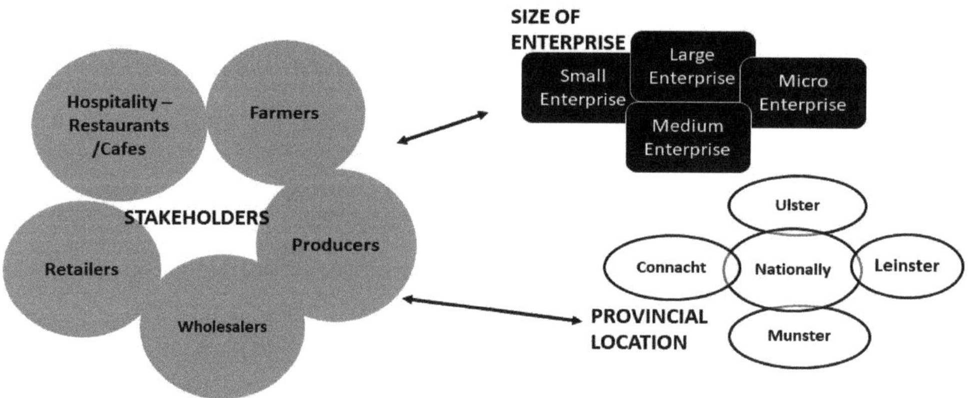

Figure 1. Sample triangulation employed in this survey to capture all stakeholders, from small and large enterprises across all provinces within Ireland.

2.3. Data Collection and Analysis

2.3.1. Survey

There were 55 responses to the survey. Within these responses, all criteria of the sample triangulation were met (Table 1). Most participants were from the Leinster region of Ireland, where a high percentage of growers are based. Most respondents (n = 22) were from large companies (>250 employees); however, all company sizes were represented in the study. The surveys were mostly completed by respondents who identified as owners or managers of the individual stakeholders. The survey results were analyzed using descriptive statistics using SPSS version 28.0 (IBM Corp. Released 2022. IBM SPSS Statistics for Windows, Version 28.0, Armonk, NY, USA, IBM Corp). Descriptive statistics were used to outline the frequencies of the responses, and cross tabulation was conducted to present the responses across the different stakeholders. The level of agreement scale was collapsed from five points—strongly disagree, disagree, neutral (neither agree nor disagree), agree, strongly agree—to three points as follows: disagree, neutral (neither agree nor disagree) and agree.

Table 1. Number of participants in each of the stakeholder groups that completed the survey and the interviews.

Stakeholder	Survey Participants (n = 55)	Interview Participants (n = 7)
Primary Producer (Farmer)	6	2 [1]
Wholesaler	9	1 (+2) [1]
Secondary Producer (Food Processor)	14	(2) [1]
Retailer	3	2
Hospitality (Restaurant/Café)	12	2
Other	11	0

[1] The two farmers interviewed also have a wholesale and food processing side to their operations.

2.3.2. Interviews

Seven interviews were conducted with participants representing each of the stakeholders identified; these participants had also completed the survey. The farmers interviewed also had a wholesale business and processed prepared products for the food service industry, thus representing more than one stakeholder category. The recorded interviews were transcribed verbatim using Otter transcription (version 3.30.0-90c819b7, US, available at https://otter.ai, accessed on 8 July 2022). To ensure accuracy, the interviews were listened back to on two separate occasions, and the transcripts amended as required. The data were coded and analyzed into themes using reflexive thematic analysis (Braun and Clarke, 2006, 2020, 2021); this was managed using Microsoft Excel. See Supplementary Material for a sample of the coding process employed (Tables S3 and S4).

3. Results and Discussion

As this was an explanatory sequential mixed methods study, the results of the survey and interviews are presented together. The interview responses inform the quantitative findings from the survey. This section begins with how the stakeholders manage fruit and vegetable surplus materials in their business right through to their views on revalorization of this material.

3.1. Measurement and Management of Surplus Material

Most survey respondents (n = 52) confirmed they do record the amount of surplus material present in their companies, with the majority reporting <10,000 tons of surplus material present per annum. Most respondents (n = 17) stated that fruit and vegetable material made up 76–100% of this tonnage, with wholesalers the highest contributor (n = 7). These volumes are in line with current reports in relation to fruit and vegetable losses from FAO's 2019 the state of food and agriculture report, where they state, "it is not surprising that fruits and vegetables incur high levels of loss given their highly perishable nature" [44].

The findings from this survey show that 20% of the stakeholders are composting their surplus fruit and vegetable material, 13% are segregating it into brown bins for collection by a third party, and 13% are using anaerobic digestion to manage the surplus material. A small percentage of survey respondents leave surplus material in the field to decompose or use if for animal feed, but unfortunately, there is still a reliance on disposal to landfill, with the results showing 15% of participants using this option. A very small percentage

(4.5%) of survey respondents stated they are currently using all of this material in a side stream (n = 2). Figure 2 shows a more detailed breakdown of food surplus management by stakeholder; for example, the stakeholders who are using all of their surplus material in the side streams are represented by a secondary producer and a company who identified as "other", explaining they are both a primary and a secondary producer.

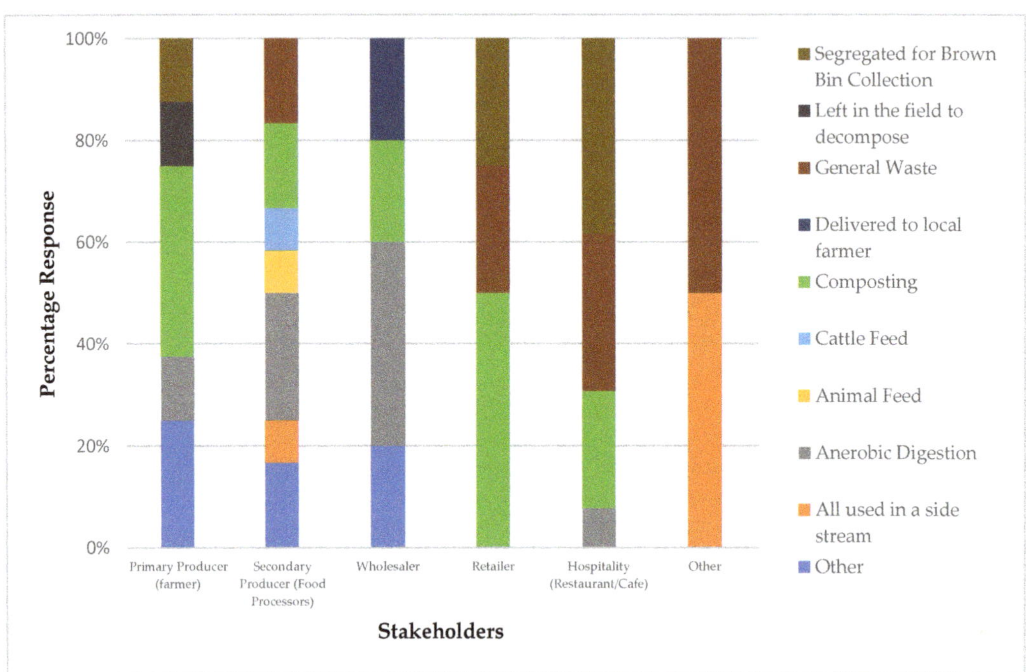

Figure 2. Current practices employed by each stakeholder to manage their surplus fruit and vegetable material (figures taken from survey respondents).

The current management of surplus material is varied. Composting is the most popular method employed by representatives from all stakeholders (n = 11). However, as mentioned, general waste with disposal to landfill is also still evident, particularly in the hospitality sector (Figure 2). As anticipated, it is the primary producers who leave surplus material in the field to decompose; this was also evident in the literature mostly to avoid excess costs associated with harvesting produce that does not meet customer specifications [18]. In the stakeholders' survey, the low number of respondents using surplus material for feeding animals was surprising, whereas other studies found this to be the most popular method for managing surplus material [9,18]. It was anticipated that secondary producers and wholesalers would be the main stakeholders using this method to manage their surplus material. Indeed, this was evident in the interviews, where the wholesalers confirmed they segregate waste material for transport to farmers for animal feed or to feed their own animals. Anaerobic digestion (n = 7) was used by primary and secondary producers and wholesalers mostly, while segregation for brown bin collection (n = 7) was mostly used by the hospitality and retail industries.

In the interviews, stakeholders acknowledged that they dispose of surplus material in ways similar to the practices found in the survey responses, as shown in Figure 2, often managing this material in a way that is convenient for the company's access to various options, such as composting or segregation for animal feed. It was suggested that measuring the surplus material is part of standard operating procedures connected with forecasting and planning. It is mostly completed for economic reasons, as waste generated leads to loss of sales, which need to be quantified. The larger businesses have built-in sustainability factors in their business models which they categorized as "sustainability credentials". These factors are linked to attaining origin green status, which is an independently audited, national program that allows the industry to incorporate measurable "sustainability targets" within their business that consider the environment and local communities [45]. Measurement of waste and surplus material would be one of the metrics they are measuring, as the quotes below explains.

"For sustainability credentials, and there is a reporting structure behind everything that we do" **Large Wholesaler**

"It's part of weekly, quarterly, and annual KPIs that we will manage food waste, it's tied to profit and loss. We would have liked a margin that the department will have to hit every week" **Retailer.**

Figure 3 details the stakeholders' perspectives on the challenges presented when managing surplus material from the fruit and vegetable sector specifically. Wholesalers and retailers mostly agree that managing surplus material is an ongoing challenge. There was, however, a difference of opinions among the primary and secondary producers as well as among the hospitality sector respondents, with some more comfortable managing this surplus material than others. All stakeholders participating in the survey agreed that it is important to manage food surplus from an environmental perspective (Figure 3). This study revealed that consumer expectations were more relevant to secondary producers, retailers and the hospitality sector. However, Göbel et al. [46] highlighted that factors such as consumer expectations can have a knock-on effect across the whole supply chain. The retailers in this survey were well-informed on how to manage food surplus effectively; however, the other stakeholders did not share this opinion and believed there was a lack of information available to them. The reason for this difference in opinion could be down to the fact that many retailers have had processes to mitigate food waste in place for some time now, such as discounting produce which is near its "best before" date [47]. However, this often moves the food waste problem onto the consumer stage. Figure 3 also shows that most stakeholders surveyed believed there was enough time within their individual day-to-day operations to manage food surplus effectively, apart from the primary producers and hospitality sector. This was further explained in the interviews when these stakeholders highlighted the short timeframes available to them in their day-to-day operations that hindered accurate management of surplus material, as the quotation below from a small café owner explains.

"I suppose the big thing would be not having time to process things, because you only have a certain window, you know, you have a fresh thing. You know, if you have to make things that are being sold immediately, you know, that that's the priority." **Hospitality (cafe owner/manager)**

The respondents from the interviews explained why food surplus generation can be challenging, alluding to time as mentioned and the impact of inconsistent labor practices, particularly among secondary producers and retailers who found it difficult to retain staff. This was mostly due to the nature of the work, as the quote below explains.

"It's getting harder to get people to do particular jobs like peeling and preparing difficult veg like turnips, it's very hard laborious work" **Family business—wholesaler/farmer.**

COVID19 was also a major disruptor for the food industry, with competent staff lost due to uncertainty in the sector at the time. The companies interviewed acknowledged they are still struggling to get back on track as a result. They also highlighted a need for investment in equipment and facilities to improve process efficiency, which is needed to manage surplus material more efficiently.

"It's either big investment or we need to stop what we're doing" **Family business—wholesaler/farmer.**

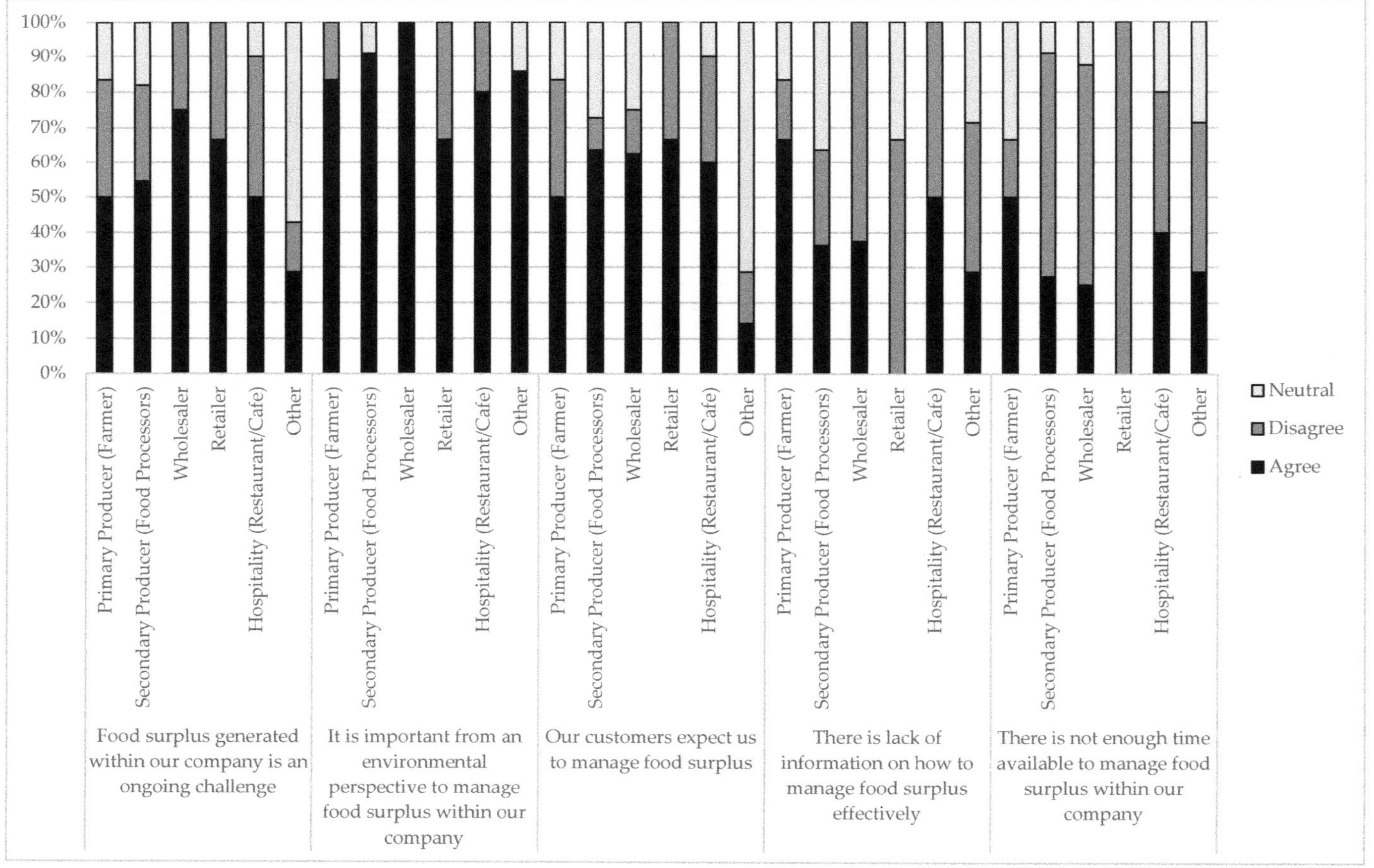

Figure 3. Results from survey focusing on statements connected with managing food surplus material generated from fruits and vegetables.

All of the stakeholders interviewed recognized this is an area that needs to be addressed, not solely from an environmental perspective but also to ensure their business models remain viable. Some of those interviewed felt they should take more of a leadership role in educating their customers. Even though they believe consumers are more aware in general, they themselves were not promoting the good work they are doing to create "zero waste kitchens", for example.

3.2. Classification of Surplus Material

The survey included a question relating to the classification of surplus fruit and vegetable material after participants were provided with the definition of food waste and food loss according to FAO [48]. This was asked to help gain a better understanding of how the stakeholders view this material, which would explain current management practices. The results showed that many respondents (n = 23) classified this material as "food waste", a high proportion also identified this material as "animal feed", and a small number (n = 5) classified this as "material for further processing". Breaking this down further to individual stakeholder sectors, the hospitality stakeholders mostly classified this material as food waste. Secondary food producers mostly classified it as material for animal feed. Primary producers (farmers) classified the material as either food losses (n = 3), food waste (n = 2) or product left in the field that did not meet specifications (n = 1). One retailer defined this material as donations. Donations would be a popular option for retailers managing surplus material across all of their produce, not only fruit and vegetables, with Irish companies like Food Cloud collecting the surplus and redirecting this valuable commodity to those in need [49].

The stakeholders' interviews conducted further explained the findings of the survey showing that stakeholders with numerous remits within their business model were able to use surplus for animal feed. For example, two of the interview participants were farmers who not only grow produce but also raise animals (cattle), making animal feed a beneficial management practice. The larger wholesaler interviewed also had an agreement with local farmers to collect surplus material for animal feed. However, composting was a common way of managing the remaining surplus material for all of those interviewed. The general view was that if another use for the surplus material was found, then it was not classified as waste. This is like the EU Commission's FUSION [6] definition, which differ from the FAO [1,48] definition in that it also considers inedible material. Oliveira et al. [50] confirmed with a review of papers from 2011–2020 that "there was no specific concept for food losses and waste which made it difficult to quantify". This was also evident in this study, with each stakeholder classifying their surplus material in different ways not necessarily following the definition of FAO. The findings from the interviews show how each stakeholder works individually, focusing on their own protocols to grow their business and remain viable. To ensure that the waste reduction targets of SDG 12.3 [4], of a 50% reduction by the year 2030, are successful, classifications or clear definitions of food loss and waste, surplus material and what is edible or not edible would need to be confirmed and communicated across stakeholders for improved uniformity.

3.3. Food Waste Reduction Targets

Most stakeholders (n = 24) in the survey disagreed that there were government incentives available to help manage food surplus, as can be seen in Figure 4. Some stakeholders acknowledged in-house incentives, although the response to this was mixed across all stakeholders, as seen in Figure 4.

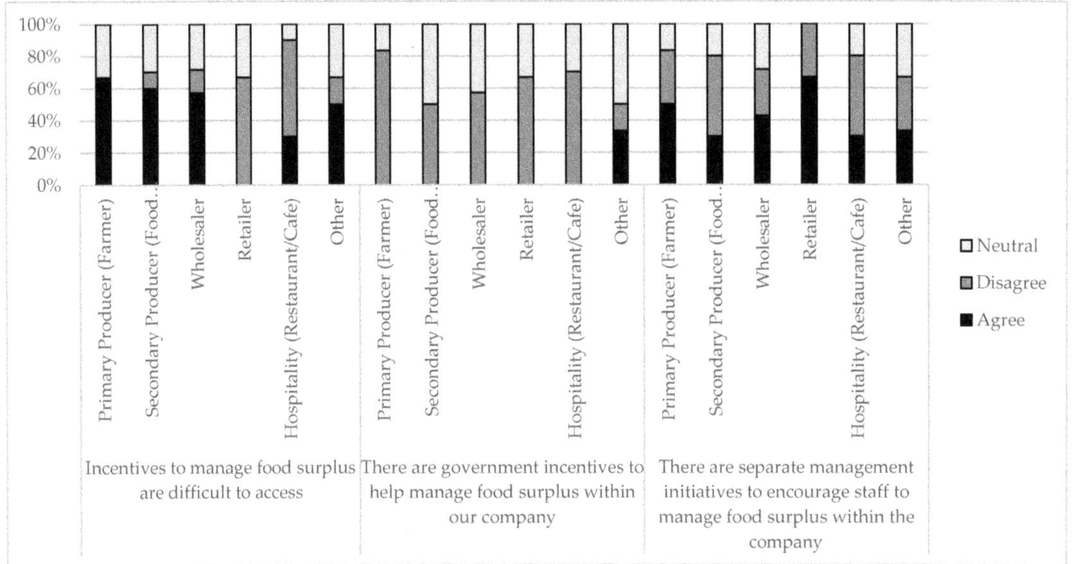

Figure 4. Survey responses to statements related to incentives for managing food surplus.

When food waste reduction targets were discussed during the interviews, most respondents noted that they were managing food waste in house anyway. Some were aware of initiatives, but most were not. The representative from retail believed the emphasis was wholly on food safety rather than food waste reduction.

However, they suggested the food waste reduction target of 50% by 2030, in line with SDG 12.3, was achievable, particularly if work has already commenced. They believed education was required to bring all stakeholders on board, consumers included, and alluded to the need for government policy to be implemented to fast track change.

3.4. Types of Surplus Material Available

The types of surplus material present varied between stakeholders, as can be seen in Figure 5. As seen in the figure, primary producers (farmers) responded that most of their surplus material comprised whole fruits and/or vegetables that do not meet retail/quality or food safety specifications. For secondary producers (food processors), it was peels, pips, cores, and other by-products generated by processing. These stakeholders also listed whole fruits or vegetables that do not meet specifications as the second highest response. Wholesalers were like primary producers in that out-of-specification whole produce was the highest response; however, they also logged products that had exceeded their expiration dates or produce that spoiled before its expiration date as other sources of surplus material. Retailers cited spoiled or contaminated products within their expiration dates as most of their surplus material. Respondents in the hospitality sector noted that their surplus material was mostly product remaining after processing, like peels, pips, and other by-products.

The reasons cited for the generation of this surplus material ranged from consumer expectations to short dates on deliveries and varied among stakeholders, as seen in Figure 6. Primary and secondary producers differed in that primary producers believed consumer expectations and retailer requirements mostly influenced the generation of surplus material. This was followed by overproduction due to forecasting or predicting orders. O'Connor et al. [14] agreed with this finding to some extent, suggesting that food waste was generated from produce that could not be sold, but also noted that pests and other production stresses also had an impact on food losses pre-harvest. However, for secondary producers, surplus material was more influenced by quality and food safety specifications and raw material quality. Retailers were impacted across the spectrum, from raw material quality to forecasting to consumer expectations. Those in hospitality listed forecasting as the main contributor, followed closely by overproduction or production inefficiencies. For wholesalers, it was the raw material quality that impacted most, followed by forecasting. Other studies, like that from Beausang et al. [18], have expressed similar findings in relation to primary producers, highlighting the need for greater awareness of the impacts of retailer specifications that are directly linked to consumer expectations. This view was echoed in the interview findings, with the primary producers in this study finding alternative ways of using the produce that is not accepted by their customers, such as for fertilizing the land or feeding their animals. Research has also shown that waste materials from primary processing have further potential as new energy sources [51]. Like this study, studies from Richards et al. [52] and Messner et al. [10] focusing on Australian horticulture found that there were many "paradoxes" connected with the generation of food surplus in primary production, with a blame game happening in terms of who is responsible for the surplus, but ultimately feeling powerless to create change with the buyers, which are often the retailers in control. The stakeholders interviewed often referred to their immediate customers, the next stage in the supply chain, as impacting their business model. For example, the wholesalers and secondary producers were at the mercy of their buyers, namely the food service/food producer sectors or retailers who require a certain product specification. The retailers were then at the mercy of auditors measuring food safety protocols to keep consumers safe and, in turn, the impact of consumer expectations in terms of product quality and specifications.

The findings from the interviews highlighted customer- and product-related factors influencing the generation of surplus material. The sale of "ugly veg" was discussed by the stakeholders during the interviews, particularly the retailers, with a strong viewpoint that consumers will not buy imperfect produce, and as a result, imperfect produce is not being offered to them for sale. One of the farmers interviewed mentioned that the appearance of vegetables needed to be acceptable to the consumer, and in order to achieve this, farmers were removing outer leaves which would have been undesirable for the consumers although perfectly edible, but which were now being used as cattle feed. Teigiserova et al. [16] suggested that educating consumers about behavior and consumption habits linked to deformed or ugly vegetables might reduce food waste and losses attributed to this material. The EU commission is also reviewing the marketing of so called "ugly fruit or vegetables" by focusing on the freshness rather than the aesthetics, which should, in turn, reduce the amount of food waste [53].

Forecasting impacted fine dining restaurants and retailers specifically. These stakeholders were also impacted by logistics connected with deliveries either of raw materials or product coming or going via central distribution. Product promotions also had a negative impact on the generation of surplus material, resulting in single produce remaining unsold in favor of the multi-packs on promotion. Aschemann-Witzel et al. [54] also explained a further negative impact of supermarket promotions creating more food waste in households, as consumers take advantage of the cheaper pricing but may end up buying more produce than will be consumed. This may solve an issue at retail, but it is ultimately only passing the problem on to the next stakeholder, in this case the consumer. External factors such as the impact of weather or local calendar events like football matches or local weddings could impact sales and create surplus material. Labor was another factor indicated; those interviewed expressed an issue in retaining staff and maintaining a level of training which keeps surplus material managed effectively.

> *"Training as well, so we will leave the ordering up to our commis chef or sous chefs or chef de parti's and they might over-order"* **Fine Dining Restaurant**

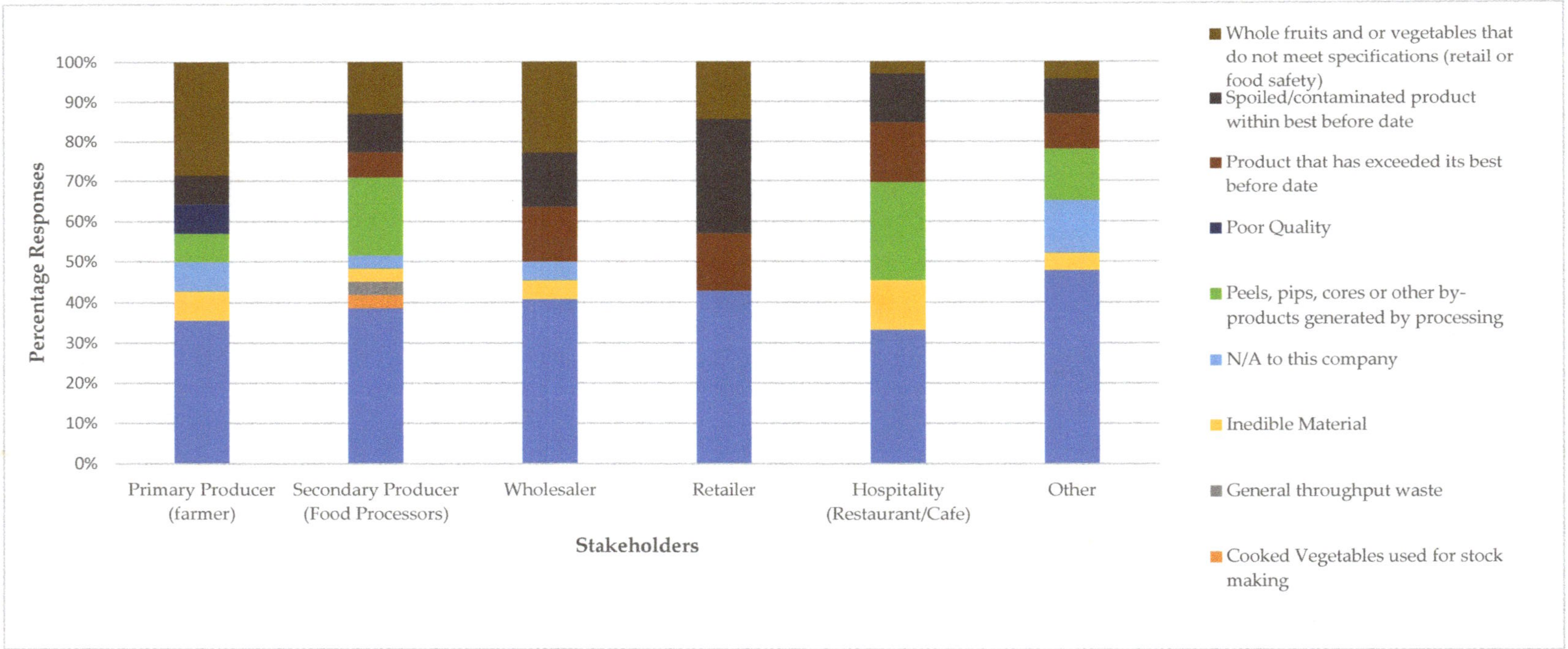

Figure 5. Types of surplus material present for each stakeholder sector (survey results).

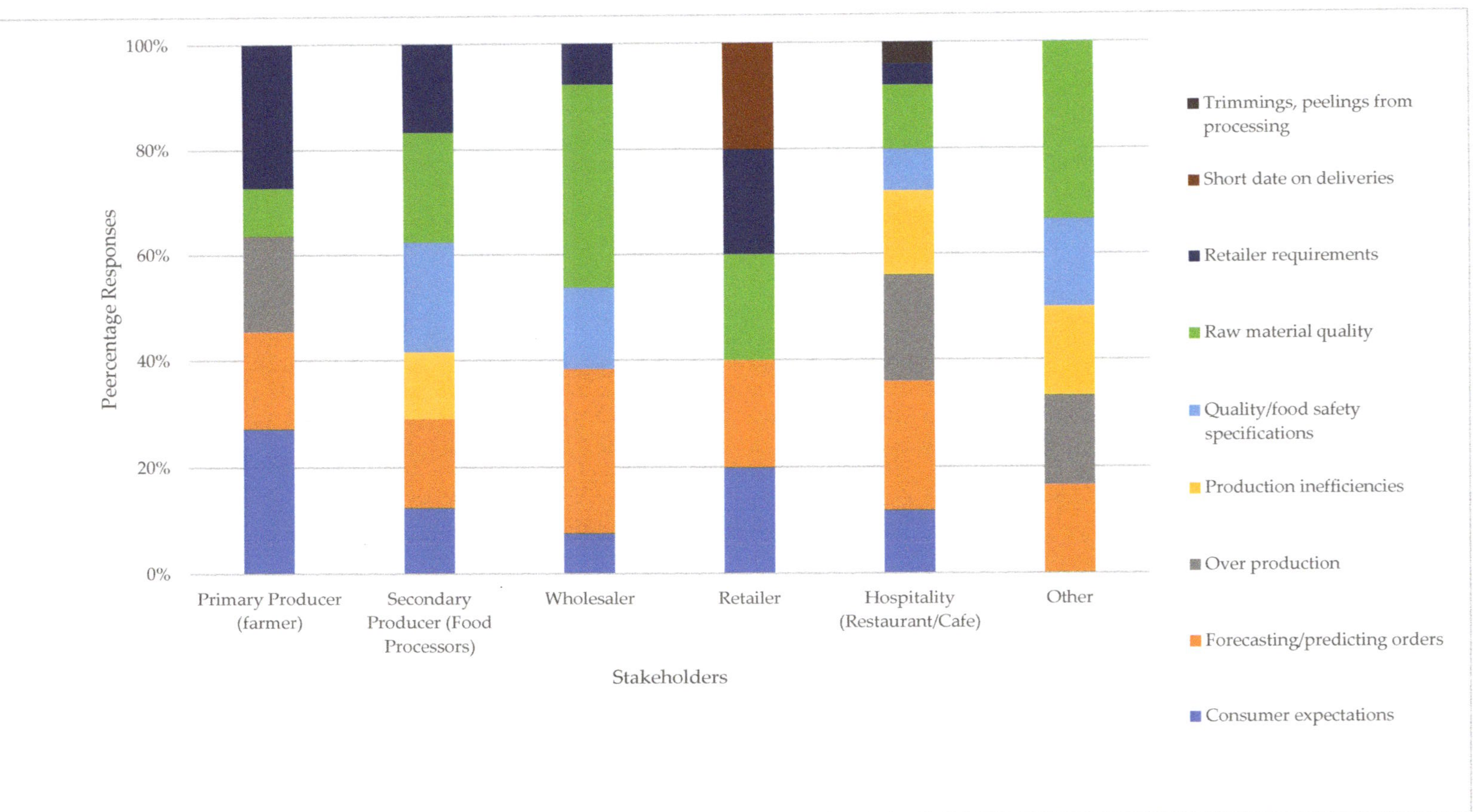

Figure 6. Reasons cited by the stakeholders in the survey for the generation of surplus fruit and vegetable material.

3.5. Valorization of Surplus Material

When investigating the barriers and drivers impacting stakeholder decisions to add value to surplus material generated (Q2), the participants were asked first, as described above, to state how they manage their surplus material. From the findings, only two respondents confirmed they are currently using surplus material in a side stream. However, almost half of the respondents (n = 21) indicated that they had identified potential side streams. The description of these side streams ranged from value-added food production (n = 7) to animal feed (n = 4), bioenergy production (n = 3), and redistribution for human consumption (n = 3). The stakeholders who had identified side streams were mostly primary or secondary producers as well as those in hospitality, but a small percentage of other stakeholders also identified side streams. Older studies stated that food producers have always considered using their by-products or side streams mostly to improve profitability [17]. What is interesting about the results in this study is the large gap between those already valorizing their surplus material and those who have not yet commenced this process. This may be due to the many barriers discussed below.

As mentioned previously, stakeholders often use this surplus material for animal feed, as it is a convenient source of feed for their own livestock. When the option to add value to this material was discussed with stakeholders in the interviews, the smaller businesses highlighted that "time" was a barrier, stating they did not have the necessary time or headspace to consider this option.

> *"We've discussed it from time to time, you know, the possibility of making broths, and stuff like that. But again, probably time, knowledge, expertise, has probably stalled us from going down that road"* **Farmer/Wholesaler**

There is also the risk of a potentially negative economic impact that businesses need to consider. Valorization of this material should ensure a viable return on their investment, otherwise time spent focusing on other revenue streams would be more appropriate. Prior research on pilot scale valorization agrees with this finding, as there is a requirement for investment to ensure the safe extraction of surplus material for use in new waste-to-value production streams [19]. The findings from this study highlight the difficulties for the smaller stakeholders. Respondents recognized the potential benefits of valorization in theory, but to date, the focus has been on minimizing food waste. As a result, businesses may have to re-structure their operations to be able to valorize surplus material.

> *"What, we've gone on is minimization of the amount of waste. And then when we get to that point, we don't have the volumes"* **Large Wholesaler**

This is in line with current recommendations, and as detailed in the food waste hierarchy, prevention is the most favorable outcome in terms of food waste management [9,18,55]. Those interviewed also preferred to donate produce to local restaurants and cafes if there were no other way of selling the produce, which is also a feature of the food waste hierarchy as a way of reducing food waste.

They also highlighted the need for external supports in terms of research and education, as well as financial support.

> *"I just don't have the facilities, like I know somebody who has facilities, and they have 0% waste in their fruit and veg. So, you know, if you had the capital investment"* **Retailer.**

Food safety legislation was believed to be restrictive, making the revalorization of food surplus, in some instances, prohibitive.

Previous review studies on food waste valorization across Europe and other jurisdictions have echoed similar barriers, particularly linked to the associated costs of starting a new valorization process, like new equipment and staffing, among other factors [19,35]. Garcia-Garcia et al. [36] also stated that it was necessary to identify a clear customer for these value-added products. However, when asked about how their customers would react to this kind of added-value product, the response was mostly positive, particularly if quality and cost factors were within customer expectations.

"There's a certain element still, that look for service and all that but the bottom line is if the price isn't lower than your competitor, they just won't get it of you. It's as simple as that" **Farmer/Wholesaler**

Ultimately, they believed it needs to make sense from a price perspective for the business model, as the quote above confirms, which is particularly relevant for the smaller businesses.

The results highlighted that transparency would be required to bring the consumer on board. McCarthy et al. [56] found that bringing positive and emotive messages linked to improving societal benefits by using surplus material that would have otherwise been wasted would improve the acceptance of these types of produce. This study also suggested that marketing and branding could be used to deliver the waste-to-value message in a user-friendly way to the consumer.

This study also recognized the importance of education in improving acceptance, with stakeholders educating their staff first, who in turn educate the customers on the lengths to which the businesses are going to reduce their food waste in creative ways. Aschemann-Witzel et al. [57] found increased acceptance of upcycled food products when frugality was highlighted. Some of the participants in this study recognized that they were missing an opportunity to demonstrate to their customers their own frugality and the sustainable practices currently in operation within their businesses.

"I suppose we should make more of a noise that the fact that it is organic, and we can, you know, therefore, we can use skins and things like that" **Cafe**

They also believed that introducing education to younger generations on where their food comes from would help to bridge the gap between industry and the consumer in terms of knowledge and ultimately acceptance of valorization of surplus material.

"I think if you showed people a basket of fruit and veg and say this is our fruit and veg, this is what you can do at home with your kids to educate them on this journey" **Retailer.**

This agrees with Rada et al. [58], who expressed the importance of educating young people on environmental issues, as this may help change behaviors within the family. It was also recognized that as the younger generation uses social media more often, this would be a good outlet to target food waste reduction campaigns [59]. Another study has shown that with more public discourse around food waste, waste-to-value products are better received by consumers, particularly if the sustainability impact of reducing food waste is expressed [39].

In terms of benefits of revalorization, they could see some potential long-term economic benefits after the initial investment period. Moreover, they recognized this as an opportunity to educate staff and customers on sustainability practices employed within their business models. It was viewed as a positive step for the company, providing the barriers alluded to above could be overcome.

3.6. Overview of Results

Although this was a small pilot study and not representative of all stakeholders, the results were in line with research findings to date, noting similar barriers to the revalorization of surplus material. Moreover, the explanatory mixed methods approach provided further validity to the findings, since the qualitative data generated from the interviews confirmed further and explained the results from the survey. The combined findings of this mixed methods study can be seen in Figure 7. These are presented under the five themes that emerged from the survey and the Interviews. The management of surplus material was viewed as important as an environmental metric by all stakeholders; however, the smaller stakeholders in this study did not have structured measurement protocols in place. The classification and the type of surplus material present are stakeholder-dependent, and most food waste reduction targets are driven by economic considerations. The findings also highlighted the lack of joined up thinking across the food supply chain on the topic of food waste management in general and the need for education and increased supports to create sustainable change in this sector.

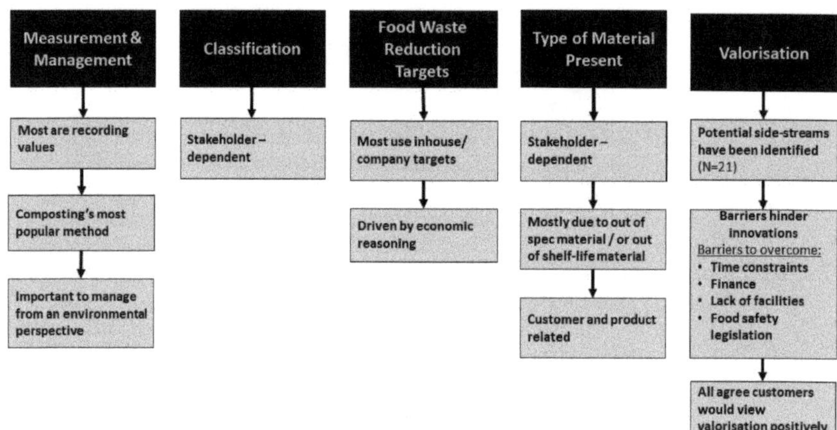

Figure 7. Summary of findings from stakeholders' survey and interviews, under the main headings investigated.

3.7. Limitations

One limitation of this pilot study was the relatively small sample size. In Ireland, several stakeholders would be classified as small or medium-sized businesses, and it proved difficult to access large numbers of these types of participants for the quantitative element of this mixed methods pilot study. Even large retailer brands are often franchised out to smaller operators. The voice of these participants was captured in the qualitative phase of this study, where a lack of resources was highlighted with the responses *"time poor"* and *"struggles maintaining a trained workforce"* mentioned frequently. However, for future studies, it would be recommended to find new avenues of recruitment to access larger numbers of these stakeholders.

4. Conclusions

This mixed methods pilot study focused on how stakeholders currently manage their fruit and vegetable surplus material, including by-products from their main production activity, and what factors impacted their decision to revalorize this material. The results confirmed that prevention is their primary motive in terms of management of their food surplus material. Trying to utilize all produce is mostly done for economic reasons. There is an interest in valorization; however, barriers such as time, lack of headspace to explore opportunities and a need for financial investment are preventing companies from taking this further, particularly small and medium-sized companies. The companies interviewed wear *"many hats"* in terms of their business models, with some being both primary and secondary producers as well as wholesalers. This means they are constantly evolving to remain profitable; therefore, policy changes or government support would be beneficial in assisting these types of companies in achieving sustainable changes.

Overall, there is a lack of joined-up thinking to bring resources together to create positive sustainable change, such as bio-refinery projects that smaller companies could feed into to help meet the food waste reduction goal of SDG 12.3. Each stakeholder interviewed showed they were doing their part to manage food waste; perhaps a new forum could be established to allow industry stakeholders to share ideas that have worked positively for them and could, in turn, help others to adapt their own practices. The disconnect between the consumer and the industry in terms of how consumer expectations impact the produce on the supermarket shelves is clear. An open, transparent dialogue between all stakeholders, including consumers, is critical if this narrative is to change. It was recognized that by sharing the current sustainability measures already in place through education with their own staff initially, followed with their customers, may start a positive chain of discourse that ultimately reaches all parties, bringing the food waste dilemma into the public arena. Then, food waste revalorization could offer products acceptable to consumers, and in turn, a viable business model for stakeholders. Education could lead to consumer understanding of the benefits of revalorization, increasing further the acceptability of the products. Future studies could explore the connections between the food supply chain links to identify ways of improving waste reduction measures and encouraging revalorization through education and improved transparency.

Supplementary Materials: The following supporting information can be downloaded at: https://www.mdpi.com/article/10.3390/su152316147/s1, Table S1: Survey Question Grid; Table S2: Interview Question Grid; Table S3: Sample of Coding File (Thematic Analysis Phase 2—descriptive loose codes); Table S4: Sample of Coding (Thematic Analysis—Phase 3—developing categories/initial themes and Phase 4—developing potential themes).

Author Contributions: Conceptualization, M.D., S.F., F.N. and O.K.; methodology, S.F., M.D., O.K. and F.N.; formal analysis, S.F. and M.D.; investigation, S.F. and M.D.; data curation, S.F. and M.D.; writing—original draft preparation, S.F.; writing—review and editing, S.F., M.D., O.K. and F.N.; visualization, S.F. and M.D.; supervision, M.D., F.N. and O.K; project administration, M.D.; funding acquisition, M.D., F.N. and O.K. All authors have read and agreed to the published version of the manuscript.

Funding: This research is part of a PhD project and was funded by the Connacht Ulster Alliance Bursary, grant number PCUAB013.

Institutional Review Board Statement: The study was conducted in accordance with the Declaration of Helsinki and approved by the Institute Review Ethics Committee (IREC) of Atlantic Technological University (IREC Reference No: 2020048, date of approval—20 December 2021).

Informed Consent Statement: Informed consent was obtained from all subjects involved in the study.

Data Availability Statement: The data presented in this study are available on request from the corresponding author. The data are not publicly available due to ethical reasons.

Acknowledgments: The authors would like to thank the respondents who participated in the piloting of the survey and the interviews. Thanks to all the anonymous stakeholders who took time to complete the survey and the individual stakeholders who took part in the interviews. Your participation is greatly appreciated. We also thank the gatekeepers from ATU for disseminating the survey.

Conflicts of Interest: The authors declare no conflict of interest. The funders had no role in the design of the study; in the collection, analyses, or interpretation of data; in the writing of the manuscript; or in the decision to publish the results.

References

1. Cederberg, C.; Sonesson, U. *Global Food Losses and Food Waste: Extent, Causes and Prevention; Study Conducted for the International Congress Save Food! At Interpack 2011, [16–17 May], Düsseldorf, Germany*; Gustavsson, J., Ed.; Food and Agriculture Organization of the United Nations: Rome, Italy, 2011; ISBN 978-92-5-107205-9.
2. Lemaire, A.; Limbourg, S. How Can Food Loss and Waste Management Achieve Sustainable Development Goals? *J. Clean. Prod.* **2019**, *234*, 1221–1234. [CrossRef]
3. Chaboud, G.; Daviron, B. Food Losses and Waste: Navigating the Inconsistencies. *Glob. Food Secur.* **2017**, *12*, 1–7. [CrossRef]
4. THE 17 GOALS | Sustainable Development. Available online: https://sdgs.un.org/goals (accessed on 11 September 2023).
5. From Farm to Fork. Available online: https://www.consilium.europa.eu/en/policies/from-farm-to-fork/ (accessed on 11 September 2023).
6. EU FUSIONS. Available online: https://www.eu-fusions.org/ (accessed on 13 September 2023).
7. Nicholes, M.J.; Quested, T.E.; Reynolds, C.; Gillick, S.; Parry, A.D. Surely You Don't Eat Parsnip Skins? Categorising the Edibility of Food Waste. *Resour. Conserv. Recycl.* **2019**, *147*, 179–188. [CrossRef]
8. Environment, U.N. UNEP Food Waste Index Report 2021. Available online: http://www.unep.org/resources/report/unep-food-waste-index-report-2021 (accessed on 11 September 2023).
9. Papargyropoulou, E.; Lozano, R.; Steinberger, J.; Wright, N.; bin Ujang, Z. The Food Waste Hierarchy as a Framework for the Management of Food Surplus and Food Waste. *J. Clean. Prod.* **2014**, *76*, 106–115. [CrossRef]
10. Messner, R.; Johnson, H.; Richards, C. From Surplus-to-Waste: A Study of Systemic Overproduction, Surplus and Food Waste in Horticultural Supply Chains. *J. Clean. Prod.* **2021**, *278*, 123952. [CrossRef]

11. Waste Framework Directive. Available online: https://environment.ec.europa.eu/topics/waste-and-recycling/waste-framework-directive_en (accessed on 19 October 2023).
12. Food Waste Charter. Available online: https://foodwastecharter.ie/ (accessed on 11 September 2023).
13. Agency, E.P. Food Waste Statistics. Available online: https://www.epa.ie/our-services/monitorin----assessment/waste/national-waste-statistics/food/ (accessed on 11 September 2023).
14. O'Connor, T.; Kleemann, R.; Attard, J. Vulnerable Vegetables and Efficient Fishers: A Study of Primary Production Food Losses and Waste in Ireland. *J. Environ. Manag.* **2022**, *307*, 114498. [CrossRef] [PubMed]
15. Directive 2008/98/EC of the European Parliament and of the Council of 19 November 2008 on Waste and Repealing Certain Directives (Text with EEA Relevance). Available online: https://eur-lex.europa.eu/legal-content/EN/TXT/?uri=celex%3A32008L0098 (accessed on 21 July 2023).
16. Teigiserova, D.A.; Hamelin, L.; Thomsen, M. Towards Transparent Valorization of Food Surplus, Waste and Loss: Clarifying Definitions, Food Waste Hierarchy, and Role in the Circular Economy. *Sci. Total Environ.* **2020**, *706*, 136033. [CrossRef]
17. Lin, C.S.K.; Pfaltzgraff, L.A.; Herrero-Davila, L.; Mubofu, E.B.; Abderrahim, S.; Clark, J.H.; Koutinas, A.A.; Kopsahelis, N.; Stamatelatou, K.; Dickson, F.; et al. Food Waste as a Valuable Resource for the Production of Chemicals, Materials and Fuels. Current Situation and Global Perspective. *Energy Environ. Sci.* **2013**, *6*, 426–464. [CrossRef]
18. Beausang, C.; Hall, C.; Toma, L. Food Waste and Losses in Primary Production: Qualitative Insights from Horticulture. *Resour. Conserv. Recycl.* **2017**, *126*, 177–185. [CrossRef]
19. Mirabella, N.; Castellani, V.; Sala, S. Current Options for the Valorization of Food Manufacturing Waste: A Review. *J. Clean. Prod.* **2014**, *65*, 28–41. [CrossRef]
20. Lange, L.; Meyer, A.S. Potentials and Possible Safety Issues of Using Biorefinery Products in Food Value Chains. *Trends Food Sci. Technol.* **2019**, *84*, 7–11. [CrossRef]
21. Caldeira, C.; Vlysidis, A.; Fiore, G.; De Laurentiis, V.; Vignali, G.; Sala, S. Sustainability of Food Waste Biorefinery: A Review on Valorisation Pathways, Techno-Economic Constraints, and Environmental Assessment. *Bioresour. Technol.* **2020**, *312*, 123575. [CrossRef] [PubMed]
22. Zelazinski, T.; Ekielski, A.; Durczak, K.; Morawska, M. Study on biodegradable materials from thermoplastic starch with the additon of nut shells. *Inmath Agric. Eng.* **2023**, *70*, 2.
23. Sasimowski, E.; Grochowicz, M.; Szajnecki, L. Preparation and Spectroscopic, Thermal and Mechanical Characterization of biocomposites of poly(butylene succinate) and onion peels or durum wheat bran. *Materials* **2023**, *16*, 6799. [CrossRef] [PubMed]
24. Pinela, J.; Prieto, M.A.; Barreiro, M.F.; Carvalho, A.M.; Oliveira, M.B.P.P.; Curran, T.P.; Ferreira, I.C.F.R. Valorisation of Tomato Wastes for Development of Nutrient-Rich Antioxidant Ingredients: A Sustainable Approach towards the Needs of the Today's Society. *Innov. Food Sci. Emerg. Technol.* **2017**, *41*, 160–171. [CrossRef]
25. Sagar, N.A.; Pareek, S.; Sharma, S.; Yahia, E.M.; Lobo, M.G. Fruit and Vegetable Waste: Bioactive Compounds, Their Extraction, and Possible Utilization. *Compr. Rev. Food Sci. Food Saf.* **2018**, *17*, 512–531. [CrossRef] [PubMed]
26. Banerjee, S.; Ranganathan, V.; Patti, A.; Arora, A. Valorisation of Pineapple Wastes for Food and Therapeutic Applications. *Trends Food Sci. Technol.* **2018**, *82*, 60–70. [CrossRef]
27. Garcia-Garcia, G.; Stone, J.; Rahimifard, S. Opportunities for Waste Valorisation in the Food Industry—A Case Study with Four UK Food Manufacturers. *J. Clean. Prod.* **2019**, *211*, 1339–1356. [CrossRef]
28. Zuin, V.G.; Segatto, M.L.; Zanotti, K. Towards a Green and Sustainable Fruit Waste Valorisation Model in Brazil: Optimisation of Homogenizer-Assisted Extraction of Bioactive Compounds from Mango Waste Using a Response Surface Methodology. *Pure Appl. Chem.* **2020**, *92*, 617–629. [CrossRef]
29. Esparza, I.; Jiménez-Moreno, N.; Bimbela, F.; Ancín-Azpilicueta, C.; Gandía, L.M. Fruit and Vegetable Waste Management: Conventional and Emerging Approaches. *J. Environ. Manag.* **2020**, *265*, 110510. [CrossRef]
30. Campos, D.A.; Gómez-García, R.; Vilas-Boas, A.A.; Madureira, A.R.; Pintado, M.M. Management of Fruit Industrial By-Products—A Case Study on Circular Economy Approach. *Molecules* **2020**, *25*, 320. [CrossRef] [PubMed]
31. Kowalski, A.; Agati, G.; Grzegorzewska, M.; Kosson, R.; Kusznierewicz, B.; Chmiel, T.; Bartoszek, A.; Tuccio, L.; Grifoni, D.; Vågen, I.M.; et al. Valorization of Waste Cabbage Leaves by Postharvest Photochemical Treatments Monitored with a Non-Destructive Fluorescence-Based Sensor. *J. Photochem. Photobiol. B Biol.* **2021**, *222*, 112263. [CrossRef] [PubMed]
32. Sabater, C.; Calvete-Torre, I.; Villamiel, M.; Moreno, F.J.; Margolles, A.; Ruiz, L. Vegetable Waste and By-Products to Feed a Healthy Gut Microbiota: Current Evidence, Machine Learning and Computational Tools to Design Novel Microbiome-Targeted Foods. *Trends Food Sci. Technol.* **2021**, *118*, 399–417. [CrossRef]
33. Srenuja, D.; Shanmugam, A.; Sinija, V.R.N. Novel Zero Waste Tactics for Commercial Vegetables—Recent Advances. *Int. J. Food Eng.* **2022**, *18*, 633–654. [CrossRef]
34. Jaouhari, Y.; Travaglia, F.; Giovannelli, L.; Picco, A.; Oz, E.; Oz, F.; Bordiga, M. From Industrial Food Waste to Bioactive Ingredients: A Review on the Sustainable Management and Transformation of Plant-Derived Food Waste. *Foods* **2023**, *12*, 2183. [CrossRef] [PubMed]
35. Sarangi, P.K.; Mishra, S.; Mohanty, P.; Singh, P.K.; Srivastava, R.K.; Pattnaik, R.; Adhya, T.K.; Das, T.; Lenka, B.; Gupta, V.K.; et al. Food and Fruit Waste Valorisation for Pectin Recovery: Recent Process Technologies and Future Prospects. *Int. J. Biol. Macromol.* **2023**, *235*, 123929. [CrossRef] [PubMed]

36. Garcia-Garcia, G.; Woolley, E.; Rahimifard, S.; Colwill, J.; White, R.; Needham, L. A Methodology for Sustainable Management of Food Waste. *Waste Biomass Valor* **2017**, *8*, 2209–2227. [CrossRef] [PubMed]
37. Khattak, K.F.; Rahman, T.U. Analysis of Vegetable Peels as a Natural Source of Vitamins and Minerals. *Int. Food Res.* **2017**, *24*, 292–297.
38. Coderoni, S.; Perito, M.A. Sustainable Consumption in the Circular Economy. An Analysis of Consumers' Purchase Intentions for Waste-to-Value Food. *J. Clean. Prod.* **2020**, *252*, 119870. [CrossRef]
39. Coderoni, S.; Perito, M.A. Approaches for Reducing Wastes in the Agricultural Sector. An Analysis of Millennials' Willingness to Buy Food with Upcycled Ingredients. *Waste Manag.* **2021**, *126*, 283–290. [CrossRef]
40. Aschemann-Witzel, J.; Peschel, A.O. How Circular Will You Eat? The Sustainability Challenge in Food and Consumer Reaction to Either Waste-to-Value or yet Underused Novel Ingredients in Food. *Food Qual. Prefer.* **2019**, *77*, 15–20. [CrossRef]
41. Grasso, S.; Asioli, D. Consumer Preferences for Upcycled Ingredients: A Case Study with Biscuits. *Food Qual. Prefer.* **2020**, *84*, 103951. [CrossRef]
42. Creswell, J.W. *Research Design: Qualitative, Quantitative and Mixed Methods Approaches*, 4th ed.; SAGE: London, UK, 2014.
43. Kumar, R. *Research Methodology: A Step-by-Step Guide for Beginners*, 4th ed.; SAGE: Los Angeles, CA, USA, 2014.
44. State of Food and Agriculture 2019. Moving Forward on Food Loss and Waste Reduction | Policy Support and Governance | Food and Agriculture Organization of the United Nations. Available online: https://www.fao.org/policy-support/tools-and-publications/resources-details/en/c/1242090/ (accessed on 11 September 2023).
45. Origin Green. Available online: https://www.origingreen.ie/ (accessed on 17 October 2023).
46. Göbel, C.; Langen, N.; Blumenthal, A.; Teitscheid, P.; Ritter, G. Cutting Food Waste through Cooperation along the Food Supply Chain. *Sustainability* **2015**, *7*, 1429–1445. [CrossRef]
47. Lidl Initiative Aims to Cut Food Waste at Irish Stores by 700,000kg. Available online: https://www.irishtimes.com/business/retail-and-services/lidl-initiative-aims-to-cut-food-waste-at-irish-stores-by-700-000kg-1.4103048 (accessed on 11 September 2023).
48. Food Loss and Food Waste | Policy Support and Governance Gateway | Food and Agriculture Organization of the United Nations | Policy Support and Governance | Food and Agriculture Organization of the United Nations. Available online: https://www.fao.org/policy-support/policy-themes/food-loss-food-waste/en/ (accessed on 13 September 2023).
49. FoodCloud: Food Waste Hurts Our Planet. Available online: https://food.cloud/ (accessed on 11 September 2023).
50. de Oliveira, M.M.; Lago, A.; Dal' Magro, G.P. Food Loss and Waste in the Context of the Circular Economy: A Systematic Review. *J. Clean. Prod.* **2021**, *294*, 126284. [CrossRef]
51. Tulej, W.; Glowacki, S. Analysis of Material-Characterization Properties of Post-Production Waste—The case of apple pomace. *Materials* **2022**, *15*, 3532. [CrossRef]
52. Richards, C.; Hurst, B.; Messner, R.; O'Connor, G. The Paradoxes of Food Waste Reduction in the Horticultural Supply Chain. *Ind. Mark. Manag.* **2021**, *93*, 482–491. [CrossRef]
53. Commission Updates Marketing Standards of Agri-Food Products to Better Address Consumer Needs and Sustainability. Available online: https://cyprus.representation.ec.europa.eu/news/commission-updates-marketing-standards-agri-food-products-better-address-consumer-needs-and-2023-04-21_en (accessed on 21 September 2023).
54. Aschemann-Witzel, J.; De Hooge, I.; Amani, P.; Bech-Larsen, T.; Oostindjer, M. Consumer-Related Food Waste: Causes and Potential for Action. *Sustainability* **2015**, *7*, 6457–6477. [CrossRef]
55. Stop Food Waste—Reduce Your Food Waste and Save Yourself Money! Available online: https://stopfoodwaste.ie/ (accessed on 11 September 2023).
56. McCarthy, B.; Kapetanaki, A.B.; Wang, P. Completing the Food Waste Management Loop: Is There Market Potential for Value-Added Surplus Products (VASP)? *J. Clean. Prod.* **2020**, *256*, 120435. [CrossRef]
57. Aschemann-Witzel, J.; Asioli, D.; Banovic, M.; Perito, M.A.; Peschel, A.O. Communicating Upcycled Foods: Frugality Framing Supports Acceptance of Sustainable Product Innovations. *Food Qual. Prefer.* **2022**, *100*, 104596. [CrossRef]
58. Rada, E.C.; Bresciani, C.; Girelli, E.; Ragazzi, M.; Schiavon, M.; Torretta, V. Analysis and Measures to Improve Waste Management in Schools. *Sustainability* **2016**, *8*, 840. [CrossRef]
59. Zamri, G.B.; Azizal, N.K.A.; Nakamura, S.; Okada, K.; Nordin, N.H.; Othman, N.; Akhir, F.N.; Sobian, A.; Kaida, N.; Hara, H. Delivery, Impact and Approach of Household Food Waste Reduction Campaigns. *J. Clean. Prod.* **2020**, *246*, 118969. [CrossRef]

 *sustainability*

Article

Consumers' Motives on Wine Tourism in Greece in the Post-COVID-19 Era

Athanasios Santorinaios, Ioanna S. Kosma and Dimitris Skalkos *

Laboratory of Food Chemistry, Department of Chemistry, University of Ioannina, 45110 Ioannina, Greece; ch06054@uoi.gr (A.S.); i.kosma@uoi.gr (I.S.K.)
* Correspondence: dskalkos@uoi.gr; Tel.: +30-2651008345

Abstract: Wine tourism is emerging as one of the most important forms of alternative, sustainable tourism in wine countries, such as Greece, in the post-COVID-19 era. In this paper, consumers' motives for wine tourism in Greece today are investigated regarding (i) their consumption habits related to wine, (ii) their experience with wine tourism, (iii) the parameters that would encourage their visit to a wine region, such as wine, the winery, and general regional characteristics, and (iv) the source of information consulted for a wine tourism experience. The questionnaire was conducted from April to May 2023, with 595 participants, via the Google Forms platform. The statistical analysis was performed with basic tools, as well as cross and chi-square tests, to analyze the data. The highlights of the results indicate that consumers (the participants of the survey) consume more wine today than before the pandemic (57%) and have previous experience in wine tourism (59.8%), with the majority of them having visited a winery more than once (67.4%). The most popular activity at the winery was found to be wine tasting (46.6%), followed by open discussion about wine (35.2%), and, at the regional level, visiting the sights (46%) and doing activities in nature (30.6%). Future participants are looking for innovation in wine tourism, with trained staff (77.5%) and organized tours (74.3%), the organization of wine festivals and other events (71.9%), opportunities to explore the local community, such as the outdoors (83.5%) and its culture and history (70.9%), during their visit, and available information on wine tourism opportunities online (73%). They also are encouraging the transition of the Greek wine tourism industry to the digital world. Based on the overall results, three types of support are proposed for the successful, sustainable development of wine tourism in wine-producing countries.

Keywords: questionnaire survey; wine tourism; post-COVID-19; Greece

check for
updates

Citation: Santorinaios, A.; Kosma, I.S.; Skalkos, D. Consumers' Motives on Wine Tourism in Greece in the Post-COVID-19 Era. *Sustainability* **2023**, *15*, 16225. https://doi.org/10.3390/su152316225

Academic Editor: Ilija Djekic

Received: 21 October 2023
Revised: 15 November 2023
Accepted: 21 November 2023
Published: 23 November 2023

1. Introduction

The COVID-19 pandemic brought significant changes to the tourism industry, with travelers reconsidering their goals and intentions for their trips [1]. Sustainability has emerged as the new trend in tourism in the post-pandemic era [2], with alternative tourism potentially replacing parts of mass tourism as part of its resurgence in the current times [3]. Wine tourism highlights its sustainability, as demonstrated by the economic and cultural considerations that have been perceived by the local communities [4]. For example, wine tourists' satisfaction is enhanced when their winery visit includes elements of local cuisine and wine culture [5]. With the discovery of grape wine being recorded from the beginning of the sixth millennium BC [6], wine production has a history of thousands of years, having survived detrimental climate changes of the past [7]. However, wine consumption has been always linked to a better quality of life, especially in old age, as well as low mortality [8]. Wine-producing countries are also now focusing more on aesthetic and experiential consumer satisfaction, rather than exclusively on knowledge-sharing about winemaking and the cellar's facilities [9]. This is a result of the similarities between wine consumption and art appreciation that have been observed from the consumer's perspective [10]. Wine tourism can promote local wine more effectively, boosting its exports and sales [11]. The aim of this research is to identify the parameters that will boost wine tourism in a wine-producing country during the post-pandemic period and beyond, supporting regional sustainable development for the local communities and wineries benefiting from this innovative, modern, and future tourism industry. To achieve this aim the relevant literature regarding wine tourism was systematically reviewed and presented, followed by the materials and method of the study, the results, and the findings.

Literature Review

The COVID-19 pandemic greatly affected all tourism sectors in Europe and Greece [12], including wine tourism. Wine tourists in the near and far future will demand safer, healthier, and more sustainable experiences [13]. After the COVID-19 pandemic, it appears that visitors prefer less popular and crowded destinations [14]. Especially for domestic travelers, tranquility and rest, elements directly linked to wine tourism through the enjoyment of the natural landscape, have been shown to be key motivators for their visit to the country's wine regions [15]. In a psychophysiology study, higher levels of satisfaction were recorded when participants were able to freely explore different areas of the winery, compared to when they were given explanations [16]. Generation Y (born between 1981 and 1994), for example, the largest group of wine tourists, have a pleasant desire to consume wine if they know its health benefits [17]. This pleasure is enhanced when combined with socialization [17].

With global competition between wine tourism regions intensifying, and the demand for wine tourism becoming more sophisticated, innovation and digital transformation in wine tourism are imperative [18]. The interest expressed in digital wine tourism arose from the fear and anxiety associated with the pandemic [19]. However, these feelings are not deterrents to the implementation of future wine tourism trips [19]. On the contrary, creating an asynchronous wine tourism experience, such as a virtual tour, could be a starting point for attracting new consumers in wine tourism [20]. In the post-pandemic era, companies and destinations are expected to redefine themselves to adapt to the digital age. The expansion of the digital presence of wineries also allows the most appropriate approach for wine tourists [21]. Wine producers can increase the value of a satisfying experience by bringing the winery closer to the customer, before, during, and after they visit the premises [22]. For example, the configuration of a winery in such a way that it gives the wine tourist the opportunity to take the "perfect Instagram photo" differentiates a winery from the rest and highlights the uniqueness of the wine destination [23].

Wine tourism in Greece is still at an early stage [14]. In a recent survey, only 22% of the surveyed Greek wineries had an online store with e-shopping [14]. At the same time, even though there is relevant national legislation for the certification of wineries that can have access to wine tourism services (Law 4276/2014) and corresponding promotions on a virtual platform, there is still no public list of these wineries accessible to visit in Greece, limiting their national support. Wine tourism in Greece, however, could be internationally supported, if requested, by the UN World Tourism Organization (UNWTO) which promotes strategies and protocols for the development of early-stage wine regions worldwide [24]. Respectively, the Global Wine Tourism Organization (GWTO) has provided scientific support towards European wine tourism by identifying, for example, the cost of living as the main determinant of tourism demand in Greece, followed by price competitiveness and income [25].

The GWTO considers wine tourism as a key contributor to the global economy with encouraging growth rates for the next decades [26]. Wine tourism addresses a niche market, which can be combined with all different types of tourism, thus promoting sustainable development for the local area [14]. Studies have shown that wine tourism can stimulate a rural local economy if small tourism businesses complement each other to form a strong local, sustainable destination image [27,28]. Therefore, the involvement of the oenological region has an important role in the sustainability of local wineries [29], thus highlighting the two-way relationship between the local wine region and wine tourism. Many wine tourists, in fact, express great interest in the variety of attractions in the area that are not directly related to wine [30]. Visiting a winery from the perspective of Generation Z, for example, seems to be less about the wine itself and more about the opportunities to enjoy the scenery, have fun, socialize, and discover local products [31]. Wine tourists in Greece usually carry out additional activities in the wine region that are not related to wine after they visit the winery [32]. Particularly for Greek youth, research has shown that there is a limited appeal for the wine itself, with young people preferring activities such as enjoying the scenery and food as well as socializing as part of their visit to a winery [33]. This appeal also results from their limited financial ability to spend during their trip [33].

The recovery process from the pandemic suggests that domestic tourism is a definite driving force of the future. Recent research analyzing the elements of the tourist behavior of Greek citizens shows their strong preference for domestic tourism [34]. Significantly higher percentages of domestic visitors indicated a strong motivation to purchase wine during their winery visit, to chat with the winemaker or oenologist, and to participate in an organized winery tour or to experience the winery and atmosphere, in contrast to international visitors who preferred to participate in a group activity and take an excursion [35].

This is the first study that identifies and analyzes the overall consumer motives regarding wine tourism in Greece in the post-COVID-19 period. The purpose of this research is to evaluate those factors or parameters related to the motivations of domestic consumers to engage in wine tourism in a wine-producing country such as Greece and to identify the main objectives that the Greek wine tourism industry should focus on to ensure development in the post-pandemic era. To achieve this goal, and based on the existing literature on the related parameters of consumer preferences for wine tourism [36], the present study examines the following factors of engagement in Greek wine tourism during the post-COVID-19 period:

(I)   Wine consumption.
(II)   Existing experience in wine tourism.
(III) Wine and winery parameters that encourage visits to a wine region.
(IV)   General characteristics of an oenological region that encourage a visit to a wine region.
(V)   Source of information consulted to encourage a visit to a wine region.

2. Materials and Methods

2.1. Data Collection and Sample Characterization

A survey was conducted based on a structured questionnaire that examined the various parameters that influence consumers' motivations for wine tourism in the post-COVID-19 era. The questionnaire was structured in six parts, formulated in such a way as to serve the purposes of the research, and was based on a previous study conducted in the past, in 2008, by Galloway et al. [36].

The first part consisted of five questions concerning the demographic data of the respondents, namely, gender, age, completed level of education, professional activity, and place of residence. The second part consisted of four questions concerning the consumption habits of the respondents regarding wine in the post-pandemic era which explores their connection with wine. The third part consisted of five questions concerning the respondents' experience with wine tourism. The fourth part consisted of 10 questions regarding the wine and winery parameters that would prompt the respondents to visit a wine region of Greece. The fifth part consisted of eight questions—parameters about the characteristics of the oenological region that encourage visits to a wine region. Finally, the sixth part consisted of six questions evaluating the importance of information sources that would encourage their involvement in wine tourism. Respondents were asked to rate each parameter from 1 (not at all important) to 5 (very important).

Table S1 presents the questionnaire, which was in electronic form (in Greek), through the Google Forms platform and was distributed via email. The email included an explanation of the purpose of the research, a brief interpretation of the concept of wine tourism, as well as a link to the digital questionnaire. The answers were anonymous, and no personal data of the respondents was collected, in accordance with GDPR regulation and data protection. The structured questionnaire was distributed to the community of the University of Ioannina (students, professors, staff, etc.) through their academic emails (@uoi.gr), justifying the high percentage of young people and students in the sample. There were no specific criteria for the target group of the study.

The high percentage of women who answered the questionnaire (73.4%) is noteworthy, an observation that has been repeated in all previous research topics we have carried out that investigate consumer behavior [37–40], even though there were no significant gender-related taste differences on wine consumption detected [41].

The survey was conducted between April and May 2023, with the total number of participants being 595 and including a wide range of demographics.

2.2. Data Analysis

The analysis of the responses was carried out through basic statistical tools. The analytical data were organized using IBM SPSS Statistics for Windows (Version 25.0, IBM Corp. Armonk, NY, USA), following the methodology outlined by Skalkos et al. [39]. Cramer's V coefficient, which ranges from 0 to 1, was utilized in the chi-squared tests. Its interpretation is as follows: V values around 0.1 indicate a weak association, around 0.3 indicate a moderate association, and approximately 0.5 or higher indicate a strong association. For all conducted tests, a significance level of 5% ($p < 0.05$) was considered.

3. Results

Table 1 presents the demographic characteristics of the participants of the questionnaire.

Table 1. Demographic characterization of the sample.

Variable	Groups	(%)
Gender	Male	26.6
	Female	73.4
Age	18–25	67.9
	26–35	6.4
	36–45	5.4
	46–55	14.5
	56+	5.9
Level of education (completed)	None/Primary School	0.2
	Secondary School	0.7
	High School	53.8
	University	45.4
Job situation	Unemployed	1.3
	Student	67.4
	Employed	29.1
	Retired	2.2
Permanent residence in Greece	North Greece (Macedonia and Thrace)	25
	West Greece (Epirus and Etoloakarnania)	27.9
	Central Greece (and Athens)	34.6
	South Greece (Peloponnese)	4.9
	Islands (Aegean, Ionian, and Crete)	7.6

Regarding the place of permanent residence, 25% of the respondents came from Northern Greece, 27.9% from Western Greece, 34.6% from Central Greece (including the capital of Greece, Athens), while there was a significantly lower representation from Peloponnese and the islands of Greece with significant lower population (4.9% and 7.6%, respectively). Regarding the level of education, the majority held a high school or university diploma (53.8% and 45.4%, respectively), which is also linked to the fact that students

(67.4%) participated the most when regarding professional activity. More than 2/3 of the sample were young people (67.9%) and currently studying (67.4%), which can draw conclusions about the new generation and highlight the prospects for attracting them to wine tourism.

Table 2 presents the consumption habits of the respondents regarding wine. Most participants purchased 1–5 bottles of wine each month (60.2%), with a significant proportion reporting zero bottles of wine consumption (33.6%). Most respondents spent up to 10 EUR (52.8%) or 10–50 EUR (42.9%) on the purchase of wine monthly while the frequency of wine consumption had a smooth variation for once, twice, and three times weekly, except for the choice of daily consumption (2.9%). Noteworthy is the observation that most of the respondents stated that they consumed more wine after the pandemic (57%) compared with the period before.

Table 2. Wine consumption habits of the participants.

Questions	Answers	(%)
How many bottles of wine do you buy per month TODAY?	Zero bottles/none	33.6
	1–5 bottles	60.2
	6–10 bottles	5.2
	11–15 bottles	0.5
	>15 bottles	0.5
How much money do you spend on wine per month TODAY?	<10 EUR	52.8
	10–50 EUR	42.9
	50–100 EUR	3.9
	>100 EUR	0.5
How often do you consume wine?	Once per month	28.1
	Once every two weeks	22
	Once per week	22.5
	Twice per week	24.5
	Everyday	2.9
Do you consume LESS or MORE wine today compared to the pre-pandemic period?	Less	43
	More	57

Table S2 depicts the significant ($p < 0.05$) associations between the respondents' consumption habits of wine based on their profile and sociodemographic variables. Specifically, many associations were found, with weak correlation on all the questions of *Part II* between age, level of education, and job situation. We also found that the question regarding the money spent per month on wine today was affected by gender as well, and the comparison regarding the consumption of wine before and after the pandemic was affected by the place of permanent residence.

Table 3 shows the results of the questions about the respondents' experience in wine tourism. Most respondents (59.8%) have visited a winery, out of which 32.6% have visited a winery once, 46.1% 2–3 times, and only 10.1% 4–5 times. Almost all the respondents who have participated in wine tourism (97.5%) have not stayed overnight at a winery (or partner accommodation). The most popular wine tourism activities they did during their visit were wine tasting (46.6%), followed by an open discussion about wine (35.2%), and buying wine at a discounted price (33%). At the same time, the most popular tourist activities carried out during their visit to the wine region were visiting the attractions of the region (46%), taking a tour of the region without a tour guide (31.5%), and activities in nature (30.3%).

Table 3. Wine tourism experience of the participants.

Question	Answers	(%)
Have you ever visited a winery?	Yes	59.8
	No	40.2
If you answered YES to the first question, how many visits in the last 12 MONTHS?	1 visit	32.6
	2–3 visits	46.1
	4–5 visits	10.1
	>5 visits	11.2
If you answered YES to the first question, did any of your visits include a stay at the winery?	Yes	2.5
	No	97.5
If you answered YES to the first question, select the oenological activities you did during your visit to a winery:	Wine tasting	46.6
	Buy wine at a discounted price	33
	Buy oenological literature	1.3
	Open discussion about wine	35.2
	Watch educational presentations	23.2
	Dinner	8.2
	Tour without a guide	26.6
	Art exhibition at the winery	13.3
	Try/buy local area products	26.3
	Picnic/BBQ	5.9
If you answered YES to the first question, select the tourist activities you did AT THE SAME TIME as your visit to a winery:	Visit area attractions	46
	Visit parks and recreation areas	21
	Dinner at local restaurants/fine dining	22.6
	Activities in nature/outdoors	30.3
	Explore the history and culture of the area	27.6
	Visit local market	27.4
	Tour of the area WITHOUT a guide	31.5
	Guided tour of the area	13.6

Table S3 presents the result of the chi-square test that revealed significant differences in the respondents' wine tourism experience based on their demographic characteristics. A medium association was found between the answers to the core question "Have you ever visited a winery" and age, with smaller ones regarding level of education (completed) and job activity. There were also many associations between age, job activity, and paid activities.

Table 4 represents the results of wine and winery parameters that would prompt participants to visit a Greek wine region as part of their wine tourism plans. Based on the positive choices (as quite and very important) ($\geq$70%), the qualified staff (77.5%), followed by the wine regional reputation (76%), the existence of a winery tour and tasting (74.3%), festivals and special events (71.9%), as well as the cost of visit (70.4%) were the preferred parameters for a visit. The range of regional wineries (35.6%) was the lowest parameter of choice for visiting a region.

Table S4 depicts the result of the chi-square test that revealed small associations between the participants' answers on parameters regarding the wine and the winery and their demographic characteristics.

Table 5 presents the results of the parameters regarding the general characteristics of the oenological region that would encourage involvement in wine tourism. Based on the positive choices (as quite and very important) ($\geq$70%), the existing natural environment (83.5%), the accessibility of the region (72.8%), and historical, cultural, and lifestyle regional opportunities (70.9%) were the most preferred parameters to encourage a visit. Surprisingly, the existence of gourmet cuisine was the least preferred parameter (37.2%). Parameters such as accommodation (58.3%), participation in social activities (55%), distance from home (53.4%), and the existence of a holiday program (50.1%) were of medium preference for visiting a region.

Table 4. Parameters regarding the wine and the winery that would encourage a participant's visit to a wine region in Greece.

Parameters Regarding the Wine and the Winery	Not Important at All (1)	Slightly Important (2)	Important (3)	Quite Important (4)	Very Important (5)
Range of wineries in one region	8.7	18.2	37.5	23.5	12.1
Educational opportunities related to wine	6.4	12.6	29.7	32.1	19.2
Wine/wine tourism festivals and special events	2.2	7.2	18.7	40	31.9
Organized winery tour and local wine tasting	1.8	7.1	16.8	38.7	35.6
The existence of quality marks such as, e.g., PDO (Protected Designation of Origin) product, etc.	2.9	9.7	19	30.1	38.3
Area grape variety	3.7	9.1	20.7	36	30.6
Reputation of the wine region and the wine	2.2	3.9	18	40	36
Cost of visit and wine tasting	2.5	5	22	32.6	37.8
Staff knowledgeable about wine and the oenological region	1.7	3.9	17	33.3	44.2
Enrichment of knowledge about wine and grape varieties	2.7	7.7	22.9	37	29.7

Table 5. Parameters regarding the general characteristics of the wine region to encourage visits to a wine region in Greece.

Parameters Regarding the General Characteristics of the Wine Region	Not Important at All (1)	Slightly Important (2)	Important (3)	Quite Important (4)	Very Important (5)
Availability and variety of types of accommodation	2	7.6	32.1	38.5	19.8
Accessibility (airport, train station, and road accessibility)	1.2	7.6	18.5	38.3	34.5
Distance from the place of permanent residence	3.2	14.1	29.2	30.4	23
Personalized and organized holiday program (individual or group)	5.4	13.4	31.1	35	15.1
Opportunity to experience the culture, history, and lifestyle of the area	1.5	6.6	21	36.3	34.6
Availability of gourmet cuisine	9.6	20.7	32.6	24.4	12.8
Attractive natural scenery and good climatic conditions	1.3	3.2	11.9	44	39.5
Participation in social activities	3.4	11.8	29.9	35	20

The results of the chi-square test that showed significant differences between consumers' opinions on the parameters regarding the general characteristics of the wine region to encourage wine tourism experience and their demographic characteristics are presented in Table S5.

Table 6 presents the results of the source of information requested by the participants in order to visit a region. Based on the quite and very important answers, information from social media (73%), and a recommendation by a third person (70.6%) were the preferred pieces of information requested for a visit, followed by the travel agent's recommendation (43.5%) and a recommendation by the local visitors' information center (42.7%). The least preferred data were from the media (35.6%) and brochures (23.6%).

Table 6. Importance of sources of information regarding wine tourism activities in Greece.

Source of Information Regarding Wine Tourism Opportunities	Not Important at All (1)	Slightly Important (2)	Important (3)	Quite Important (4)	Very Important (5)
Brochures	16	24.4	36.1	16	7.6
Recommendation from a familiar person	1.8	6.4	21.2	43.2	27.4
Media (TV, and radio)	8.1	18	38.5	25.7	9.7
Local visitor information centers	8.1	13.8	35.5	30.3	12.4
Travel agencies—group visits	8.9	15.1	32.4	29.9	13.6
Internet/social media (Facebook, Instagram, TikTok,) etc.	2.9	5	19.2	31.3	41.7

Table S6 presents the results of the chi-square test showing the significant differences between the participants' preference for the tool of information regarding wine tourism opportunities and their demographic characteristics.

4. Discussion

This study investigates, for the first time, the effect of the COVID-19 pandemic on consumer's motivations for wine tourism in a wine-producing region, such as Greece, to examine its development prospects in the post-pandemic period. The demographic characteristics of the participants depicted an increased percentage of young people, aged 18–25, like our other previous survey regarding quality wine consumption [40], which may highlight the desires and expectations of the next generation of wine tourists, as well as the path that wine tourism in Greece must take in order to attract them.

Regarding wine consumption habits, the participants, although they are frequent wine consumers, are not spending much on bottled wine, as they probably prefer bulk wine, which can be interpreted as a desire to buy locally produced wines, as proven by Di Vita et al. [42]. However, after the pandemic, there has been an increased consumption of wine compared to before, an observation that verifies a previous study by Pytell et al. [43].

Regarding the previous wine tourism experience of the participants, our findings indicate that the increased satisfaction of the consumer's wine tourism experience may increase their loyalty and encourage further engagement with wine tourism services (e.g., more than one winery visit), as recorded by other researchers [5,44]. The participants did not choose to use the accommodation options of the winery and instead were more driven to purchase wine after they visited the winery, as previously stated, and to invest in the local community, stimulating the economy of rural areas [45].

Regarding the motivations of the participants related to wine and the winery to keep them engaged with wine tourism, our results show that they were eager after COVID-19 to experience knowledge-sharing regarding the wine by qualified staff, rather than pleasure, as recorded by Bruwer et al. as well in 2019 (the pre-pandemic period) [46]. The young audience still seeks sustainable visit costs in the post-pandemic period, as suggested in 2018 by Stergiou et al. [33], with a primary interest in wine tourism festivals and special events. Such events can create new target groups for the wine tourism industry, as suggested by Yuan et al. first [47].

In terms of motivation regarding the wine region's amenities, we have shown that consumers seek a more sustainable wine tourism experience in the post-pandemic period, as shown already by Ouvrard et al. [48], with easy accessibility and nature-based activities. They are also willing to discover the history and culture of the region during their wine visit.

Regarding the sources of information for wine tourism opportunities, our results show that most participants consider it very important for the wineries to form a digital identity in the wine tourism industry, through the Internet and social media, highlighting the need for the inclusion of wine tourism in the digital world. Survey participants dodid not consider brochures or mass media to be important sources of information. These findings, regarding the source of information sought for wine tourism selection and considering the high percentage of young (students of Generation Z) participants, are fully expected since the younger generations worldwide, more than other generations, have stopped reading brochures, maps, or watching TV as part of their daily habits.

5. Conclusions

This research paper investigates the motivations of consumers regarding their participation in wine tourism initiatives in wine-producing countries such as Greece, highlighting the parameters that will attract more domestic wine tourism in the post-pandemic period. The survey showed that although the participants consume more wine today compared to the period before the COVID-19 pandemic, the domestic demand for wine remains low (1–5 bottles, up to 10 EUR per month). Regarding the parameters for engagement in wine tourism, participants look for an organized winery experience, with trained staff and participation in events. Young Greeks are also looking for affordable wine tourism packages. The participants, moreover, showed that they are interested in exploring the local community when participating in wine tourism. Finally, they expect the wine industry to have all the proper information accessible online.

With the COVID-19 pandemic officially reaching its end, it is very important to analyze how it has affected human activity, what priorities it has now set, and how the wine tourism industry can ultimately adapt to today's needs. The wine tourism industry benefits from tourists' strong desire for sustainable tourism away from mass crowds. However, it is being tested with the need to enter the digital age and the competition between other forms of alternative tourism. The early stage of wine tourism in Greece is also an opportunity to adapt and evolve to new trends while ensuring its identity.

Wine tourism in wine-producing countries, based on the results of our work, needs three types of support to be able to develop further and successfully:

The first type is institutional support from the state, such as an institutional framework that ensures the quality of wine tourism services, highlighting wine wealth both in the context of representation and the context of digital promotion accessible to the public, and the creation of training programs for education and information about financing opportunities (e.g., "Leader" program of the EU).

The second type is support from the local community, including a collective effort made to be able to offer a complete wine tourism experience.

The third type is individual support from every winery in the country to create a comprehensive strategic plan to attract new consumers, to offer an enriched wine tourism experience, training, or hiring qualified staff, as well as innovation and the maintenance of a digital identity to be able to cope with the competition and evolve in the new trends that are created.

This study involved more women as well as an increased number of young students. Although it can be considered a limitation of the research, young Greeks will shape the future trends of domestic wine tourism and the results of this research can be used for this prospect. Another limitation is the participation of only Greek residents and not foreigners, who also shape the sustainability of wine tourism in Greece. A study based on an audience over 30 years old can highlight the desires needing to be satisfied by Greek wine tourism in the near future.

More studies, such as the impression that Greeks have of wine tourism today, their opinion on its future prospects, as well as a comparison with similar surveys in other EU states, will allow the formation of a more comprehensive view of how the pandemic has affected wine tourism, both in Greece and globally, and what challenges it has to face.

Supplementary Materials: The following supporting information can be downloaded at: https://www.mdpi.com/article/10.3390/su152316225/s1, Table S1: Questionnaire about wine tourism in Greece and its prospects of sustainable development in the post-pandemic period, Table S2: Associations between the consumption habits of wine and demographic variables, Table S3: Associations between participants' previous wine experience and demographic variables, Table S4: Associations between respondents' answers on parameters regarding wine and wineries and their demographic variables, Table S5: Associations between the parameters regarding the general characteristics of the wine region to encourage participation in tourism activities in Greece and demographic variables, and Table S6: Associations between preference on source of information regarding wine tourism activities in Greece and demographic characteristics.

Author Contributions: Conceptualization and methodology, D.S. and A.S. writing—original draft preparation, A.S. and I.S.K. supervision and editing, D.S. All authors have read and agreed to the published version of the manuscript.

Funding: This research received no external funding.

Institutional Review Board Statement: Not applicable.

Informed Consent Statement: Not applicable.

Data Availability Statement: The data presented in this study are available on request from the corresponding author.

Conflicts of Interest: The authors declare no conflict of interest.

References

1. Rokni, L. The Psychological Consequences of COVID-19 Pandemic in Tourism Sector: A Systematic Review. *Iran. J. Public. Health* **2021**, *50*, 1743–1756. [CrossRef]
2. Bhatt, K.; Seabra, C.; Kabia, S.K.; Ashutosh, K.; Gangotia, A. COVID Crisis and Tourism Sustainability: An Insightful Bibliometric Analysis. *Sustainability* **2022**, *14*, 12151. [CrossRef]
3. Gamage, D.; Samarathunga, M. Alternative Tourism as an Alternate to Mass Tourism during the Post-COVID-19 Recovery Phase. *Sunday Times* **2020**, *2*. Available online: https://www.dailymirror.lk/opinion/Alternative-Tourism-as-Alternate-to-Mass-Tourism/172-188860 (accessed on 20 October 2023).
4. Andrade-Suárez, M.; Caamaño-Franco, I. The Relationship between Industrial Heritage, Wine Tourism, and Sustainability: A Case of Local Community Perspective. *Sustainability* **2020**, *12*, 7453. [CrossRef]
5. López-Guzmán, T.; Vieira-Rodríguez, A.; Rodríguez-García, J. Profile and Motivations of European Tourists on the Sherry Wine Route of Spain. *Tour. Manag. Perspect.* **2014**, *11*, 63–68. [CrossRef]
6. McGovern, P.; Jalabadze, M.; Batiuk, S.; Callahan, M.P.; Smith, K.E.; Hall, G.R.; Kvavadze, E.; Maghradze, D.; Rusishvili, N.; Bouby, L.; et al. Early Neolithic Wine of Georgia in the South Caucasus. *Proc. Natl. Acad. Sci. USA* **2017**, *114*, E10309–E10318. [CrossRef]
7. Ashenfelter, O.; Storchmann, K. The Economics of Wine, Weather, and Climate Change. *Rev. Environ. Econ. Policy* **2016**, *10*, 25–46. [CrossRef]
8. Strandberg, T.E.; Strandberg, A.Y.; Salomaa, V.V.; Pitkala, K.; Tilvis, R.S.; Miettinen, T.A. Alcoholic Beverage Preference, 29-Year Mortality, and Quality of Life in Men in Old Age. *J. Gerontol. A Biol. Sci. Med. Sci.* **2007**, *62*, 213–218. [CrossRef]
9. Williams, P. The Evolving Images of Wine Tourism Destinations. *Tour. Recreat. Res.* **2001**, *26*, 3–10. [CrossRef]
10. Charters, S.; Pettigrew, S. Is Wine Consumption an Aesthetic Experience? *J. Wine Res.* **2005**, *16*, 121–136. [CrossRef]
11. Almeida, D.; Massuça, J.; Fialho, A.; Dionisio, A. Sustainable Wine Tourism as a Diversification Strategy: A Different Approach in a Rural Cooperative. *CASE J.* **2023**, *19*, 204–231. [CrossRef]
12. Korinth, B.; Wendt, J.A. The Impact of COVID-19 Pandemic on Foreign Tourism in European Countries. *Stud. Ind. Geogr. Comm. Pol. Geogr. Soc.* **2021**, *35*, 186–204. [CrossRef]
13. Festa, G.; Cuomo, M.T.; Genovino, C.; Alam, G.M.; Rossi, M. Digitalization as a Driver of Transformation towards Sustainable Performance in Wine Tourism—The Italian Case. *Br. Food J.* **2023**, *125*, 3456–3467. [CrossRef]
14. Karagiannis, D.; Metaxas, T. Sustainable Wine Tourism Development: Case Studies from the Greek Region of Peloponnese. *Sustainability* **2020**, *12*, 5223. [CrossRef]
15. Gaetjens, A.; Corsi, A.M.; Plewa, C. Customer Engagement in Domestic Wine Tourism: The Role of Motivations. *J. Destin. Mark. Manag.* **2023**, *27*, 100761. [CrossRef]
16. Bem-Haja, P.; Santos, I.M.; Cunha, D.; Kastenholz, E. Wine Tourism Market Research: Bringing the Psychophysiology Lab to the Field. In Proceedings of the 21st European Conference on Research Methodology for Business and Management Studies, University of, Aveiro, Aveiro, Portugal, 2–3 June 2022; Academic Conferences International Limited: Reading, UK, 2022.
17. Thiwachaleampong, R.; Maneekobkulwong, S.; Yimcharoen, P. An Empirical Study of Factors Influencing Behavioral Intention to Purchase Wine in Generation Y. In Proceedings of the 2022 7th International Conference on Business and Industrial Research (ICBIR), Bangkok, Thailand, 19 May 2022; pp. 573–576.
18. Sigala, M. Thriving in Wine Tourism through Technology and Innovation: A Survival or a Competitiveness Need? In *Technology Advances and Innovation in Wine Tourism*; Springer Nature Singapore: Singapore, 2023; pp. 3–11.
19. Gastaldello, G.; Giampietri, E.; Zaghini, E.; Rossetto, L. Virtual Wine Experiences: Is COVID Extending the Boundaries of Wine Tourism? *Wine Econ. Policy* **2022**, *11*, 5–18. [CrossRef]

20. Gómez-Carmona, D.; Paramio, A.; Cruces-Montes, S.; Marín-Dueñas, P.P.; Aguirre Montero, A.; Romero-Moreno, A. The Effect of the Wine Tourism Experience. *J. Destin. Mark. Manag.* **2023**, *29*, 100793. [CrossRef]
21. Alebaki, M.; Psimouli, M.; Kladou, S.; Anastasiadis, F. Digital Winescape and Online Wine Tourism: Comparative Insights from Crete and Santorini. *Sustainability* **2022**, *14*, 8396. [CrossRef]
22. Zamarreño Aramendia, G.; Cruz Ruiz, E.; Hernando Nieto, C. La Digitalización de La Experiencia Enoturística: Una Revisión de La Literatura y Aplicaciones Prácticas. *Doxa Comun. Rev. Interdiscip. Estud. Comun. Cienc. Soc.* **2021**, *33*, 257–283. [CrossRef]
23. Esau, D.; Senese, D.M. The Sensory Experience of Wine Tourism: Creating Memorable Associations with a Wine Destination. *Food Qual. Prefer.* **2022**, *101*, 104635. [CrossRef]
24. Andaç, F. UN World Tourism Organization's Contributions to World Tourism. July 2014, pp. 993–996. Available online: https://www.unwto.org/archive/global/annualreport2014 (accessed on 15 October 2023).
25. Barros, C.P.; Botti, L.; Peypoch, N.; Solonandrasana, B. Editorial. *Tour. Hosp. Res.* **2009**, *9*, 1–2. [CrossRef]
26. Smed, K.M.; Bislev, A. New Tourists at Old Destinations: Chinese Tourists in Europe. In *Emerging Powers, Emerging Markets, Emerging Societies*; Palgrave Macmillan UK: London, UK, 2016; pp. 235–255.
27. Güzel, Ö.; Ehtiyar, R.; Ryan, C. The Success Factors of Wine Tourism Entrepreneurship for Rural Area: A Thematic Biographical Narrative Analysis in Turkey. *J. Rural. Stud.* **2021**, *84*, 230–239. [CrossRef]
28. Tomay, K.; Tuboly, E. The Role of Social Capital and Trust in the Success of Local Wine Tourism and Rural Development. *Sociol. Ruralis* **2023**, *63*, 200–222. [CrossRef]
29. Dimitrakaki, I. "Managing Small Wine Firms–Wine Tourism-a Lever for the Development of a Wider Area" The Case of Domaine Hatzimichalis. *Int. J. Soc. Sci. Hum. Res.* **2022**, *5*, 1461–1470. [CrossRef]
30. Cunha, D.; Kastenholz, E.; Silva, C. Analyzing Diversity amongst Visitors of Portuguese Wine Routes Based on Their Wine Involvement. *Int. J. Wine Bus. Res.* **2023**, *35*, 121–141. [CrossRef]
31. Stergiou, D.P.; Airey, D.; Apostolakis, A. The Winery Experience from the Perspective of Generation Z. *Int. J. Wine Bus. Res.* **2018**, *30*, 169–184. [CrossRef]
32. Fytopoulou, E.; Karasmanaki, E.; Galatsidas, S.; Andrea, V.; Tsantopoulos, G. Enhancing Wine Tourism Experience through Developing Wine Tourist Typology and Providing Complementary Activities. *Int. J. Agric. Resour. Gov. Ecol.* **2022**, *18*, 145. [CrossRef]
33. Stergiou, D.P. An Importance-Performance Analysis of Young People's Response to a Wine Tourism Situation in Greece. *J. Wine Res.* **2018**, *29*, 229–242. [CrossRef]
34. Poulaki, I.; Nikas, I.A. Measuring Tourist Behavioral Intentions after the First Outbreak of COVID-19 Pandemic Crisis. Prima Facie Evidence from the Greek Market. *Int. J. Tour. Cities* **2021**, *7*, 845–860. [CrossRef]
35. Nella, A.; Christou, E. Market Segmentation for Wine Tourism: Identifying Sub-Groups of Winery Visitors. *Eur. J. Tour. Res.* **2021**, *29*, 2903. [CrossRef]
36. Galloway, G.; Mitchell, R.; Getz, D.; Crouch, G.; Ong, B. Sensation Seeking and the Prediction of Attitudes and Behaviours of Wine Tourists. *Tour. Manag.* **2008**, *29*, 950–966. [CrossRef]
37. Skalkos, D.; Kosma, I.S.; Vasiliou, A.; Guine, R.P.F. Consumers' Trust in Greek Traditional Foods in the Post COVID-19 Era. *Sustainability* **2021**, *13*, 9975. [CrossRef]
38. Skalkos, D.; Kosma, I.S.; Chasioti, E.; Bintsis, T.; Karantonis, H.C. Consumers' Perception on Traceability of Greek Traditional Foods in the Post-COVID-19 Era. *Sustainability* **2021**, *13*, 12687. [CrossRef]
39. Skalkos, D.; Kosma, I.S.; Chasioti, E.; Skendi, A.; Papageorgiou, M.; Guiné, R.P.F. Consumers' Attitude and Perception toward Traditional Foods of Northwest Greece during the COVID-19 Pandemic. *Appl. Sci.* **2021**, *11*, 4080. [CrossRef]
40. Skalkos, D.; Roumeliotis, N.; Kosma, I.S.; Yiakoumettis, C.; Karantonis, H.C. The Impact of COVID-19 on Consumers' Motives in Purchasing and Consuming Quality Greek Wine. *Sustainability* **2022**, *14*, 7769. [CrossRef]
41. Bodington, J.C. Wine, Women, Men, and Type II Error. *J. Wine Econ.* **2017**, *12*, 161–172. [CrossRef]
42. Di Vita, G.; Caracciolo, F.; Brun, F.; D'Amico, M. Picking out a Wine: Consumer Motivation behind Different Quality Wines Choice. *Wine Econ. Policy* **2019**, *8*, 16–27. [CrossRef]
43. Pytell, J.D.; Thakrar, A.P.; Chander, G.; Colantuoni, E. Changes in Alcohol Consumption by Beverage Type Attributable to the COVID-19 Pandemic for 10 States, March 2020 to November 2020: An Ecological Simulation-Based Analysis. *J. Addict. Med.* **2022**, *16*, e412–e416. [CrossRef]
44. Leri, I.; Theodoridis, P. The Effects of the Winery Visitor Experience on Emotions, Satisfaction and on Post-Visit Behaviour Intentions. *Tour. Rev.* **2019**, *74*, 480–502. [CrossRef]
45. Trišić, I.; Štetić, S.; Privitera, D.; Nedelcu, A. Wine Routes in Vojvodina Province, Northern Serbia: A Tool for Sustainable Tourism Development. *Sustainability* **2019**, *12*, 82. [CrossRef]
46. Bruwer, J.; Rueger-Muck, E. Wine Tourism and Hedonic Experience: A Motivation-Based Experiential View. *Tour. Hosp. Res.* **2019**, *19*, 488–502. [CrossRef]

47. Yuan, J.J.; Cai, L.A.; Morrison, A.M.; Linton, S. An Analysis of Wine Festival Attendees' Motivations: A Synergy of Wine, Travel and Special Events? *J. Vacat. Mark.* **2005**, *11*, 41–58. [CrossRef]
48. Ouvrard, S.; Jasimuddin, S.M.; Spiga, A. Does Sustainability Push to Reshape Business Models? Evidence from the European Wine Industry. *Sustainability* **2020**, *12*, 2561. [CrossRef]

sustainability

Article

Consumers' Attitudes towards Differentiated Agricultural Products: The Case of Reduced-Salt Green Table Olives

Aikaterini Paltaki [1,*], Fani Th Mantzouridou [2], Efstratios Loizou [3], Fotios Chatzitheodoridis [3], Panagiota Alvanoudi [2], Stelios Choutas [4] and Anastasios Michailidis [1]

[1] Department of Agricultural Economics, School of Agriculture, Aristotle University of Thessaloniki, 52124 Thessaloniki, Greece; tassosm@auth.gr
[2] Laboratory of Food Chemistry and Technology, School of Chemistry, Aristotle University of Thessaloniki, 52124 Thessaloniki, Greece; fmantz@chem.auth.gr (F.T.M.); alvanoudi@chem.auth.gr (P.A.)
[3] Department of Regional Development and Cross Border Studies, University of Western Macedonia, Koila Campus, 50100 Kozani, Greece; eloizou@uowm.gr (E.L.); fxtheodoridis@uowm.gr (F.C.)
[4] CHOUTAS-MARDAS S.A., 63017 Chalkidiki, Greece; stelios1065@gmail.com
* Correspondence: apaltaki@agro.auth.gr

Abstract: Contemporary healthy food issues and food safety concerns induce consumers to become more interested in a healthier diet such as foods reduced in salt. This study explores consumers' behaviour, attitude, and expectations for the development of a new reduced-salt table olive product from Chalkidiki, an area of Greek. In this context, the main aim of this paper is to investigate the knowledge and attitudes of consumers about health and nutrition, reduced salt consumption, and consumption of Chalkidiki reduced-salt green table olives. Summary statistics and multivariate analysis were performed to examine consumers' perceptions. The results of the research highlight a remarkable consumer interest in products with reduced salt content. Furthermore, the majority are willing to purchase such foods which is a possible action that can be taken to reduce salt intake. These outcomes emphasise that producing a new reduced-salt table olive product is promising, as the interest of consumers, industries, and the research community has turned to innovative actions that add nutritional value and meet the consumers' expectations.

Keywords: consumers; Greece; multivariate statistical analysis; salt; table olive

Citation: Paltaki, A.; Mantzouridou, F.T.; Loizou, E.; Chatzitheodoridis, F.; Alvanoudi, P.; Choutas, S.; Michailidis, A. Consumers' Attitudes towards Differentiated Agricultural Products: The Case of Reduced-Salt Green Table Olives. *Sustainability* **2024**, *16*, 2392. https://doi.org/10.3390/su16062392

Academic Editor: Dimitris Skalkos

Received: 25 January 2024
Revised: 1 March 2024
Accepted: 11 March 2024
Published: 13 March 2024

1. Introduction

Table olive production has made a decisive contribution to the agricultural and processed sector of several countries around the Mediterranean basin. Table olives are important parts of the Mediterranean diet [1]. Olive fruit contains two bitter phenolic compounds, oleuropein and ligstroside, in high levels, which render the fruit inedible, and some form of processing (like fermentation) is necessary to reduce the concentration of these compounds [2].

Olive processing is one of the main agro-industrial sectors in Mediterranean countries such as Spain, Italy, and Greece [3]. Greece has a long tradition in table olive production and processing, while, recently, the table olive industry has become a significant sector of the Greek economy [4]. Greece is the second-largest producing and exporting country in the European Union [5]. One of the most important varieties cultivated in Greece, from an economic point of view, is cv. Chalkidiki for green olives [6,7].

There are three main ways of processing table olives, namely (a) the Spanish style, consisting of green olives in brine, (b) the California style, consisting of black olives in brine, and (c) the Greek style, consisting of natural black olives in brine [8]. The most widespread is the Spanish-style, in which olives are treated with a NaOH solution (salt) [9]. This process is used almost exclusively for the processing of green olives and has been studied extensively on a laboratory scale. However, large-scale studies are rather limited for each type of table olive processing [1].

Salt addition is necessary for table olives' fermentation both for safety and flavour [10–12]. It is a key ingredient to delay the growth of new bacteria, control the rate of fermentation, and as a flavouring agent in the processing of table olives [13]. Table olives are fruits with high nutritional value and are proposed as functional foods [14,15]. More specifically, the concentration of a variety of lipophilic and non-lipophilic compounds affects their quality and defines them as a functional food [16].

Salt consumption is among the higher concerns of the population. This is not a new trend, as health agencies have been constantly warning people about excess salt, sugar, and carbohydrate consumption in their diet [17]. According to the WHO [18], the daily recommended salt intake should be reduced to <5 g/day of salt for adults. Salt intake at this level helps reduce blood pressure and the risk of cardiovascular disease, stroke, and coronary heart disease. The main benefit of reducing salt intake is a corresponding reduction in high blood pressure [19].

High salt concentration is a matter that needs to be addressed. A new functional product, a reduced-salt green table olive, is very promising. Salt reduction is indeed a crucial public health concern; however, research on salt intake awareness and behaviour might not be as extensive as one might expect [20]. The problem of salt overconsumption greatly affects the entire human society. Various policies have been formulated to promote informational campaigns aimed at changing inappropriate eating habits and highlighting the need for healthy food choices and diets [21]. A strategy to achieve population-level salt reduction is the reformulation of food in markets. Predicting consumers' reactions to reformulated foods is critical to this issue [22].

The development of healthier products, with reduced-salt content is an innovation that would improve consumers' opinions about such products and could contribute to the reduction in daily salt intake [23]. One of the most studied variables in marketing research is consumers' knowledge levels [24]. Consumer knowledge is an important predictor due to its strong influence on consumer behaviour when evaluating a product [25]. Other important variables regarding consumers' segmentation are gender and age [21]. Food labels are linked with positive dietary outcomes [26] and consumers' willingness to buy reformulated foods (e.g., reduced-salt content) at a premium price [27]. Salt reduction programs as well as awareness campaigns should take into account the level of knowledge, attitudes, and behaviour of consumers. Measures taken before the implementation of a program provide an opportunity to assess the impact and effectiveness of an intervention [28].

The goal of a healthier diet is to develop strategies that can help and support consumers' choices. The benefits of a healthier diet will occur only if consumers are willing to make long-term eating changes and if markets increase the offered variety of reduced-salt foods [19]. Some strategies to reduce salt intake in the general population may be public education, nutritional counselling, food labelling, coordinated voluntary actions, and, finally, regulation of salt levels in foods [29]. At the same time, all actions and campaigns should encourage and educate consumers to think about their salt intake [30]. Consumers' awareness of salt intake risks would affect their decision to buy reduced-salt products [31]. Thus, the issue of salt intake with food is of great concern among the scientific community and policy makers engaged in public health strategies. Many measures are proposed in this direction by policy makers, both at the European Union and world level. Since the promotion of reduced salt in food attracts the interest of so many people and experts, it is very important to have information about the consumers' perceptions regarding reduced salt foods and demand about products such as olive oils. Within this context lies the aim and the scientific contribution of the current study; it aims to support, with valuable info, both policy makers and market stakeholders dealing with the production and supply of table olives.

Specifically, this paper explores consumer behaviour, attitude, and expectations for developing a new reduced-salt table olive product from Chalkidiki olive varieties. The main aim of this paper is to investigate the knowledge and attitudes of consumers about health and nutrition, reduced salt consumption, and consumption of Chalkidiki reduced-salt green olives.

The remainder of this paper is organized as follows. The material and methods section describes the research on data collection, the formulation of the questionnaire, validity, and reliability tests, the methodology followed for data analysis as well as research hypotheses. The next section presents the results of multivariate statistical analysis and hypothesis testing. Finally, we present the conclusions of the research.

2. Materials and Methods

The questionnaire that was developed was formulated on a five-point Likert scale and was structured in five sections:

PART I. Demographic—personal characteristics.

PART II. Respondent's personal information.

PART III. Knowledge and attitudes about health and nutrition.

PART IV. Consumers' attitudes about reducing salt consumption.

PART V. Consumers' attitudes regarding the consumption of Chalkidiki reduced-salt green table olives.

Pilot research was first conducted with the above-formulated questionnaire on a small sample of 50 consumers. The purpose of this research was to finalize the questionnaire, modify, add, or delete questions. To calculate the minimum required sample size [32], the following Functions (1), (2) and (3) were used:

$$n = \frac{Nx}{(N-1)E^2 + x} \tag{1}$$

$$E = \sqrt{\frac{(N-n)x}{n(n-1)}} \tag{2}$$

$$x = Z\left(\frac{c}{100}\right)^2 r(100 - r) \tag{3}$$

where n = sample size, N = population size of the two largest cities in Greece, Athens and Thessaloniki, r = fraction of responses, c = confidence level, and $Z(c/100)$ is the critical value for the c.

The maximum acceptable margin of error was defined as $E = 5\%$, the critical value of the normal distribution at the required confidence level was $c = 95\%$ and the response distribution was considered equal to $r = 70\%$. The minimum sample size that emerged is 323 consumers. Data collection started from 25 September to 30 November 2022. The sample was from the two largest cities in Greece, Athens and Thessaloniki, where most consumers gather, being a representative part of the Greek population. The sample was randomly selected. An effort was made for the good geographical distribution of the questionnaires. The questionnaires were completed in person by 331 consumers.

Validity and reliability tests were performed. Validity represents the extent to which research findings accurately reflect reality [33]. It is considered a measure of the quality of the measurement process, indicates the value of the research, and is accepted and respected by researchers and users of research [34]. According to Punch [35], validity is about the accuracy and representativeness of the data. Equally important in analysis is reliability, especially in the case of multivariate statistical analysis. Reliability seeks to minimize errors and biases in the results of research instruments [33,34].

First, a validity test was conducted. Specifically, 5 experts in questionnaire research checked the structure of the questionnaire before it was distributed to consumers. The procedure was carried out using a five-point Likert scale. The result of this process was that, after 3 rounds, the group of experts agreed on all the statements and questions, with an average score equal to or greater than 4.

Then, to ensure the reliability of this research, Cronbach's alpha was used to determine the consistency, accuracy, stability, and objectivity of the research instruments. In total, 67 variables were included and analysed to determine the degree to which these variables were related to each other and to identify cases that had to be excluded. The coefficient value of Cronbach's alpha was found to be equal to 0.93, showing a reliable scale [36–38]. It is worth emphasizing that none of the 331 cases were rejected from the analysis.

From a methodological point of view, this research includes both descriptive statistics (frequencies, percentages, and mean values) and multivariate analyses (using the statistical program SPSS-Version 28). In particular, Two-Step Cluster Analysis (TSCA) was used to classify the population of consumers with common characteristics. Bivariate Cluster Analysis is a scalable cluster analysis algorithm designed to handle large data sets and categorical variables as well as attributes [37]. Categorical Regression (CATREG) was performed to highlight the possible relationships between the purchase of Chalkidiki reduced-salt green table olives, and a set of other variables. In particular, the CATREG model is a modern regression technique, more holistic and efficient than the most commonly used models [36], and figures relationships between the dependent variable and a set of independent variables [37]. Finally, a normality test was initially performed for hypothesis testing. In all the relevant checks, the distribution of the observations satisfactorily approached the normal standard distribution, and we did not need to proceed with "normalization". In addition, a statistical test of residuals was carried out and, in all tests, no statistical residuals were found; therefore, they did not need to be excluded from further testing of the research hypotheses.

Hypotheses emerged through the literature review. Hypothesis testing was performed using a Chi-square statistical test and a descriptive test (explore means). Through hypothesis testing, we understand the interaction of the variables under consideration. The research hypotheses are as follows:

- H1 Consumers who are more concerned about their health are more willing to buy reduced-salt foods [31].
- H2 Consumers who know more about a product's benefits, tend to have a greater purchase intention towards it [24,25].
- H3a Young people are more willing to buy products with reduced-salt content [21].
- H3b Male consumers are more willing to buy products with reduced-salt content [21].
- H4 The existence of a food label is associated with positive dietary outcomes such as reduced-salt products [26].
- H5 Consumers are willing to buy a product at a premium price when it has a label on the package, compared to another similar product without a label [27].

3. Results

3.1. Consumers' Demographic and Personal Characteristics

Table 1 presents consumers' demographic and personal characteristics, regarding gender, age, marital status, family members and minors, profession, educational level, annual household income, health status, following a special diet for health reasons, and whether it is advisory to reduce salt intake.

Table 1. Consumers' demographic and personal characteristics (*n* = 331 consumers).

Demographic—Personal Characteristics	Value
Gender	
Male	48.34%
Female	51.66%
Age	
18–30 years old	39.90%
31–43 years old	19.60%
44–56 years old	29.60%
>57 years old	10.90%
Marital status	
Married, in a cohabitation agreement, or in a long-term relationship	48.40%
Single	46.20%
Separated/divorced	2.40%
Widowed	2.40%
Do not wish to answer	0.60%
Family members	
One member	17.80%
Two members	22.40%
Three members	16.30%
Four members	33.20%
More than four members	10.30%
Minor members	
No minors	68.00%
$\leq$2 minors	29.30%
>2 minors	2.70%
Profession	
Private employees	30.80%
Public/municipal employees	18.70%
Single professionals, self-employed, business owners	15.10%
Students	22.70%
Retired	6.90%
Domestic	3.30%
Unemployed	2.50%
Educational level	
Some years of basic education	0.90%
Basic education of Primary and High School	3.00%
High school	17.50%
Higher education (University)	48.30%
Master's degree or a PhD	30.30%
Annual household income	
EUR 0–18,000	48.60%
EUR 18,001–30,000	26.90%
EUR > 30,001	24.50%
Health status	
Very burdened	0.00%
Burdened	6.40%
Neutral	16.00%
Good	46.20%
Very good	31.40%
Special diet for health reasons	
Yes	8.80%
No	91.20%
Advised to reduce salt intake	
Yes	20.20%
No	79.80%

3.2. Knowledge and Attitudes about Health and Nutrition

Table 2 presents the self-declaration attitudes of consumers towards health and nutrition; 61.60% of consumers believe they have good knowledge about healthy eating.

Furthermore, 45.00% prefer products that do not endanger health (products with a reduced content of salt, sugar, and preservatives), and 46.50% buy products that help prevent diseases (rich in fibre, vitamins, antioxidants, minerals such as calcium, omega-3 fatty acids, etc.). In addition, consumers take into account the nutritional value of the products they buy (52.00%) and put the quality of the products as the first selection criterion (47.10%). Finally, some consumers prefer to buy organic products (20.80%), and others believe that organic food is better for health than conventionally grown food (52.00%).

Table 2. Consumers' self-declaration attitudes towards health and nutrition (*n* = 331 consumers).

	Strongly Disagree [1]	Disagree [2]	Neither Agree nor Disagree [3]	Agree [4]	Strongly Agree [5]	Mean Value and St. Dev.
I have good knowledge about healthy eating	8 (2.40%)	4 (1.20%)	50 (15.10%)	204 (61.60%)	65 (19.60%)	3.95 0.78
I put the quality of the products as the first selection criterion	7 (2.10%)	24 (7.30%)	74 (22.40%)	156 (47.10%)	70 (21.10%)	3.78 0.93
I consider the nutritional value of the products	8 (2.40%)	39 (11.80%)	70 (21.10%)	172 (52.00%)	42 (12.70%)	3.61 0.94
I buy products that help prevent disease (rich in fibre, vitamins, antioxidants, minerals such as calcium, omega-3, fatty acids, etc.)	10 (3.00%)	39 (11.80%)	90 (27.20%)	154 (46.50%)	38 (11.50%)	3.52 0.95
I prefer products that do not endanger my health (products with a reduced content of salt, sugar, and preservatives)	11 (3.30%)	58 (17.50%)	69 (20.80%)	149 (45.00%)	44 (13.30%)	3.47 1.03
I buy products that help control my health problems (high cholesterol, high blood pressure, osteoporosis, etc.)	20 (6.00%)	67 (20.20%)	129 (39.00%)	81 (24.50%)	34 (10.30%)	3.13 1.04
I put the price of the products as the first selection criterion	24 (7.30%)	76 (23.00%)	106 (32.00%)	92 (27.80%)	33 (10.00%)	3.10 1.09
I always follow a healthy and balanced diet	17 (5.10%)	68 (20.50%)	132 (39.90%)	101 (30.50%)	13 (3.90%)	3.08 0.93
I am concerned about the amount of salt in my diet	22 (6.60%)	73 (22.10%)	118 (35.60%)	93 (28.10%)	25 (7.60%)	3.08 1.03
I avoid foods that contain additives	16 (4.80%)	83 (25.10%)	120 (36.30%)	94 (28.40%)	18 (5.40%)	3.05 0.97
I prefer to buy organic products	40 (12.10%)	108 (32.60%)	114 (34.40%)	62 (18.70%)	7 (2.10%)	2.66 0.99
Organic food is not better for my health than conventionally grown food	46 (13.90%)	126 (38.10%)	112 (33.80%)	41 (12.40%)	6 (1.80%)	2.50 0.94

3.3. Consumers' Attitudes about Reducing Salt Consumption

The level of consumers' knowledge about the risks of increased salt consumption is presented in Table 3. We observe that, in general, consumers are highly aware of the potential risks. There is an agreement that high salt consumption can cause health problems (M = 4.38), fluid retention (M = 4.39), and lead to the development of hypertension (M = 4.34).

Table 3. Consumers' knowledge about the health risks of increased salt consumption (n = 331 consumers).

	Strongly Disagree [1]	Disagree [2]	Neither Agree nor Disagree [3]	Agree [4]	Strongly Agree [5]	Mean Value and St. Dev.
High salt consumption can cause fluid retention	2 (0.60%)	6 (1.80%)	7 (2.10%)	163 (49.20%)	153 (46.20%)	4.39 0.68
High salt consumption can cause health problems	4 (1.20%)	6 (1.80%)	6 (1.80%)	159 (48.00%)	156 (47.10%)	4.38 0.73
Increased salt intake can lead to the development of hypertension	6 (1.80%)	2 (0.60%)	15 (4.50%)	160 (48.30%)	148 (44.70%)	4.34 0.76
My health might improve if I reduced my salt intake	4 (1.20%)	14 (4.20%)	78 (23.60%)	150 (45.30%)	85 (25.70%)	3.90 0.87
High salt consumption can cause a heart attack	3 (0.90%)	10 (3.00%)	96 (29.00%)	165 (49.80%)	57 (17.20%)	3.79 0.79
High salt consumption can cause kidney disease	7 (2.10%)	13 (3.90%)	113 (34.10%)	132 (39.90%)	66 (19.90%)	3.72 0.90
High salt consumption can cause a stroke	4 (1.20%)	15 (4.50%)	117 (35.30%)	144 (43.50%)	51 (15.40%)	3.67 0.83
High salt consumption can cause osteoporosis	6 (1.80%)	29 (8.80%)	144 (43.50%)	116 (35.00%)	36 (10.90%)	3.44 0.87
High salt consumption can cause stomach cancer	12 (3.60%)	41 (12.40%)	172 (52.00%)	76 (23.00%)	30 (9.10%)	3.21 0.90

Consumers were then asked about the likelihood of implementing actions to reduce their salt intake (Table 4). The two actions with a higher probability of application are reducing the amount of salt added to food (M = 3.86) and avoiding the frequent consumption of highly salted foods (e.g., pasta, pickles, ready-to-eat snacks) (M = 3.86). However, stopping salt consumption appears to be the least likely (M = 2.74). Additionally, the largest percentage of consumers, 63.10%, have already bought a reduced-salt product.

Table 4. Implementation of actions to reduce the intake of salt (n = 331 consumers).

	Strongly Disagree [1]	Disagree [2]	Neither Agree nor Disagree [3]	Agree [4]	Strongly Agree [5]	Mean Value and St. Dev.
Avoiding frequent consumption of highly salted foods (e.g., pastes, pickles, ready-to-eat snacks)	9 (2.70%)	25 (7.60%)	56 (16.90%)	153 (46.20%)	88 (26.60%)	3.86 0.98
Reducing the amount of salt added to food	8 (2.40%)	31 (9.40%)	42 (12.70%)	169 (51.10%)	81 (24.50%)	3.86 0.97
Reducing the consumption of processed foods	8 (2.40%)	32 (9.70%)	69 (20.80%)	179 (54.10%)	43 (13.00%)	3.66 0.91
Buy reduced-salt alternatives	12 (3.60%)	40 (12.10%)	64 (19.30%)	167 (50.50%)	48 (14.50%)	3.60 1.00
Check product labels for salt content	13 (3.90%)	45 (13.60%)	71 (21.50%)	142 (42.90%)	60 (18.10%)	3.58 1.06
Stop eating salt	78 (23.60%)	73 (22.10%)	68 (20.50%)	81 (24.50%)	31 (9.40%)	2.74 1.31

3.4. Consumers' Attitudes Regarding the Consumption of Chalkidiki Reduced-Salt Green Table Olives

Figure 1 shows the possibility of buying Chalkidiki reduced-salt green table olives. The majority, 50.80%, said "likely" (168 consumers), 22.10% "neither unlikely nor likely" (73 consumers), 18.40% "extremely likely" (61 consumers), 7.30% "extremely unlikely" (24 consumers), and 1.50% "unlikely" (5 consumers). Afterward, consumers were asked to rate their willingness to pay a premium price for a Chalkidiki reduced-salt green table olive compared to a similar product with normal salt content. As we can see in Figure 1, the majority of consumers (174) showed a high probability with a rate of 52.60%.

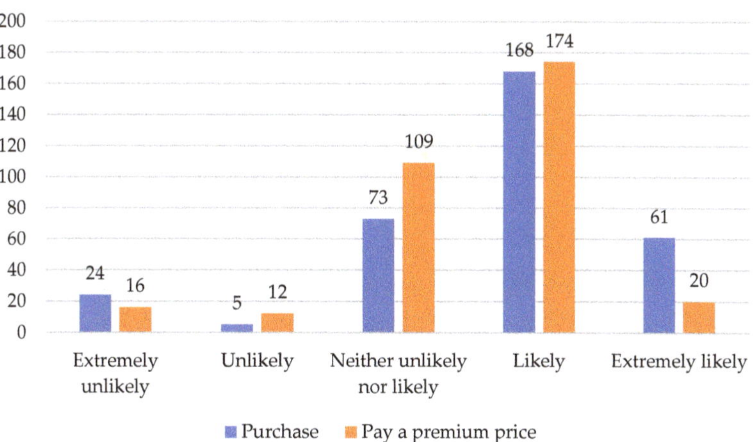

Figure 1. Chalkidiki reduced-salt green table olive purchase; Probability of paying a premium price to buy Chalkidiki reduced-salt green table olives compared to a similar product with normal salt content (*y*-axis: number of responders) (*n* = 331 consumers).

Regarding the premium price, 39.30% of the consumers were willing to pay EUR 0.50–1.00 more for the purchase of Chalkidiki reduced-salt green table olives (200 g packaging), and 30.80% would pay EUR 0–0.50 more. It is worth noting that 23.60% of consumers would pay above EUR > 1.00 for this new product. The remaining 6.30% were not interested in buying a premium price product.

3.5. Multivariate Statistical Analysis

TSCA was applied to segment the population into groups of consumers with common characteristics regarding the "Purchase of Chalkidiki reduced-salt green table olives". From the analysis, three clusters were created using eight variables. According to the silhouette cohesion and separation measure, the clustering process is considered satisfactory. Table 5 lists the average values of the variables of each cluster.

The first cluster consists of 124 consumers (37.50%). The majority are men (56.50%), 18–30 years old (39.50%), married (46.00%), with a family of two (25.80%), and no minors (75.80%). They are private employees (37.10%), with a university degree (48.40%), and low income, EUR 0–18,000 (47.60%). These are consumers whose health status is neutral; they have been advised to reduce their salt intake and may follow a healthy and balanced diet. They also argue that high salt consumption can cause problems for their health. They are willing to purchase alternative products with reduced-salt content but have not purchased such a product in the past. Finally, they are willing to buy Chalkidiki reduced-salt green table olives, while testing this specific product will positively influence their purchase decision.

Table 5. Results of Bivariate Analysis in clusters—Characteristics of each cluster (*n* = 331 consumers).

	Cluster 1 Mean Value	Cluster 2 Mean Value	Cluster 3 Mean Value	Total Sample Mean Value
Health status	3.47	4.38	4.36	4.03
Salt reduction recommendation *	1.46	2.00	2.00	1.80
I follow a healthy diet	2.69	2.62	3.47	3.08
High salt consumption causes health problems	4.36	3.77	4.54	4.38
Buy alternative products with reduced-salt content	3.48	2.21	4.02	3.60
Previous purchase of reduced-salt product *	1.66	1.49	1.20	1.41
Buy Chalkidiki reduced-salt green olives content	3.80	3.41	3.73	3.72
Change in purchase decision after product trial	4.10	2.36	4.22	3.95

All variables are on a 5-point Likert scale (1 = strongly disagree, 5 = strongly agree) with 2 exceptions *. * 1 = Yes, 2 = No.

The second cluster, which is the smallest, has 39 consumers (11.80%). The majority are women (51.30%), 44–56 years old (41.00%), married (48.70%), with a family of four (53.80%), and no minors (64.10%). They are public and private employees at the same percentage (28.20%), with a university degree (69.20%), and low income, EUR 0–18,000 (64.10%). These are consumers whose health status is good; they have not been recommended to reduce their salt intake and do not follow a healthy and balanced diet. They also argue that high salt consumption can cause problems for their health. They are not willing to purchase alternative products with reduced-salt content but may have purchased such a product. Finally, they are indifferent to buying Chalkidiki reduced-salt green table olives, while testing this particular product is unlikely to positively influence their purchase decision.

The third cluster is the most numerous, as it has 168 consumers (50.70%). The majority are women (57.70%), 18–30 years old (41.70%), married (50.00%) with a family of four (34.50%), and no minors (63.10%). They are private employees and students at the same percentage (26.80%), with a university degree (43.50%), and low income, EUR 0–18,000 (45.80%). These are consumers who are in good health, have not been recommended to reduce their salt intake and follow a healthy and balanced diet. They also argue that high salt consumption can cause problems for their health. They are willing to purchase alternative products with reduced-salt content and have already purchased such a product. Finally, they are willing to buy Chalkidiki reduced-salt green table olives, while testing this specific product will positively influence their purchase decision.

As far as consumers' willingness to buy Chalkidiki reduced-salt green table olives is concerned, we can assign the three clusters to three groups of consumers. Thus, the first cluster includes consumers who would buy the product under certain conditions, the second cluster includes consumers who would not buy the product, and, finally, the third cluster consumers who would buy the product.

To further analyse the variable from the TSCA "Purchase of Chalkidiki reduced-salt green table olives", it was used as the dependent variable for the CATREG model to identify those factors that influence consumer attitudes towards this product. The independent variables of the CATREG analysis are presented in Table 6.

Table 6. CATREG results ($n = 331$ consumers).

Independent Variables	Sig.	F	Standard Error	Importance
1. Recommendation to reduce salt intake	0.000	84.654	0.055	0.424
2. Health status	<0.001	28.761	0.058	0.215
3. To reduce salt intake I would buy reduced-salt alternatives	0.000	24.349	0.053	0.092
4. I always follow a healthy and balanced diet	<0.001	10.857	0.057	0.061
5. The quality of the products is the first selection criterion	<0.001	12.616	0.072	0.052
6. Marital status	<0.001	7.855	0.062	0.028
7. Annual household income	<0.001	7.994	0.036	0.028
8. Number of minor household members	<0.001	8.147	0.047	0.025
9. Profession/Employment	<0.001	10.104	0.051	0.024
10. High salt consumption can cause health problems	<0.001	8.191	0.055	0.020
11. The price of the products is the first selection criterion	<0.001	9.246	0.042	0.018
12. Purchase of products that help control health problems (high cholesterol, high blood pressure, osteoporosis, etc.)	<0.001	9.008	0.051	0.017
13. Level of education	<0.001	6.729	0.042	0.016
14. Gender	0.064	3.472	0.042	0.013
15. Number of household members	<0.001	13.641	0.041	0.006
16. The reduction in salt intake might improve health	0.082	2.093	0.057	0.006
17. Concern about the amount of salt in the diet	<0.001	8.022	0.051	0.004
18. Age	0.025	3.169	0.072	0.002
19. Preference for products that do not endanger health (products with a reduced content of salt, sugar, and preservatives)	<0.001	13.582	0.040	0.000

Dependent variable: Purchase of Chalkidiki reduced-salt green table olives.

The multiple correlation coefficient value was $R^2 = 0.77$, indicating that 77.0% of the variance in the transformed values of the dependent variable is explained by the transformed values of the independent variables participating in the regression equation.

The results of the categorical regression are presented in Table 6. The statistically significant variables are those with sig. < 0.05. The relative importance of the independent variables appears greater for the variable "Recommendation to reduce salt intake", followed in order by the variable "Health status". Cumulatively, these variables explain 63.90% of the total significance.

3.6. Hypothesis Testing

All the results are strictly related to the actual dataset, and a significantly bigger dataset may change it.

To test H1, a Chi-square statistical test was performed between the variables "interest in health" and "willingness to purchase reduced-salt foods". The value of Pearson's Chi-square index was found to be $\alpha < 0.05$ at a 5% level of statistical significance. Then, to check whether or not our H1 is accepted, we proceeded with a descriptive test (explore means). Specifically, the test showed that consumers who are more willing to buy food with reduced-salt content are the ones who answered "likely" and "extremely likely" to follow a healthy and balanced diet, with a mean value of 3.23 and 3.33, respectively. All the above observations lead to the acceptance of H1.

To test H2, a Chi-square statistical test was performed between the variables "reduced-salt table olive purchase" and "benefits of salt intake reduction". The value of Pearson's Chi-square index value was found to be $\alpha < 0.05$ at a 5% level of statistical significance. Then, to check whether or not our H2 is accepted, we proceeded with a descriptive test (explore means). Specifically, it was found that the level of knowledge significantly affects consumers in terms of buying table olives of reduced salinity. Consumers who are more likely to buy the product are those who answered that they have a higher knowledge of the benefits of reduced-salt consumption with a mean value of 4.50. Taking into account all the above observations, H2 is accepted.

H3 was split into H3a and H3b. To test H3a, a Chi-square statistical test was performed between the variables "age" and "willingness to buy reduced-salt products". The value of Pearson's Chi-square index was found $\alpha < 0.05$ at a 5% level of statistical significance. Then, to check whether or not our H3a is accepted, we proceeded with a descriptive test (explore means). Specifically, the test showed that age is among the most important variables in consumers' willingness to buy reduced-salt products, with a mean of 3.78. All the above observations lead to the acceptance of Hypothesis 3a. To test H3b, a Chi-square statistical test was performed between the variables of "gender" and "willingness to buy reduced-salt products". The value of Pearson's Chi-square index was found $\alpha < 0.05$ at a 5% level of statistical significance. Then, to check whether or not our H3b is accepted, we proceeded with a descriptive test (explore means). Specifically, the test showed that gender is among the most important variables in willingness to buy reduced-salt products, with a mean of 3.69. All the above observations lead to the acceptance of Hypothesis 3b.

To test H4, a Chi-square statistical test was performed between the variables "food label" and "purchase of reduced-salt products". The value of Pearson's Chi-square index was found $\alpha < 0.05$, so at a 5% level of statistical significance. Then, to check whether or not our H4 is accepted, we proceeded with a descriptive test (explore means). Specifically, the test indicated that consumers who would buy the reduced-salt products are those who answered that they "strongly agree" that having a food label has a positive effect, with a mean value of 3.81. All the above observations lead to the acceptance of Hypothesis 4.

To test H5, a Chi-square statistical test was performed between the variables "food label" and "buying products at a premium price". The value of Pearson's Chi-square index was found $\alpha < 0.05$ at a 5% level of statistical significance. Then, to check whether or not our H5 is accepted, we proceeded with a descriptive test (explore means). Specifically, the test indicated that consumers who are likely to buy products with a premium price are those who answered that having a food label on the package is "very important", with a mean value of 3.67. All these observations lead to the **acceptance of Hypothesis 5**.

4. Discussion

Research results revealed a remarkable consumer interest in reduced-salt products, which is consistent with another study that showed the high willingness of consumers to purchase reduced-salt food products even without having a salt-reduction goal [30]. Additionally, all participants recognize the potential health risks of increased salt consumption. Furthermore, the majority agree that purchasing reconstituted foods with reduced-salt content is a possible action that can be taken to reduce salt intake. Efforts should be focused on reformulating table olives to reduce their salt content [22,39]. Table olives with a reduced salt content are a matter of great interest for the establishment of table olives as a functional food, according to international nutritional guidelines [1]. However, consumers have the perception that reduced-salt foods would not be tasty and thus do not want to purchase them, especially if they have a front-pack label emphasizing this reduction [40,41]. A proposition could be to add health logos instead of salt reduction labels which indicate the healthiness of reduced-salt products and are less likely to affect consumers [29]. On the contrary, Goodman et al. [42] mentioned that front-package labels with content descriptors positively affect consumers in selecting reduced-salt products. Moreover, completely removing salt can lead to an increased risk of survival or growth of pathogenic microorganisms which can alter the taste or colour, causing significant economic losses [43]. An alternative solution to this could be the possible substitution of salt with other minerals with a mineral with a more favourable effect on human health which does not alter the taste of food [24]. Such minerals could be potassium chloride (KCl), calcium chloride (CaCl$_2$), and magnesium chloride (MgCl$_2$) [24]. However, the effect of NaCl replacement on the organoleptic characteristics of the product is still controversial [43].

According to the analysis of the data, the majority of the participating consumers are willing to buy reduced-salt table olives from cv. Chalkidiki olives, while the majority is even willing to pay a premium price for the product in question. Research by Sánchez-Rodríguez et al. [44] showed that 88.00% of participants are willing to pay more than the usual price by EUR 1.35 per 200 g. In addition, health reasons are among the factors that have the greatest positive impact on the market for reduced-salt green table olives. Consumers' interest in health is a very positively influential factor in willingness to pay [27]. Furthermore, the higher the concern of consumers about their health, the greater their willingness to pay for safer and healthier foods [45], and the more attention that will be paid to salt content information on foods [32].

Plenty successful actions regarding salt consumption were implemented through awareness campaigns, voluntary support, regulations, and intervention programs [46]. An effective tool for promoting behaviour change and reduction in salt consumption could be mass media campaigns [47]. Awareness campaigns towards reduced-salt products and their health benefits will influence consumers. This is also supported by Mork et al. [30], who mentioned that health authorities should educate consumers about salt intake through public campaigns. The importance of education is highlighted by Tsimitri et al. [48], who proposed media and social media as potential education tools. The lack of knowledge and understanding of the negative effects of salt excessive consumption, and the lack of knowledge of the salt content of food are among the barriers to salt reduction [49]. The food manufacturing industry could be the key to food reformulation. A proposed strategy referred to as "health by stealth" includes gradually lowering salt content by small percentages, without consumers noticing it. This action will take decades to make any impact on public health, but it can train consumer palates into salt-reduced foods. Efforts of the food industry should be widespread to adjust consumers' sensory acceptance. Food labelling and social marketing campaigns allow manufacturers to promote the healthy and nutritious characteristics of products and encourage consumers towards their consumption [50]. From an academic point of view, the results of this research could be a useful guide in designing training and awareness programs. The cooperation of science, technology, and health organisations is the best way to understand and create foods that can benefit people's health [50].

5. Conclusions

This paper's scientific contribution is to provide knowledge about consumers' behaviour towards new products with specific characteristics, as well as to propose possible actions to raise their awareness. The present research focuses on investigating the demand/market for a new table olive product with reduced-salt content from cv. Chalkidiki olives. At the same time, the level of knowledge and the attitude of consumers regarding health issues and eating habits as well as their behaviour towards reduced-salt foods are evaluated. In terms of content, the present research helps to segment the population into groups of consumers with common characteristics (similar consumption behaviour) and identify the main variables influencing consumers' willingness to purchase the product of this research. The emerged results of the study, through the application of a sound and adequate statistical methodology, can be found to be useful to both policy makers and market stakeholders. Important messages regarding consumers preferences about salt can be transferred to the producers of products such as table olives.

The statistical analysis performed allowed the discovery of statistical relationships that were not apparent through the descriptive statistical analysis initially performed. TSCA was applied to segment the population into groups of consumers with common characteristics regarding the "Purchase of Chalkidiki reduced-salt green table olives", creating three clusters using eight variables. It is worth mentioning that the classification of consumers into three different groups, those who would buy the product under certain conditions (first cluster), those who would not buy the product (second cluster), and those who would buy the product (third cluster), is really important for food industries to choose the right way to attract them into this market. Segmenting consumers into clusters with similar behaviour is the first step to more successful promotional and marketing strategies. Separate actions could be designed based on the specific characteristics and needs of each cluster. Furthermore, CATREG revealed the two most important factors, "health status" and "recommendation on reducing salt intake", influencing consumers' willingness to purchase this new product. Informational tools such as health campaigns could raise consumers' awareness towards reduced-salt products and their health benefits while reflecting positively on the new product "Chalkidiki reduced-salt green table olive".

To sum up, the results of this research conclude that the prospect of producing a new reduced-salt table olive product is promising, as the interest of consumers has turned to foods concerning health and safety. A limitation of the research is that we cannot generalize the results at the level of the country's population due to the relatively small sample, nevertheless, it shows a trend with which we can draw important conclusions about consumers' attitudes and behaviour. The size of the sample follows the specifications of the project that this research was carried out. Another weakness is the order and the way the questions were worded as there is a risk of biasing the responses. Moreover, the demographic questions should have been raised at the end of the questionnaire rather than at the beginning. Future research can use larger samples to uncover potential differences with the findings of the present research. Comparing results of similar studies, but in different areas or countries, presents a promising future research path.

Author Contributions: Conceptualization, A.P. and A.M.; methodology, A.P., A.M. and E.L.; validation, F.C., F.T.M. and S.C.; formal analysis, A.P. and A.M.; investigation, P.A.; writing—original draft preparation, A.P., A.M. and P.A.; writing—review and editing, A.P., A.M., E.L., F.C., F.T.M. and P.A.; project administration, A.M. and F.T.M. All authors have read and agreed to the published version of the manuscript.

Funding: This research was funded by the project "FILELIA—Development of edible olives friendly to a salt-reduced diet" (Project code: KMP6–0079456) under the framework of the Action "Investment Plans of Innovation" of the Operational Program "Central Macedonia 2014 2020", that is co-funded by the "European Regional Development Fund and Greece".

Institutional Review Board Statement: The study which was part of the project FILELIA which was conducted in accordance with the Declaration of Helsinki, and approved by the Research Ethics and Deontology Committee Board of the Aristotle University of Thessaloniki (151062/2022, 7 June 2022).

Informed Consent Statement: Not applicable.

Data Availability Statement: Data are available upon request.

Acknowledgments: The present study has been presented at the 15th International Conference "Economies of the Balkan and Eastern European Countries", EBEEC 2023, Chios, Greece (http://ebeec.ihu.gr/) (accessed on 25 January 2024).

Conflicts of Interest: The authors declare no conflict of interest. Author Stelios Choutas was employed by the company CHOUTAS-MARDAS S.A. The remaining authors declare that the research was conducted in the absence of any commercial or financial relationships that could be construed as a potential conflict of interest.

References

1. Mantzouridou, F.T.M.; Mastralexi, A.; Filippidou, M.; Tsimidou, M.Z. Challenges in the Processing Line of Spanish Style cv. Chalkidiki Green Table Olives Spontaneously Fermented in Reduced NaCl Content Brines. *Eur. J. Lipid Sci. Technol.* **2020**, *122*, 1900453. [CrossRef]
2. Johnson, R.L.; Mitchell, A.E. Reducing Phenolics Related to Bitterness in Table Olives. *J. Food Qual.* **2018**, *2018*, 3193185. [CrossRef]
3. Tsapatsaris, S.; Kotzekidou, P. Application of central composite design and response surface methodology to the fermentation of olive juice by *Lactobacillus plantarum* and *Debaryomyces hansenii*. *Int. J. Food Microbiol.* **2004**, *95*, 157–168. [CrossRef] [PubMed]
4. Tzamourani, A.P.; Di Napoli, E.; Paramithiotis, S.; Economou-Petrovits, G.; Panagiotidis, S.; Panagou, E.Z. Microbiological and physicochemical characterisation of green table olives of Halkidiki and Conservolea varieties processed by the Spanish method on industrial scale. *Int. J. Food Sci. Technol.* **2021**, *56*, 3845–3857. [CrossRef]
5. International Olive Oil Council. World Table Olive Figures. 2022. Available online: https://www.Internationaloliveoil.Org/What-We-Do/Economic-Affairs-Promotion-Unit/#figures (accessed on 17 March 2023).
6. Argyri, K.; Doulgeraki, A.I.; Manthou, E.; Grounta, A.; Argyri, A.A.; Nychas, G.J.E.; Tassou, C.C. Microbial Diversity of Fermented Greek Table Olives of Halkidiki and Konservolia Varieties from Different Regions as Revealed by Metagenomic Analysis. *Microorganisms* **2020**, *8*, 1241. [CrossRef]
7. Kazou, M.; Tzamourani, A.; Panagou, E.Z.; Tsakalidou, E. Unraveling the Microbiota of Natural Black cv. Kalamata Fermented Olives through 16S and ITS Metataxonomic Analysis. *Microorganisms* **2020**, *8*, 672. [CrossRef]
8. Kiai, H.; Raiti, J.; El Abbassi, A.; Hafidi, A. Chemical Profiles of Moroccan Picholine Olives and Its Brines during Spontaneous Fermentation. *Int. J. Fruit Sci.* **2020**, *20*, 1297–1312. [CrossRef]
9. Mikrou, T.; Kasimati, K.; Doufexi, I.; Kapsokefalou, M.; Gardeli, C.; Mallouchos, A. Volatile composition of industrially fermented table olives from Greece. *Foods* **2021**, *10*, 1000. [CrossRef]
10. Bautista-Gallego, J.; Arroyo-López, F.N.; López-López, A.; Garrido-Fernández, A. Effect of chloride salt mixtures on selected attributes and mineral content of fermented cracked Aloreña olives. *LWT-Food Sci. Technol.* **2011**, *44*, 120–129. [CrossRef]
11. Panagou, E.Z.; Tassou, C.C. Changes in volatile compounds and related biochemical profile during controlled fermentation of cv. Conservolea green olives. *Food Microbiol.* **2006**, *23*, 738–746. [CrossRef]
12. Panagou, E.Z.; Hondrodimou, O.; Mallouchos, A.; Nychas, G.J.E. A study on the implications of NaCl reduction in the fermentation profile of Conservolea natural black olives. *Food Microbiol.* **2011**, *28*, 1301–1307. [CrossRef]
13. Güngör, F.; Ocak, Ö.; Ünal, M. Effects of different preservation methods and storage on Spanish-style domat olives fermented with different chloride salts. *J. Food Process. Preserv.* **2021**, *46*, 15236. [CrossRef]
14. Bautista-Gallego, J.; Arroyo-López, F.N.; Bordons, A.; Jiménez-Díaz, R. Editorial: New Trends in Table Olive Fermentation. *Front. Microbiol.* **2019**, *10*, 1880. [CrossRef] [PubMed]
15. Grounta, A.; Tassou, C.C.; Panagou, E.Z. Greek-Style Table Olives and their Functional Value. In *Olives and Olive Oil as Functional Foods: Bioactivity, Chemistry and Processing*, 1st ed.; Shahidi, F., Kiritsakis, A., Eds.; John Wiley & Sons: Hoboken, NJ, USA, 2017; pp. 325–342. [CrossRef]
16. Paiva-Martins, F.; Barbosa, S.; Silva, M.; Monteiro, D.; Pinheiro, V.; Mourão, J.L.; Fernandes, J.; Rocha, S.; Belo, L.; Santos-Silva, A. The effect of olive leaf supplementation on the constituents of blood and oxidative stability of red blood cells. *J. Funct. Foods* **2014**, *9*, 271–279. [CrossRef]
17. Bautista-Gallego, J.; Arroyo-López, F.N.; Durán-Quintana, M.C.; Garrido-Fernández, A. Fermentation profiles of Manzanilla-Aloreña cracked green table olives in different chloride salt mixtures. *Food Microbiol.* **2010**, *27*, 403–412. [CrossRef]
18. WHO. *Guideline: Sodium Intake for Adults and Children*; World Health Organization, Department of Nutrition for Health and Development: Geneva, Switzerland, 2012. Available online: https://www.who.int/publications/i/item/9789241504836 (accessed on 17 March 2023).
19. WHO. Salt Reduction. 2020. Available online: https://www.who.int/news-room/fact-sheets/detail/salt-reduction (accessed on 17 March 2023).
20. Zandstra, E.H.; Lion, R.; Newson, R.S. Salt reduction: Moving from consumer awareness to action. *Food Qual. Prefer.* **2016**, *48*, 376–381. [CrossRef]
21. Di Vita, G.; D'Amico, M.; Lombardi, A.; Pecorino, B. Evaluating low sodium content in food: The willingness to pay for salt-reduced bread. *Agric. Econ. Rev.* **2016**, *17*, 82–99.
22. Regan, Á.; Kent, M.P.; Raats, M.M.; McConnon, Á.; Wall, P.; Dubois, L. Applying a Consumer Behavior Lens to Salt Reduction Initiatives. *Nutrients* **2017**, *9*, 901. [CrossRef] [PubMed]
23. López-López, A.; Bautista-Gallego, J.; Moreno-Baquero, J.; Garrido-Fernández, A. Fermentation in nutrient salt mixtures affects green Spanish-style Manzanilla table olive characteristics. *Food Chem.* **2016**, *211*, 415–422. [CrossRef] [PubMed]
24. Torres-Ruiz, F.J.; Garrido-Castro, E.; Gutiérrez-Salcedo, M. Exploring consumer non-knowledge in the agrifood context and its effects on behaviour. *Br. Food J.* **2022**, *124*, 3624–3643. [CrossRef]
25. Espejel, J.; Fandos, C.; Flavián, C. The Influence of Consumer Degree of Knowledge on Consumer Behavior: The Case of Spanish Olive Oil. *J. Food Prod. Mark.* **2009**, *15*, 15–37. [CrossRef]
26. Grimes, C.A.; Riddell, L.J.; Nowson, C.A. Consumer knowledge and attitudes to salt intake and labelled salt information. *Appetite* **2009**, *53*, 189–194. [CrossRef] [PubMed]

27. Rizzo, G.; Borrello, M.; Guccione, G.D.; Schifani, G.; Cembalo, L. Organic Food Consumption: The Relevance of the Health Attribute. *Sustainability* **2020**, *12*, 595. [CrossRef]

28. Bhana, N.; Utter, J.; Eyles, H. Knowledge, Attitudes and Behaviours Related to Dietary Salt Intake in High-Income Countries: A Systematic Review. *Curr. Nutr. Rep.* **2018**, *7*, 183–197. [CrossRef]

29. Cobb, L.; Appel, L.; Anderson, C. Strategies to Reduce Dietary Sodium Intake. *Curr. Treat. Options Cardiovasc. Med.* **2012**, *14*, 425–434. [CrossRef]

30. Mørk, T.; Lähteenmäki, L.; Grunert, K.G. Determinants of intention to reduce salt intake and willingness to purchase salt-reduced food products: Evidence from a web survey. *Appetite* **2019**, *139*, 110–118. [CrossRef] [PubMed]

31. Bryła, P. Selected Predictors of the Importance Attached to Salt Content Information on the Food Packaging (a Study among Polish Consumers). *Nutrients* **2020**, *12*, 293. [CrossRef]

32. Hamburg, M. *Basic Statistics: A Modern Approach*, 3rd ed.; Harcourt Brace Jovanovich: San Diego, CA, USA, 1985.

33. Neuman, W.L. *Social Research Methods: Qualitative and Quantitative Approaches*, 7th ed.; Pearson: London, UK, 2014.

34. Sarantakos, S. *Social Research*, 3rd ed.; Palgrave Mac-Millan: New York, NY, USA, 2005.

35. Punch, K.F. *Introduction to Social Research: Quantitative and Qualitative Approaches*, 3rd ed.; Sage: Los Angeles, CA, USA, 2013.

36. Bournaris, T.; Correia, M.; Guadagni, A.; Karouta, J.; Krus, A.; Lombardo, S.; Lazaridou, D.; Loizou, E.; da Silva, J.R.M.; Martínez-Guanter, J.; et al. Current Skills of Students and Their Expected Future Training Needs on Precision Agriculture: Evidence from Euro-Mediterranean Higher Education Institutes. *Agronomy* **2022**, *12*, 269. [CrossRef]

37. Michailidis, A.; Partalidou, M.; Nastis, S.A.; Papadaki-Klavdianou, A.; Charatsari, C. Who goes online? Evidence of internet use patterns from rural Greece. *Telecommun. Policy* **2011**, *35*, 333–343. [CrossRef]

38. Paltaki, A.; Loizou, E.; Chatzitheodoridis, F.; Partalidou, M.; Nastis, S.; Michailidis, A. Farmers' knowledge, training needs and skills in the Bioeconomy: Evidence from the Region of Western Macedonia. *Proceedings* **2024**, *94*, 7. [CrossRef]

39. Rocha, J.; Borges, N.; Pinho, O. Table olives and health: A review. *J. Nutr. Sci.* **2020**, *9*, e57. [CrossRef]

40. Liem, D.G.; Miremadi, F.; Zandstra, E.H.; Keast, R.S.J. Health labelling can influence taste perception and use of table salt for reduced-sodium products. *Public Health Nutr.* **2012**, *15*, 2340–2347. [CrossRef] [PubMed]

41. Feunekes, G.I.J.; Gortemaker, A.I.; Willems, A.A.; Lion, R.; van den Kommer, M. Front-of-pack nutrition labelling: Testing effectiveness of different nutrition labelling formats front-of-pack in four European countries. *Appetite* **2008**, *50*, 57–70. [CrossRef]

42. Goodman, S.; Hammond, D.; Hanning, R.; Sheeshka, J. The impact of adding front-of-package sodium content labels to grocery products: An experimental study. *Public Health Nutr.* **2013**, *16*, 383–391. [CrossRef]

43. Pino, A.; Vaccalluzzo, A.; Solieri, L.; Romeo, F.V.; Todaro, A.; Caggia, C.; Arroyo-López, F.N.; Bautista-Gallego, J.; Randazzo, C.L. Effect of sequential inoculum of beta-glucosidase positive and probiotic strains on brine fermentation to obtain low salt sicilian table olives. *Front. Microbiol.* **2019**, *10*, 174. [CrossRef]

44. Sánchez-Rodríguez, L.; Cano-Lamadrid, M.; Carbonell-Barrachina, Á.A.; Sendra, E.; Hernández, F. Volatile Composition, Sensory Profile and Consumer Acceptability of HydroSOStainable Table Olives. *Foods* **2019**, *8*, 470. [CrossRef] [PubMed]

45. Li, R.; Lee, H.Y.; Lin, Y.T.; Liu, C.W.; Tsai, P.F. Consumers' Willingness to Pay for Organic Foods in China: Bibliometric Review for an Emerging Literature. *Int. J. Environ. Res. Public Health* **2019**, *16*, 1713. [CrossRef] [PubMed]

46. Pouraram, H.; Afshani, F.; Ladaninejad, M.; Siassi, F. What do we Need to Start a Multimedia Salt Reduction Campaign? *Int. J. Prev. Med.* **2023**, *14*, 28. [CrossRef]

47. Gupta, A.K.; Carroll, T.E.; Chen, Y.; Liang, W.; Cobb, L.K.; Wang, Y.; Zhang, J.; Chen, Y.; Guo, X.; Mullin, S.; et al. 'Love with Less Salt': Evaluation of a sodium reduction mass media campaign in China. *BMJ Open* **2022**, *12*, 056725. [CrossRef]

48. Tsimitri, P.; Michailidis, A.; Loizou, E.; Mantzouridou, F.T.; Gkatzionis, K.; Mugampoza, E.; Nastis, S.A. Novel Foods and Neophobia: Evidence from Greece, Cyprus, and Uganda. *Resources* **2022**, *11*, 2. [CrossRef]

49. Sánchez, G.; Peña, L.; Varea, S.; Mogrovejo, P.; Goetschel, M.L.; De los Ángeles Montero-Campos, M.; Mejía, R.; Blanco-Metzler, A. Knowledge, perceptions, and behavior related to salt consumption, health, and nutritional labeling in Argentina, Costa Rica, and Ecuador. *Pan Am. J. Public Health* **2012**, *32*, 259–264. [CrossRef] [PubMed]

50. Granato, D.; Barba, F.J.; Bursać Kovačević, D.; Lorenzo, J.M.; Cruz, A.G.; Putnik, P. Functional Foods: Product Development, Technological Trends, Efficacy Testing, and Safety. *Annu. Rev. Food Sci. Technol.* **2020**, *11*, 93–118. [CrossRef] [PubMed]

 *sustainability*

Article

The Global Growth of 'Sustainable Diet' during Recent Decades, a Bibliometric Analysis

Maria Gialeli [1,*], Andreas Y. Troumbis [2], Constantinos Giaginis [1], Sousana K. Papadopoulou [3,*], Ioannis Antoniadis [4] and Georgios K. Vasios [1]

1 Department of Food Science and Nutrition, School of Environment, University of the Aegean, 81400 Myrina, Greece
2 Department of Environment, School of Environment, University of the Aegean, 81100 Mytilene, Greece; atro@aegean.gr
3 Department of Nutritional Sciences and Dietetics, School of Health Sciences, International Hellenic University, 57001 Thessaloniki, Greece
4 Department of Management Science and Technology, School of Economic Sciences, University of Western Macedonia, Koila, 50100 Kozani, Greece
* Correspondence: gialeli.m@aegean.gr (M.G.); souzpapa@gmail.com (S.K.P.)

Abstract: The term 'sustainable diets' (SDs) was first introduced in the scientific literature in 1986 and later defined in detail by the Food and Agriculture Organization (FAO) as pertaining to those diets that can promote environmental health ad effectively ensure food and nutrition security as well as a healthy lifestyle in humans, combining the notion of sustainability with dietary patterns and their beneficial impacts. Since then, various international events have been held promoting sustainability as a significant component of food production, nutrition, and human health. These events have enhanced the knowledge transition and awareness between the scientific community and policymakers concerning the importance of SDs. In this aspect, this is the first study that aims to identify trends and turning points over time concerning the research on SDs. We performed a comprehensive bibliometric analysis of 1407 scientific documents published in Scopus during the period 1986–2022. The documents were screened following the PRISMA guidelines, and bibliometric analysis was conducted using the Bibliometrix R-package and VOSviewer and the detection of Sustainable Development Goals with the text2sdg R-package. Overall, there was an exponential growth in the literature on SDs that followed international events from 2009 onward. Among the most impactful journals were *Sustainability*, *Nutrients*, and *Frontiers in Nutrition*. The leading countries in research were pointed out, as well as the high rate of collaborations and partnerships between them. The research interest was mainly focused on (a) climate change, greenhouse gas emissions, and environmental impact; (b) food systems, security, and consumption; and (c) health, Mediterranean Diet (MD), and dietary guidelines. The significance of these keywords changed over time, following the evolution of SDs concepts from the planetary environmental impact of food production to the healthier dietary habits of individuals. Among several dietary patterns, MD was identified as the most popular among the local SDs, with synergies among scientists in the Mediterranean region. Overall, the novelty of this study is the mapping of the expansion of knowledge over the last 36 years regarding the term SDs while taking into consideration international events and their impact on scientific research.

Keywords: sustainable diets; bibliometric analysis; Scopus; bibliometrix; VOSviewer; text2sdg; Sustainable Development Goals

Citation: Gialeli, M.; Troumbis, A.Y.; Giaginis, C.; Papadopoulou, S.K.; Antoniadis, I.; Vasios, G.K. The Global Growth of 'Sustainable Diet' during Recent Decades, a Bibliometric Analysis. *Sustainability* 2023, 15, 11957. https://doi.org/10.3390/su151511957

Academic Editor: Dimitris Skalkos

Received: 3 July 2023
Revised: 28 July 2023
Accepted: 29 July 2023
Published: 3 August 2023

1. Introduction

1.1. Definition of the Term

As mentioned by the Lancet Commission in 2019: "Food is the single strongest lever to optimize human health and environmental sustainability on the Planet. However, food is currently threatening both people and the planet" [1]. Many terms have been considered to describe the efforts of humankind to produce and consume food in a more sustainable manner. Terms such as: 'sustainable diet' [2], 'econutrition' [3], 'sustainable food consumption' [4], 'ecological diet', 'resilient diet', 'biodiverse diets' [5], 'climate friendly diets' [6], 'sustainable nutrition' and 'nutritional sustainability' [7,8] are those more reported in the international scientific literature.

The most commonly used term is 'Sustainable Diets' (SDs) which was introduced by Gussow and Clancy in 1986 [2]. Twenty-five years later, the term SDs was defined by the Declaration of the World Summit of Food and Agriculture Organization (FAO) [9] as "those diets with low environmental impacts that contribute to food and nutrition security and to healthy life for present and future generations. Sustainable diets are protective and respectful of biodiversity and ecosystems, culturally acceptable, accessible, economically fair and affordable; nutritionally adequate, safe, and healthy, while optimizing natural and human resources.". This definition is widely accepted by the scientific community and is usually mentioned in most papers published on SDs [10–12].

From the SDs definition by FAO [9], six 'key components' are derived: (1) well-being and health; (2) biodiversity, environment, and climate; (3) equity and fair trade; (4) eco-friendly, local, and seasonal foods; (5) cultural and heritage skills; and (6) food and nutrient requirements, food security, and accessibility [9,13,14]. The components were represented in the literature as flower-like figures [13,14]. These crucial elements were developed taking into consideration the three intersecting pillars of Sustainable Development: the biological (and other resource) system, the economic system, and the social system as described by a Venn diagram from Barbier in 1987 [15].

1.2. Background on Key International Events

It should be noted that sustainability and sustainable development have been a focus of the United Nations (UN) for more than three decades. These terms were undoubtedly implied in the Brundtland Report in 1987 and articulated later in 1992, during the 'Earth Summit' in Rio de Janeiro when Agenda 21 was adopted [16]. Ten years later, in 2002, in Johannesburg, 'The World Summit on Sustainable Development', strategies were outlined to obtain a more efficient and effective implementation of Agenda 21 [17]. Hence, sustainable agriculture has gradually become a significant aim at a global level [18].

Due to the rise in awareness about the inequality between world hunger and obesity [9], International and European Organizations have adopted several measures (Table 1). In 1996, during the World Food Summit held by the FAO in Rome, the Rome Declaration on World Food Security and the World Food Summit Plan of Action were signed to promote food security. Four years later, in 2000, at the Millennium Summit, the Millennium Development Goals (MDGs) were set.

An important link between biodiversity, food, and nutrition was established in 2006 by the Convention of Biological Diversity, creating an early definition of SDs [12]. Food security was underlined not only in this initiative but also in the World Food Summit on Food Security, where the Declaration of the World Summit on Food Security was adopted in 2009 with the commitment that all countries in the world will work to minimize or even end hunger.

Evolving from the MDGs, after the Conference on Sustainable Development in 2012 that took place in Rio de Janeiro, Brazil, with the outcome document 'The Future We Want', and the Open Working Group in SDGs, we were led to the 2030 Agenda for Sustainable Development, adopted by all United Nations Member States in 2015 and the Sustainable Development Goals (SDGs) [19]. In reference to the 2030 Agenda, 17 individual SDGs aim as an 'indivisible whole' to solve severe issues such as poverty, education, and climate change [20]. Out of the 17 goals, SDG 2 aims to 'end hunger, achieve food security, improve nutrition, and promote sustainable agriculture'. This goal is characterized by multiple dimensions (social, economic, and environmental) that go much further than food security [21].

Table 1. International Events, Conferences, Summits, and Treaties related to Sustainable Development Goals and Sustainable Diets.

Year		Event	Abbr.	Place	Notes
1992	UN	Earth Summit	UNCED	Rio de Janeiro, Brazil	Conference on Environment and Development, Agenda 21 [1][**]
1996	FAO	World Food Summit	WFS	Rome, Italy	Rome Declaration [*]
2000	UN	Millennium Summit	MDGs	New York, NY, USA	Millennium Development Goals (MDGs) [2][**]
2002	UN	World Summit on Sustainable Development	WSSD	Johannesburg, South Africa	The Johannesburg Declaration on Sustainable Development and the Plan of Implementation [**]
2006	UN/FAO	Convention of Biological Diversity	CBD	Rome, Italy	Initiative on Biodiversity for Food and Nutrition [*]
2009	FAO	World Summit on Food Security	WSFS	Rome, Italy	Declaration of the World Summit on Food Security [*]
2012	UN	Conference on Sustainable Development	UNCSD	Rio de Janeiro, Brazil	Rio+20, 'The Future We Want' [**]
2013		Open Working Group on SDGs	OPWG SDGs	-	Proposal on the SDGs [**]
2015	UN	Sustainable Development Summit	SDGs	New York, NY, USA	2030 Agenda for Sustainable Development [3], SDGs [**]
2015	UN	Climate Change Conference	COP21	Paris, France	Paris Agreement on Climate Change [**]
2019	EU	European Green Deal [4]	EGD	European Union	'Farm to Fork' Strategy [*]
2021	FAO	Food Systems Summit	UNFSS	New York, NY, USA	Food Systems Summit of FAO [*]

[1] Agenda 21: Program of action for sustainable development; Rio declaration on environment and development; statement of forest principles; the final text of agreements negotiated by governments at the United Nations Conference on Environment and Development (UNCED), 3–14 June 1992, Rio de Janeiro, Brazil (2. print). (1994). Conference on Environment and Development, New York, NY, USA. Department of Public Information, United Nations. [2] General Assembly resolution 55/2, United Nations Millennium Declaration, A/RES/55/2 (18 September 2000), available online: undocs.org/en/A/RES/55/2 (accessed on 2 July 2023). [3] General Assembly resolution 70/1, Transforming our world: the 2030 Agenda for Sustainable Development, A/RES/70/1 (21 October 2015), available online: undocs.org/A/RES/70/1 (accessed on 2 July 2023). [4] European Commission, Communication from the Commission The European Green Deal, (11 December 2019) COM (2019) 640 final, available online: eur-lex.europa.eu/legal-content/EN/TXT/?uri=COM%3A2019%3A640%3AFIN (accessed on 2 July 2023). [*] SDs Related International Events. [**] SDGs Related International Events and Treaties.

Additionally, all SDGs were meant to ensure homogeneity between different policy sectors [22]. Therefore, there is a link to nutrition among them; for instance, SDG 1 is related to nutrition as poverty makes it harder to meet nutritional recommendations and restricts access to proper food intake, and SDG 2 promotes that undernourishment is brought on by unsustainable food production and SDG 3 proposes that healthy and sustainable nutrition promotes health [23]. In another example, regarding food systems, SDG 12 promotes sustainable consumption and production patterns [24]. Furthermore, in the 'Fixing Food 2021' report from the Barilla Center for Food and Nutrition [25], it has been underlined that food sustainability is crucial for all SDGs and not only in the two mentioned as mainly food related.

There are other important initiatives linked to the SDGs at both international and European levels. At the international level, the Paris Agreement on Climate Change was finalized in 2015 during the Climate Change Conference in Paris, France, unraveling strong connections to the 2030 Agenda. At the European level, the European Green Deal, adopted by the European Member States in 2019 [26], promotes initiatives for the end of hunger, the achievement of food security, the improvement of nutrition, and the promotion of sustainable agriculture, especially with the 'Farm to Fork' Strategy.

Apart from International and European organizations, several other scientific events took place regarding SDs, and certain scientific documents were published [12] (see

Supplementary Table S1 for details). As a result of an international 'conversation' regarding the choice of indicators of a sustainable diet that began after 2009 [27], the Mediterranean Diet (MD) was selected as a characteristic example of SD [28]. The indicators were selected using strict scientific criteria and were categorized into four thematic areas: (1) nutrition and human health, (2) environmental health, (3) economy, and (4) favorable socio-cultural factors [29].

Furthermore, in addition to highlighting the importance of local traditional diets, there has been an effort to identify the total of the indicators that measure effectively and critically assess SDs [5,8]. In the same context and embracing the SDGs, more than 20 experts from the EAT-Lancet Commission have proposed a universal healthy reference diet [1].

In the present study, the bibliographic data are studied alongside international and European Organizations' activities regarding SDs specifically and SDGs, Climate Change, and Food Security generally, in a quantitative approach. The focus of the research did not remain solely to SDs related events as the disciplinarity of the term allows a more comprehensive approach. Furthermore, SDGs have become a framework widely accepted by United Nations Members and other International and Regional Organizations, creating partnerships forming policies, and becoming a driving force of research and innovation.

1.3. Objectives

This research's objective is to determine the underlying international 'dialogue' about Sustainable Diets through a methodical review and assessment of the international scientific literature. The goal of our research is to step on past analyses and define trends and progression of the term 'Sustainable Diets' for the previous 36 years, from 1986 to 2022. Furthermore, the dataset was critically analyzed and scrutinized to effectively detect the collaboration among the countries, the relationship between the corresponding authors' nationality and the type of diet they are studying, and the role of international organizations, international treaties, and funding in scientific production regarding the term Sustainable Diets. Finally, six query systems were used to detect the SDGs in our dataset.

2. Materials and Methods

2.1. Bibliometric Analysis and General Workflow

Defined as a field in 1934 by the librarian Paul Otlet, termed 'bibliométrie', bibliometry was described to measure all the factors concerning both publication and reading of books and documents [30]. Since the release of online databases that provide access to bibliographic metadata and the fast growth of scientific publications, the necessity of using bibliometric methods to assess the latest literature has highly increased [31].

The method explored in this study focuses on bibliometric mapping and content analysis by monitoring the SDGs in the selected papers. A bibliometric analysis was methodologically designed and carefully performed to investigate the literature addressing the terms 'sustainable diets' and 'sustainable nutrition' in the bibliographic database Scopus using a five-stage standard workflow [31,32]. Firstly, the study was designed; secondly, the data were collected; thirdly, the obtained were analyzed; fourthly, the derived data were visualized; then, in the fifth and final stage they were critically interpreted. In the data analysis stage, data networks were created for collaboration and co-occurrence analyses, in addition to the descriptive analysis of the bibliographic data. Moreover, network mapping was performed during the data visualization stage.

2.2. Data Sources

After posing the research questions, the bibliographic database Scopus was selected for the present study in order to identify potentially relevant documents. This database was selected because it indexes more journals than other popular electronic databases such as PubMed and Web of Science and provides approximately 20% greater coverage for citation analysis than Web of Science [33–35]. Scopus was comprehensively searched for peer-reviewed articles published between the periods of 1986–2022 in the English language until April 2023. Titles and abstracts were carefully read and critically screened by two authors (M.G. and G.K.V.), who carefully identified and eliminated duplicates.

2.3. Protocol

The protocol was drafted using the PRISMA extension for Scoping Reviews (PRISMA-ScR) guidelines [36], which is an extension of the PRISMA 2020 statement [37], the updated version of the Preferred Reporting Items for Systematic reviews and Meta-Analyses [38].

2.4. Software and Analyses

The bibliometrics analysis was performed using the Bibliometrix R package [32]. Moreover, we used the text2sdg R package [39] to detect SDGs in the initial collection. We also used VOSviewer [40], a software tool for constructing and visualizing bibliometric networks, to perform the co-occurrences analysis.

For the bibliometric analysis, the shiny app Biblioshiny was used. This application provides a web interface for Bibliometrix [32]. We used Biblioshiny to import the final dataset and convert it to a dataframe collection. Furthermore, analyses and plots for sources, authors, and documents were performed, and the Conceptual and the Social Structure of the bibliographic collection were analyzed.

In particular, the main information of the collection was presented. Next, using Microsoft Excel Spreadsheet Software, the annual scientific production was visualized. In the same figure, the literature findings about major events related to SDs were added. The sources that made the publications were studied, and the country's production and collaboration were presented. Moreover, as for the keywords, the most frequently used keywords by the authors were retrieved, and the trending topics were identified. In this analysis, the words that were used in the search queries were not included by the authors (sustainable diet, sustainable diets, diet, diets, nutrition, sustainable nutrition, sustainable) in order to highlight the main trends related to SDs. By doing that, we focused mainly on the other keywords and themes that emerged in the bibliographic dataset studied than our original searches. Additionally, a thematic evolution chart was created using Sankey diagrams. Four cutting points were set by splitting the studied collection into five time slices: 1986–2006, 2007–2011, 2012–2015, 2016–2019, and 2020–2022. These cutting points derived from some of the major turning points of the upward trend of the scientific production of this collection. They were strategically selected to identify the new themes introduced in the literature after every increase in the number of published documents.

A co-occurrence network of the authors' keywords was visualized using VOSviewer software. VOSviewer creates maps using a co-occurrence matrix in three steps; (a) based on the co-occurrence matrix, a similarity matrix is created in the first stage, (b) using the VOS mapping technique, a map is created from the similarity matrix (c) the map is then translated, rotated, and mirrored [40].

Keyword co-occurrence revealed clusters in the research discipline providing auxiliary support to scientific research [41]. Out of the keywords in all articles, 49 with a minimum threshold of five co-occurrences. The threshold was chosen to ensure that enough research topics were obtained while effectively avoiding overcrowding. These sets of author keywords were separated into six clusters using the VOSviewer keyword co-occurrence algorithm. Each cluster contained nodes (circles) of the same color and included links or relationships (lines) between nodes. Nodes with similar colors belong to the same cluster. The size of each node and word in each cluster was proportional to the node's weight/frequency, and clusters with the most keywords were more centered on the study areas [42].

Lastly, apart from the bibliometric analysis, SDGs were also detected in the titles of the present bibliographic collection using the open source, multi-system analysis R package text2sdg [39] was used to identify SDGs in the titles of the documents included in the final dataset. A stacked barplot of absolute frequencies was visualized using six different individual query systems used for the identification (Auckland, Aurora, Elsevier, SDGO,

SDSN, SIRIS). This package allows to use text data in order to detect SDGs and compare the outcomes quantitively.

2.5. Data Collection, Eligibility Criteria and Screening

First of all, the eligibility criteria, both inclusion and exclusion criteria, were identified. Articles were eligible for inclusion in this review if they included the terms 'sustainable diets' or 'sustainable nutrition' or other derivatives of the terms in their titles, abstracts, or keywords, and referred to the sustainability of human diet and nutrition. Only articles (articles, data papers, reviews, and short surveys) published after 1986 in English were eligible for inclusion [43]. Articles were excluded if they did not fit the meaning of the search terms. Double citations were also excluded.

We also performed several searches in the Scopus database (the last search was conducted in April 2023) (Table 2). The first two presented considered the term 'Sustainable Diets' (SD1, SD2) and the other two the term 'Sustainable Nutrition' (SN1, SN2). In these queries, we looked for the adjective 'sustainable' and the nouns 'diet' or 'nutrition' in the titles of the scientific documents (SD1, SN1). We also located the word 'sustainable' and the word 'diet' or 'nutrition' in the title, abstract, and keywords of the documents (SD2, SN2) within one word proximity using the proximity operator within W/n.

Table 2. Search strategy on Scopus database, where four different queries were conducted and the resulting number of articles before and after filtering.

Topic	Search	Search Query	Dataset [1]	No. of Docs
Sustainable Diets	SD1	TITLE (*sustainab* AND *diet*)	SD1—(SD2 or SN1 or SN2)	247
			(SD1 and SD2)—(SN1 or SN2)	462
	SD2	TITLE-ABS-KEY (*sustainab* W/1 *diet*)	SD2—(SD1 or SN1 or SN2)	841
			(SD1 or SD2) and (SN1 or SN2)	197
Sustainable Nutrition	SN1	TITLE (*sustainab* AND *nutritio*)	SN1—(SN2 or SD1 or SD2)	390
			(SN1 and SN2)—(SD1 or SD2)	185
	SN2	TITLE-ABS-KEY (*sustainanb* W/1 *nutritio*)	SN2—(SN1 or SD1 or SD2)	551
Total Dataset			SD1 or SD2 or SN1 or SN2	2873

[1] The column 'dataset' is displaying the logic behind the different sub searches that were performed in Scopus database using Boolean operators (AND, OR, NOT) according to set theory in order to identify the unique publications across the searches and duplicates.

The asterisk wildcard (*) was used because we wanted to include all the grammatical forms of the words diet and nutrition. After some trial tests, we noted that, except for the noun diet, other derived terms appeared in the results: the term appeared both in singular and plural (diet and diets) as an adjective (dietary, dietetic) as a noun (dietetics, dietitians, dieter, dietotherapy, dietand), and as a gerund (dieting, non-dieting). We excluded papers that in their title contained terms other than diet (e.g., diethyl, dietro, etc.).

To avoid the duplicated entries in the total dataset, several sub-searches were performed using the three Boolean operators; AND, OR, AND NOT. Explaining the column Dataset in Table 2, the word 'and' represented AND Boolean operator and the intersection of the two searches. The symbol '-' was applied for AND NOT excluding specific searches. Finally, the word 'or' was for OR; this Boolean operator expanded the search showing the union of two searches. These searches were made to identify the unique documents of the final dataset and eliminate the results that were recognized as common (Table 2). For example, we found 247 documents using the query string: (TITLE (*sustainab* AND *diet*) AND NOT (TITLE-ABS-KEY (*sustainab* W/1 *diet*) OR TI-TLE (*sustainab* AND *nutritio*) OR TITLE-ABS-KEY (*sustainab* W/1 *nutritio*))).

In order to perform a consistent review, three of the authors (M.G., A.Y.T., and G.K.V.) systematically screened the titles and abstracts of 20 same publications and discussed the results. The goal was to establish a common view and understanding. Subsequently, each author searched a subset of databases. All discrepancies between the authors were resolved by discussion. Three reviewers (M.G., A.Y.T., and G.K.V.) screened the title, abstract, and keywords and graded the relevancy of each entry of the database in a spreadsheet using a modified Likert scale (1–5). This scale was termed as Bibliometric Scale, where grade 1 was assigned to the highly irrelevant documents and grade 5 the highly relevant. The papers included in the final dataset should score 3–5 to be considered relevant.

Documents were excluded if they did not fit the meaning of the search terms and did not have a direct connection to human food consumption, as reported below.

- Nutritional sustainability of animal/sustainable animal nutrition feeding (e.g., camels, pigs, ruminants);
- Sustainability in cultivation methods (e.g., maize, corn, soybean);
- Sustainability in aquaculture methods;
- Animal breeding, Embryo development;
- Sustain weight or body mass (e.g., sustainable weight loss, sustainably preserve muscle mass);
- Weight loss sustainability;
- Use of the word sustainability in terms of continuity, viability, and feasibility.

As a result of the data charting process, the data collected were as follows: citation information (authors, year, doi number), bibliographic information (affiliation, correspondence address, language of original document), abstract and keywords, and funding details (number, acronym, sponsor, funding text).

A total of 2873 documents were initially identified through multiple searches (Figure 1). Duplicates were then removed ($n = 4$). Subsequently, 2869 documents were finally assessed for eligibility. Documents that did not fit the time period studied ($n = 124$), language ($n = 93$), and document type ($n = 493$) were then excluded. In the Screening phase, reports with no data available in the abstract and keywords domain were considered as not retrieved ($n = 78$). Finally, after the screening process, the papers were graded by using the 1–5 scale, which had been standardized by the three researchers (M.G., A.Y.T., and G.K.V.) as mentioned above. More specifically, 124 papers were graded with 1 as highly irrelevant, 480 papers were graded with 2, as likely to be irrelevant, and 70 papers were graded with 3 as more or less relevant. A total of 1407 studies were finally included in the bibliometric and content analyses.

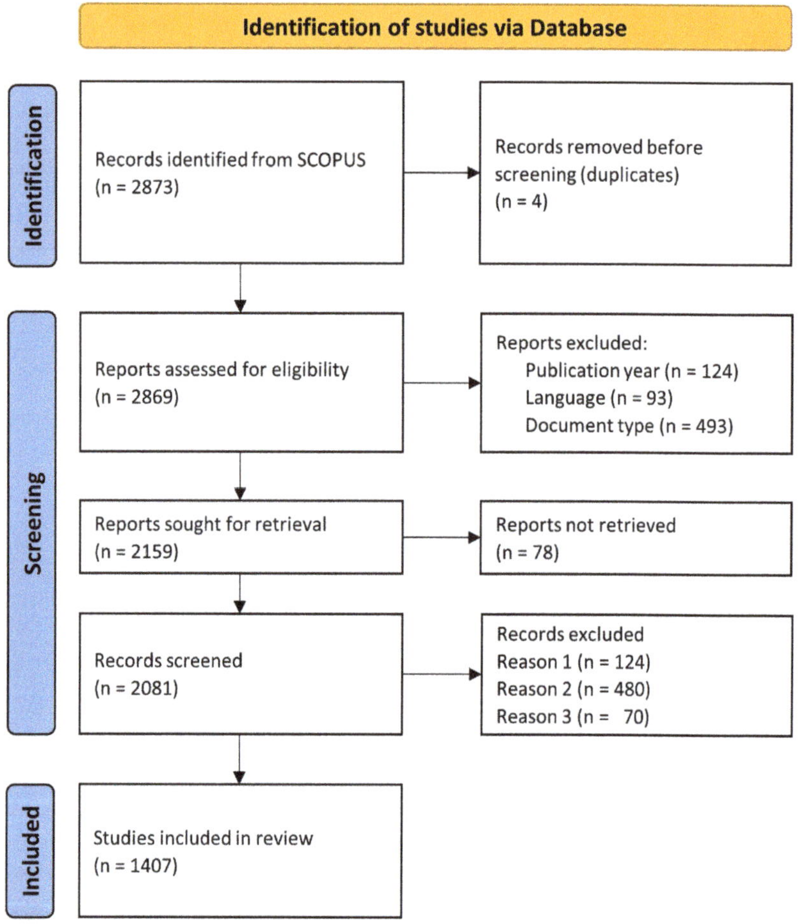

Figure 1. Flowchart indicating identification and selection of studies used in the present analysis following PRISMA statement guidelines.

3. Results

In this section, we demonstrate the main results of the descriptive bibliometric analysis, the annual scientific production of this collection, the main journals that published the scientific papers of the collection, the countries involved, the main keywords, their connection in a co-occurrence network, their thematic evolution, and the connection between the SDGs and the titles of the papers of the final dataset.

3.1. Main Information

Table 3 summarizes the main information of the present bibliographic collection. The total number of examined documents (papers) examined was 1407. Most documents were articles (1116 documents), and reviews (283 documents). Data papers (1 document) and short surveys (7 documents) were also included. The 5191 authors used a total of 3397 keywords. Furthermore, the number of single-authored documents was 122, while the number of co-authors per document, that is, the mean number of authors' appearances per document, was 4.99. Both indicators (authors per document and co-authors per document) evaluated the level of author collaboration [44]. Finally, the level of international co-authorship reached the percentage of 34.68%.

Table 3. Main information regarding the final collection of documents.

Description	Results
MAIN INFORMATION ABOUT DATA	
Timespan	1986–2022
Sources (Journals, Books, etc.)	428
Documents	1407
Annual Growth Rate %	17.65
Document Average Age	3.89
Average citations per doc	27
References	93,027
DOCUMENT TYPES	
Article	1116
data paper	1
Review	283
short survey	7
DOCUMENT CONTENTS	
Keywords Plus (ID)	4590
Author's Keywords (DE)	3397
AUTHORS	
Authors	5191
Authors of single-authored docs	109
AUTHORS COLLABORATION	
Single-authored docs	122
Co-Authors per Doc	4.99
International co-authorships %	34.68

3.2. Annual Scientific Production

The publication trends of the 1407 documents were systematically studied from 1986 to 2022. Figure 2 shows the number of publications on SDs for this period and the major international events related to the SDGs and SDs that took place. The number of publications began to increase in 2007 (5 documents) and significantly increased over the examined period of time. Figure 2 shows an escalation in publications from one document in 1986 to 348 documents in 2022, which is noted as the peak number of publications. The annual production since 1986 fits an exponential trend line with an R^2 value of 0.9843, while the annual growth rate was 17.65%. Taking the above under consideration, research on SDs appears to be at the stage of exponential growth, implying that SDs are highly increasingly receiving the attention of the scientific community, particularly in the last few years.

Additionally, we compared annual scientific production with literature findings and major international events, such as summits and treaties related to sustainability, food, nutrition, and SDGs. Approximately 5% of the articles were published between 1986 and 2012 (27 out of 1407). After 2012, there was an upward trend in the number of publications. This happened after the official definition of SDs by the FAO during the International Scientific Symposium 'Biodiversity and Sustainable Diets: United against Hunger' in 2010 [45]. Moreover, almost 95% of the documents (1332 out of 1407) were published in the last ten years and especially after 2015, when the Paris Agreement on Climate Change took place and the 2030 Agenda for Sustainable Development was set, promoting the 17 SDGs.

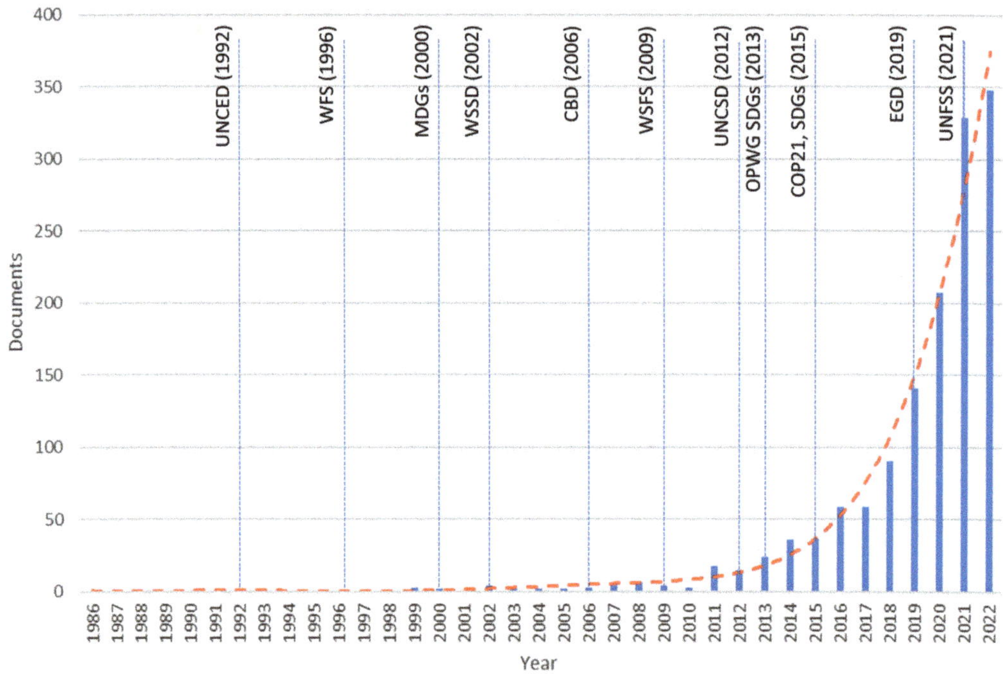

Figure 2. Annual scientific production of articles on the terms 'sustainable diets' and 'sustainable nutrition' and international events occurred during the studied period. The annual production since 1986 fits an exponential trend line (red dash line) with an R^2 value of 0.9843, while the annual growth rate was 17.65%.(UNCED: UN Earth Summit, WFS: World Food Summit, MDGs: Millennium Development Goals, WSSD: World Summit on Sustainable Development, CBD: Convention of Biological Diversity, WSFS: World Summit on Food Security, UNCSD: UN Conference on Sustainable Development, OPWG SDGs: Open Working Group on SDGs, SDGs: Sustainable Development Goals, COP21: Climate Change Conference, EGD: European Green Deal, UNFSS: UN Food Systems Summit).

3.3. Sources

In our collection, we found 428 sources (journals) and cited sources included in the 1407 documents' bibliographies. According to Table 4, the five most relevant sources for the terms under study are *Sustainability* (111 publications), *Nutrients* (91 publications), *Frontiers in Nutrition* (58 publications), *Public Health Nutrition* (44 publications), and *Journal of Cleaner Production* (39 publications) (Table 4). The main subject areas covered by these sources are agricultural and biological sciences, environmental sciences, social sciences, nursing, and medicine. According to Bradford's Law, these ten journals (2.34% of the total number of journals) were also the core–nucleus journals that were particularly devoted to the subject. These ten journals have published 495 articles in total, representing 35% of the entire collection.

Table 4. Most relevant sources by the number of articles they published.

Sources	Articles
Sustainability	111
Nutrients	91
Frontiers in Nutrition	58
Public Health Nutrition	44
Frontiers in Sustainable Food Systems	39
Journal of Cleaner Production	39
Foods	32
Appetite	30
Global Food Security	29
International Journal of Environmental Research and Public Health	22

3.4. Country Production and Collaboration

Between 1986 and 2022, 110 countries published publications on the studied subject. In Figure 3, the color intensity is proportional to the number of publications. The map shows the country based on the authors' affiliation. Countries in the Mediterranean basin have published a lot of scientific documents, with Italy in the first place, followed by other countries of the region, such as Spain, Greece, and Portugal (Supplementary Table S2). Nordic countries (Sweden, Denmark, Norway, and Finland) that could relate to the New Nordic Diet were also at the top places. The countries with the largest scientific production were the USA ($n = 777$), the UK ($n = 437$), Italy ($n = 337$), Australia ($n = 274$), and France ($n = 251$).

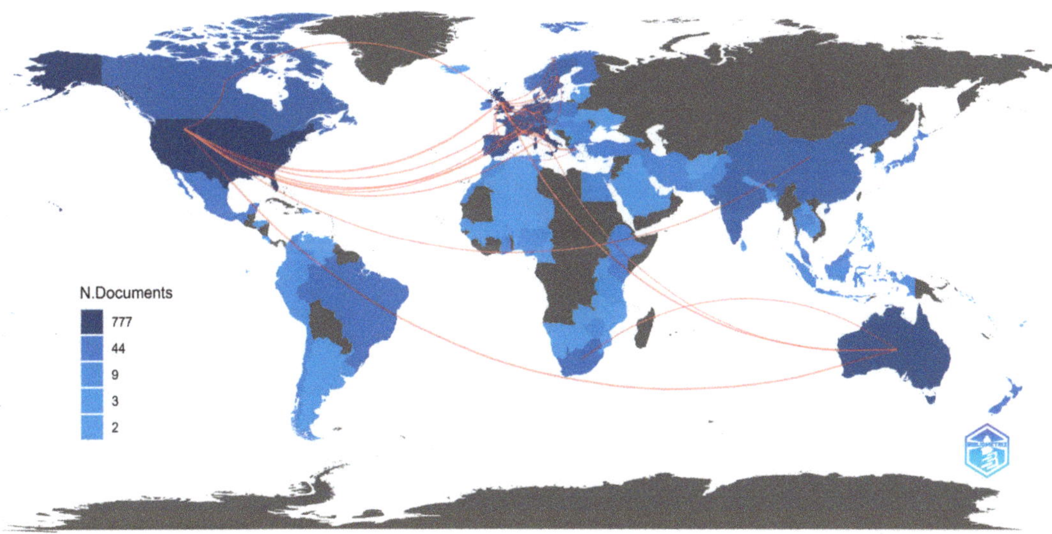

Figure 3. World maps showing the number of documents per country; the color intensity is proportional to the number of publications. The red lines indicate the collaboration among countries.

The research collaborations among countries are also presented in Figure 3, where countries with equal or more than ten shared papers are shown by connectors. The countries with the strongest linkages were the USA with the UK (40 links), the UK with Italy (27 links), the USA with Australia (27 links), and Italy with France (23 links). Most of them were from English-speaking countries. Italy is a Mediterranean country with a great tradition in the Mediterranean Diet. Moreover, the FAO headquarters are located in Rome, Italy.

Finally, there was also close collaboration between countries in the Mediterranean basin; for example, Italy and Spain shared 20 collaborations.

3.5. Keywords and Trend Topics

Table 5 highlights the most frequently used keywords by the authors in their publications. Some of the first results (sustainable diet, sustainable, nutrition, nutrition and diet) are trivial because they were included in the set of terms used to be built up the search query in the bibliographic database. However, they reveal that diet as a term is used more often than the term nutrition in combination with the term sustainable. In addition, the food system had the highest number of occurrences (99 occurrences), followed by food security (87 occurrences). The next three authors' keywords, greenhouse gas emission (74 occurrences), climate change (65 occurrences), and environmental impact (60 occurrences), connect global warming with dietary choices. The next word was health (60 occurrences). Mediterranean diet (60 occurrences) and Sustainable Development Goals (43 occurrences) were also in the top twenty list.

Table 5. Most frequent words (Author's Keywords).

Words	Occurrences
sustainable diet	326
sustainable	293
nutrition	141
food system	99
food security	87
diet	85
greenhouse gas emissions	74
climate change	65
environmental impact	60
health	60
Mediterranean Diet	60
environment	47
food	45
Sustainable Development Goals	43
carbon footprint	42
life cycle assessment	42
food consumption	40
sustainable food system	40
sustainable nutrition	40
water footprint	40

Alarmingly enough, climate change and greenhouse gas emissions became trending topics in 2017 (Figure 4), soon after the Climate Change Conference (COP21) in Paris in 2015, where the Paris Agreement on Climate Change was adopted by 196 Parties. Even though the Mediterranean Diet has been identified studied as a Sustainable Diet Model since 2009 in the collection. It mainly appears to be a trending topic 10 years later, in 2019. In the last few years (2019–2021), they have become popular topics regarding vegetarianism and plant-based foods. Lastly, food and nutrition security has become a well–studied topic since 2016, as mentioned in the second Goal of SDGs (SDG 2: Zero Hunger).

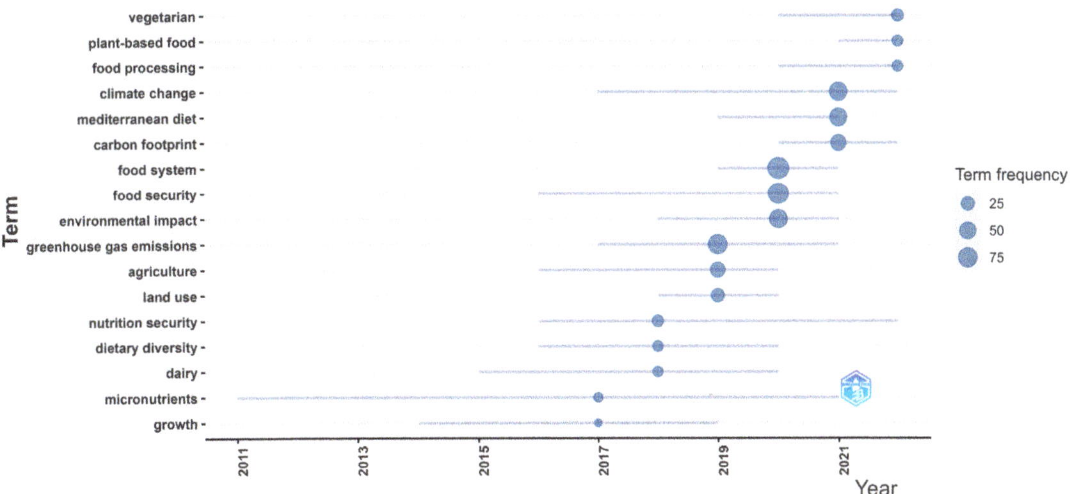

Figure 4. Trend Topics for the period 2011–2022 representing the evolution of the authors' keywords.

3.6. Thematic Evolution

Moreover, the evolution of the main themes in SDs research was systematically explored by the use of a timeline view analysis of the authors' keywords (Figure 5). The first period (1986–2016) begins with the first definition of the term sustainable diet and ends the year of the Convention of Biological Diversity. There have been merely a few publications throughout these 30 years; therefore, there are two themes: sustainable and food systems. In the next period (2007–2011), these two themes remained, but food security emerged as a new theme. It should be mentioned that in 2009, FAO held the World Summit on Food Security. During the following period (2012–2015) there was an appearance of eight new themes ((1) climate change, (2) environmental impact, (3) environmental sustainability, (4) sustainable food system, (5) food consumption, (6) food production, (7) nutrition security and (8) biodiversity). During these years, the Conference on Sustainable Development took place in Rio, Brazil (RIO+20) with the outcome document 'The Future We Want', where the importance of sustainable food production, food, and nutrition security was underlined. Beginning in 2013, and as a result of RIO+20, an Open Working Group on SDGs started a dialogue about the sustainability of food production and consumption, among others. Subsequently, between 2016–2019, SDGs were identified as a new theme. Sustainable diets play a major part in this time slice, but sustainable nutrition also makes an appearance as a novel theme. The newly emerged themes are strongly related to the production and consumption of meat ('livestock' and 'meat reduction') and other sustainable protein sources (edible insects, plant proteins, legumes). In the last two years (2020–2022), 'sustainable diet' has prevailed as a term. Several branches of the previous period were brought together in the 'carbon footprint' theme. Lastly, 'vegetarian' is one of the new themes that have risen in popularity (see Supplementary Figure S1 for a detailed thematic evolution).

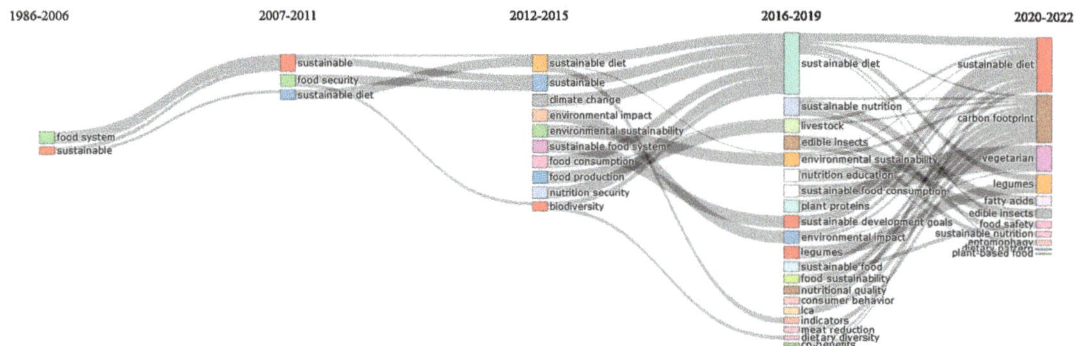

Figure 5. Evolution of the main themes in research using authors' keywords in five time slices: 1986–2006, 2007–2011, 2012–2015, 2016–2019, and 2020–2022.

3.7. Authors' Keyword Co-Occurrence

As shown in Figure 6, there were 49 items and 633 links with a total link strength of 27,612, assigned into six clusters by the VOSviewer algorithm. The name of each cluster noted in bold letters is our interpretation:

1. General overview of SDs with a focus on the environmental pillar: (red color) sustainable diet, carbon footprint, COVID-19, dietary change, environmental footprint, environmental sustainability, food choice, food consumption, food sustainability, greenhouse gas emissions, healthy diet, Mediterranean diet, nutritional quality, sustainable consumption, sustainable food system, sustainable healthy diets, and water footprint;
2. Linkage SDs with health: (green color) dietary patterns, children, food safety, food waste, healthy diets, obesity, public health, sustainable nutrition;
3. Sustainable Nutrition and meat consumption: (blue color) sustainable, nutrition, diet, environment, food, health, meat, protein;
4. SDGs with a focus on SDG 2 (Zero Hunger): (yellow color) food system, agriculture, climate change, food and nutrition security, food security, malnutrition, Sustainable Development Goals;
5. Environmental impact of dietary choices: (purple color) environmental impact, biodiversity, diet quality, dietary guidelines, land use, life cycle assessment;
6. Consumer behavior and plant-based food choices: (bright blue color) food policy, consumer behavior, vegetarian.

The keyword with the largest node is 'sustainable' (with 12,091 total link strength) and follows 'sustainable diet' (with 9219). The strongest link is between the keywords 'sustainable' and 'nutrition' with 3721 link strength. However, these keywords were expected to be in high places as they were included in the keywords searched in the Scopus database. The first word after them was 'food system', which belonged in the fourth cluster and has a total link strength of 3403.

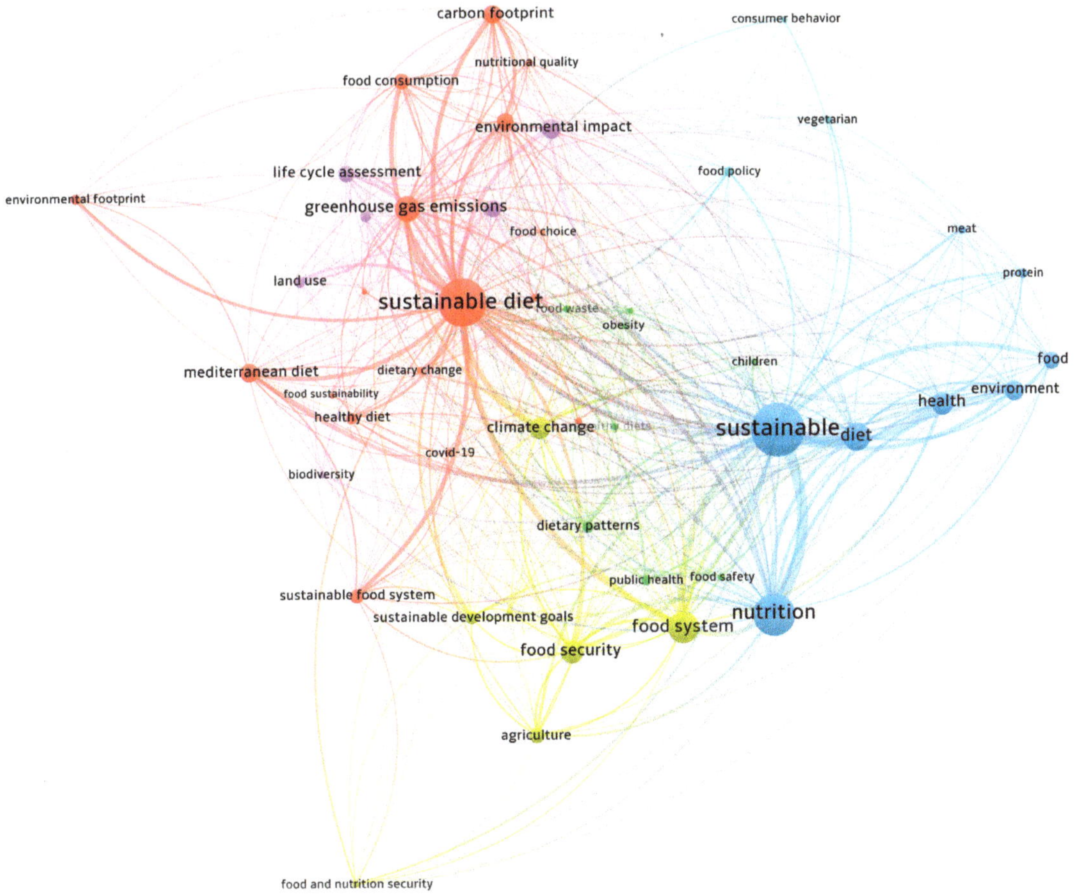

Figure 6. Co-occurrences network of Author keywords of Sustainable Diets and Sustainable nutrition in the final dataset (*n* = 1407) using the VOSviewer software.

3.8. Detecting UN Sustainable Development Goals in Text

As depicted in Figure 7, the papers included in the final dataset were mostly related to SDG 2 (Zero Hunger), with almost 2500 hits, SDG 12 (Responsible Consumption and Production) followed by almost 1500 hits, SDG 3 (Good Health and Well-being), and SDG 11 (Sustainable Cities and Communities), followed by approximately 1000 hits each. However, all SDGs were represented in the titles of this collection, some more and some less, revealing a clear interconnection among them.

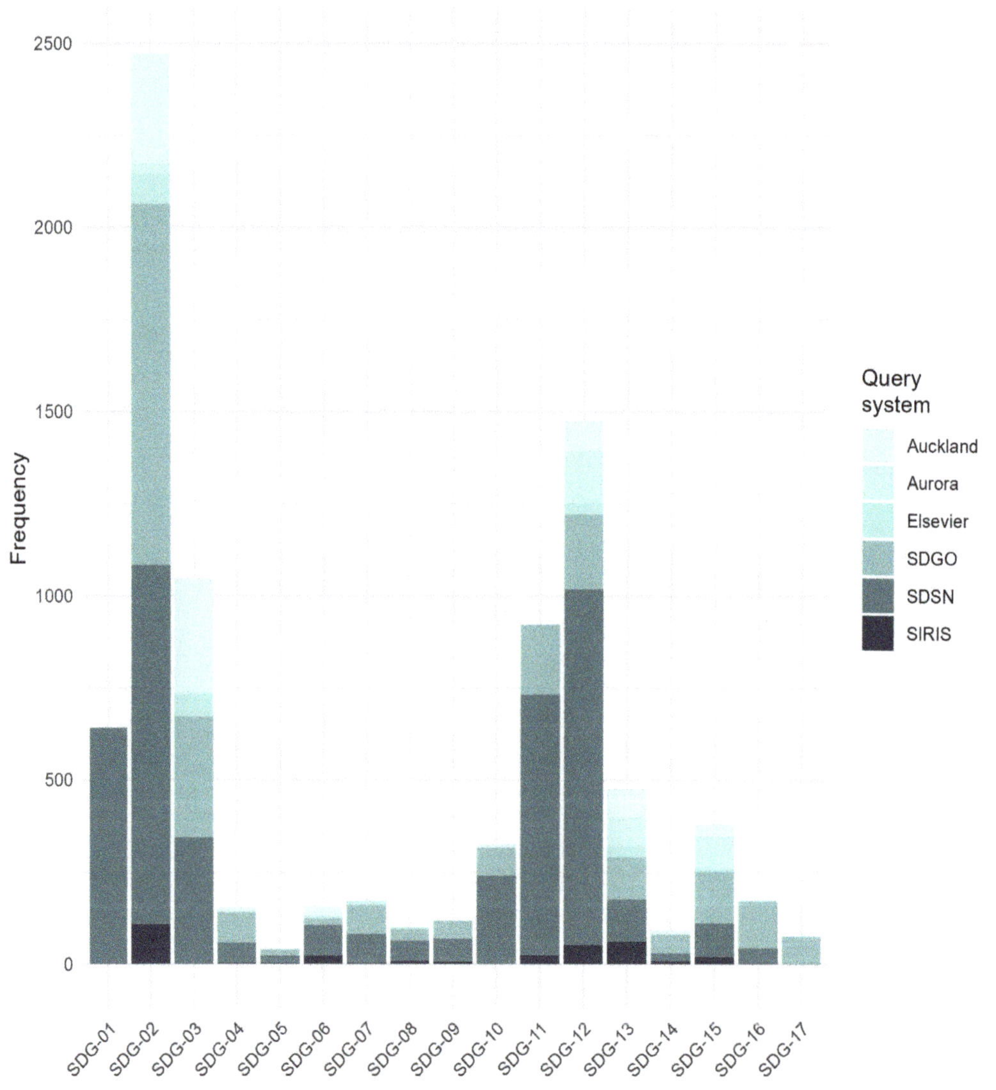

Figure 7. Relative number of hits per SDG from the papers' titles cumulatively across 6 query systems using R package text2sdg in the final dataset (*n* = 1407).

4. Discussion

4.1. Summary of the Findings

The international events that took place in recent years exerted a considerable impact on scientific knowledge and production soon after 2012 when there was an increase in the production of scientific documents. The definition of SDs by FAO played an important role in this issue. In 2015, there were two important key-point events: the Paris Agreement on Climate Change and the 2030 Agenda for Sustainable Development, which was set to promote the 17 SDGs. Therefore, the turning points in the production of scientific literature corresponded to some of the major international events, revealing an interaction between policymaking and scientific research.

The core journals publishing about SDs have as main subject areas the agricultural and the biological sciences, the environmental sciences, the social sciences, nursing, and medicine, confirming the interdisciplinary nature of this field and the connections between nutrition, health, food systems, agricultural production, and environmental protection as main subject areas.

Regarding the two terms 'sustainable diet' and 'sustainable nutrition', the first one was more highly related to environmental issues compared to the latter one, which was directly connected with health. The term 'sustainable diets' is broadly used as a phrase and is often characterized by additional adjectives (e.g., healthy). Furthermore, the geographic distribution of studies could be related to the different dietary patterns studied for their sustainability (e.g., the Mediterranean Diet and New Nordic Diet).

Notably, the countries belonging to the Mediterranean Region (Italy, Greece, Spain, and Portugal) had multiple publications in the relevant bibliographic collection. MD is a topic that has been highly studied by researchers in these countries. There was also identified a close collaboration between countries in the Mediterranean area (e.g., Italy and Spain). Mediterranean Diet was set as an example of SDs [28]. In this collection, 95 articles and reviews contained the term Mediterranean in their title, abstract, or keywords. Apart from its health benefits and economic viability, it has been studied as a dietary pattern with a low environmental impact [46–48], as its environmental footprints (energy consumption, water consumption, GHG emissions, and agricultural land use) and its ecological footprint are lower compared to that of other Western diets [49,50]. Interestingly, as mentioned by Gaforio et al., several studies have highlighted the beneficial healthy effects of extra virgin olive oil consumption for the prevention of several chronic diseases are highlighted making it a good choice both for human health and sustainable agriculture as verified by numerous clinical studies [51]. We also identified papers with different updates to the classic Mediterranean Diet Pyramid, the double pyramid model developed by the Barilla Center [49], and the updated Mediterranean Diet Pyramid [52]. Another considerable aspect of the Mediterranean Diet was the relevant indicators of its sustainability [27].

Alongside the MD, the New Nordic Diet (NND) was characterized as a sustainable diet in this collection ($n = 13$). It is a newly created diet with strong roots in Nordic culinary traditions, which principally aims at improving both human health and environmental health. Noticeably, it was founded in 2004 by a group of well-recognized Nordic chefs who set out to give prominence to regional, local products [53]. Although the NND mainly focuses on Denmark, Finland, Iceland, Sweden, and Norway, a new variant of the NND, called the Southern New Nordic Diet, has recently been discussed as a nutritional scenario in southern Europe (northwestern Spain and northern Portugal) [54].

In comparison with the Average Danish Diet, NND exerts significant environmental and social benefits [55]; however, it has been estimated as more expensive than other dietary patterns [56]; NND has been examined for potential effects on various aspects of childhood health and development [57,58], promoting healthy eating habits formed throughout pregnancy and infancy [59].

According to our findings, COVID-19 was found in the red cluster linked with MD (link strength = 9, total link strength = 109) (Figure 6). Given the various health advantages of the Mediterranean Dietary Pattern, it may be considered the most optimal dietary pattern for preventing COVID-19 infection in non-infected persons, as well as it could be rendered as an adjuvant therapy for improving disease outcomes while simultaneously favoring recovery in COVID-19 patients [60]. COVID-19 exerted a detrimental impact on food security, but it has also provided a 'window of opportunity' to initiate a shift to more sustainable diets in Palestine [61]. In this territory, the multiple crises alongside with COVID-19 pandemic influenced dietary and eating behaviors. Agricultural production and consumption were affected and limited on a local scale, and thus a policy for growing vegetables by households has highly been promoted for growing vegetables in spaces available close to peoples' homes. In Italy, during the first and second phases of the COVID-19 pandemic, there was a growing desire to consume certified sustainable food products [62]. On the other hand, the lockdown caused significant changes in family dietary and eating patterns [63]. However, there are no adequate published studies so far to provide reliable and conclusive evidence about the potential benefits of MD against COVID-19. In this aspect, forthcoming studies are currently in progress, and they will publish their results in the next few years in order for more precise conclusion to be drawn in this topic.

Food security has widely been researched in 294 publications. As shown in Figure 5, a whole cluster is dedicated to this term, as it has extensively been studied due to the even more rapidly growing world population. Several human activities, such as the way governing food production and consumption, put the natural resources of our planet in danger, whereas climate change and environmental negative impacts are gradually being accelerated by unsustainable consumption choices [12]. Importantly, sustainability is linked to the viability of the food system, public health, and prosperity. Additionally, SDGs Agenda has increasingly promoted food security mainly by eliminating malnutrition [64–66]. Goal 2 of the SDGs is actually set to 'end hunger, achieve food security and improve nutrition'. FAO in Rome Declaration on World Food Security (1996) and in Declaration of the World Summit on Food Security (2009) set clear goals to eradicate hunger.

Climate change and greenhouse gas emissions constitute two terms that have been studied in relation to SDs since 2017, showing the impact of the Paris Agreement and the SDGs [67]. Alarmingly enough, climate change could also affect agricultural and livestock production [68,69]. Furthermore, greenhouse gas emissions could be associated with a reduction in agriculture and food production yields [70]. Therefore, it has been proposed that dietary changes may positively affect both human health and environmental health [71,72].

Meat production and consumption have attracted considerable scientific interest. At the same time, alternative sources of protein, such as edible insects, plant proteins, and legumes, have also appeared in the international scientific literature. Entomophagy is considered a controversial and alternative topic proposed mainly as a sustainable solution for protein intake in the fight against malnutrition [73–75]. Despite its long history in some cultures, the willingness to consume these novel food products remains still disputable [76,77]. Vegetarianism has also been studied as a plant-based diet with a low environmental footprint and a more accepted dietary pattern [75,78,79], whereas home cooking is studied as being more environmentally friendly and beneficial to health [80].

By observing the clusters emerging from the co-occurrence network, six separate categories have appeared: (1) a general overview of SDs with a focus on the environmental pillar, (2) a linkage SDs with human health, (3) sustainable nutrition and meat consumption, (4) SDGs with a focus on SDG 2 (Zero Hunger), (5) the environmental impact of dietary choices, and (6) consumer behaviors and plant-based food choices.

SDGs focusing on areas such as hunger, human health, and climate were identified as the most obvious beneficiaries of a shift toward sustainable food systems. However, there were found important linkages concerning food systems and less obvious linkages regarding SDGs, such as SDG 12 (Responsible Consumption and Production), SDG 3 (Good Health and Well-being), and SDG 11 (Sustainable Cities and Communities). The above findings support substantial evidence that food could be considered as a common thread linking all 17 SDGs. Earlier policymakers realized that reforming food systems may provide a potentially powerful lever for sustainable development, which may allow us to meet the SDGs.

4.2. Limitations and Future Research

There are certain limitations, mainly regarding the bibliometric analysis approach itself. We set a time limit for this study to the last 35 years by the first introduction of the term in the international scientific literature. However, it could be interesting to explore how the terms were used before that time limit and, in parallel, to study the thematic evolution throughout the previous decades if the documents before 1985 were digitally available. Second, in this study, the Scopus bibliographic database was used. However, comparing the findings in other databases (for example, Web of Science, Dimensions) could provide additional information about the relevant study area. Beyond the limitations, the data in this study provide significant insights into the editorial activity of international scientific journals and the corresponding topics that have been discussed over the last few years. Moreover, this study clearly presents a quantitative analysis of the international scientific literature. Moreover, the present findings reinforce the fact that future research is strongly recommended to adopt a more qualitative approach.

5. Conclusions

This is the first bibliometric analysis study that provides a thorough overview of the international research concerning the currently available scientific papers on sustainable diets and sustainable nutrition published in peer-reviewed journals between 1986 and 2022. The research topics were answered by using two alternative methodologies. Firstly, a bibliometric analysis answered the questions regarding the article production growth in the field within almost 35 years, the trending topics, the journals, and their subject areas, as well as the geographic distribution of the studied papers. A co-occurrence network was derived to reliably identify the core author keywords and their relationships, while the trend topics and their evolution over time were explored. Secondly, a content analysis of the dataset was efficiently performed by using R-language to detect which SDGs were related to the terms under study.

Therefore, we have drawn the following main conclusions: The findings of this study clearly confirm the exponential growth of scientific interest in the undersigned subject area. Some of the turning points in the evolution of publications follow the international timeline of events, revealing whether they can impact scientific production. Considering the key components of SDs, the greatest emphasis of the scientific community in this bibliographic analysis was highly ascribed to food security, climate change, environmental impact (water footprint, carbon footprint, greenhouse gas emissions), food systems, and their sustainability, Mediterranean Diet, and Sustainable Development Goals. This study first presents a thorough overview of the current trends of the scientific community concerning the global interest in the SDs, which could be highly considered as an effective strategy to promote environmental sustainability and planet health.

Supplementary Materials: The following supporting information can be downloaded at: https://www.mdpi.com/article/10.3390/su151511957/s1, Table S1: Main Scientific Events and publications (Burlingame et al., 2022 [12]); Table S2: The top 25 publishing countries in the SDs topic during 1986–2022.; Figure S1: Thematic Evolution using authors' keywords in in five time slices: 1986–2006, 2007–2011, 2012–2015, 2016–2019, 2020–2022.

Author Contributions: Conceptualization, M.G., A.Y.T., C.G. and G.K.V.; methodology, M.G., A.Y.T., C.G. and G.K.V.; software, M.G. and G.K.V.; validation, M.G., C.G. and G.K.V.; formal analysis, M.G., A.Y.T. and I.A.; investigation, M.G., S.K.P. and G.K.V.; resources, M.G., S.K.P. and I.A.; data curation, M.G. and G.K.V.; writing—original draft preparation, M.G.; writing—review and editing, M.G., A.Y.T., C.G., S.K.P., I.A. and G.K.V.; visualization, M.G.; supervision, G.K.V. All authors have read and agreed to the published version of the manuscript.

Funding: This research received no external funding.

Institutional Review Board Statement: Not applicable.

Informed Consent Statement: Not applicable.

Data Availability Statement: The data collection was made from the Scopus bibliographic database in April 2023.

Acknowledgments: The authors greatly appreciate the valuable comments and suggestions of the anonymous reviewers that helped improve the quality of this paper.

Conflicts of Interest: The authors declare no conflict of interests.

References

1. Willett, W.; Rockström, J.; Loken, B.; Springmann, M.; Lang, T.; Vermeulen, S.; Garnett, T.; Tilman, D.; DeClerck, F.; Wood, A.; et al. Food in the Anthropocene: The EAT–Lancet Commission on Healthy Diets from Sustainable Food Systems. *Lancet* **2019**, *393*, 447–492. [CrossRef]
2. Gussow, J.D.; Clancy, K.L. Dietary Guidelines for Sustainability. *J. Nutr. Educ.* **1986**, *18*, 1–5. [CrossRef]
3. Declerck, F.A.J.; Fanzo, J.; Palm, C.; Remans, R. Ecological Approaches to Human Nutrition. *Food Nutr. Bull.* **2011**, *32*, S41–S50. [CrossRef]
4. Friel, S.; Barosh, L.J.; Lawrence, M. Towards Healthy and Sustainable Food Consumption: An Australian Case Study. *Public Health Nutr.* **2014**, *17*, 1156–1166. [CrossRef]
5. Jones, A.D.; Hoey, L.; Blesh, J.; Miller, L.; Green, A.; Shapiro, L.F. A Systematic Review of the Measurement of Sustainable Diets. *Adv. Nutr.* **2016**, *7*, 641–664. [CrossRef]
6. Eustachio Colombo, P.; Elinder, L.S.; Lindroos, A.K.; Parlesak, A. Designing Nutritionally Adequate and Climate-Friendly Diets for Omnivorous, Pescatarian, Vegetarian and Vegan Adolescents in Sweden Using Linear Optimization. *Nutrients* **2021**, *13*, 2507. [CrossRef]
7. Smetana, S.M.; Bornkessel, S.; Heinz, V. A Path From Sustainable Nutrition to Nutritional Sustainability of Complex Food Systems. *Front. Nutr.* **2019**, *6*, 39. [CrossRef]
8. Portugal-Nunes, C.; Nunes, F.M.; Fraga, I.; Saraiva, C.; Gonçalves, C. Assessment of the Methodology That Is Used to Determine the Nutritional Sustainability of the Mediterranean Diet-A Scoping Review. *Front. Nutr.* **2021**, *8*, 772133. [CrossRef]
9. Burlingame, B.; Charrondiere, U.R.; Dernini, S.; Stadlmayr, B.; Mondovì, S. Food Biodiversity and Sustainable Diets: Implications of Applications for Food Production and Processing. In *Green Technologies in Food Production and Processing*; Boye, J.I., Arcand, Y., Eds.; Food Engineering Series; Springer: Boston, MA, USA, 2012; pp. 643–657, ISBN 978-1-4614-1586-2.
10. Allen, T.; Prosperi, P.; Cogill, B.; Flichman, G. Agricultural Biodiversity, Social-Ecological Systems and Sustainable Diets. *Proc. Nutr. Soc.* **2014**, *73*, 498–508. [CrossRef]
11. Burlingame, B.; Dernini, S. Sustainable Diets: The Mediterranean Diet as an Example. *Public Health Nutr.* **2011**, *14*, 2285–2287. [CrossRef]
12. Burlingame, B.; Lawrence, M.; Macdiarmid, J.; Dernini, S.; Oenema, S. IUNS Task Force on Sustainable Diets—LINKING NUTRITION AND FOOD SYSTEMS. *Trends Food Sci. Technol.* **2022**, *130*, 42–50. [CrossRef]
13. Alsaffar, A.A. Sustainable Diets: The Interaction between Food Industry, Nutrition, Health and the Environment. *Food Sci. Technol. Int.* **2016**, *22*, 102–111. [CrossRef]
14. Johnston, J.L.; Fanzo, J.C.; Cogill, B. Understanding Sustainable Diets: A Descriptive Analysis of the Determinants and Processes That Influence Diets and Their Impact on Health, Food Security, and Environmental Sustainability. *Adv. Nutr.* **2014**, *5*, 418–429. [CrossRef]
15. Barbier, E.B. The Concept of Sustainable Economic Development. *Environ. Conserv.* **1987**, *14*, 101–110. [CrossRef]
16. Purvis, B.; Mao, Y.; Robinson, D. Three Pillars of Sustainability: In Search of Conceptual Origins. *Sustain. Sci.* **2019**, *14*, 681–695. [CrossRef]
17. Hens, L.; Nath, B. The Johannesburg Conference. *Environ. Dev. Sustain.* **2003**, *5*, 7–39. [CrossRef]
18. Johnson, R.B. Sustainable Agriculture: Competing Visions and Policy Avenues. *Int. J. Sustain. Dev. World Ecol.* **2006**, *13*, 469–480. [CrossRef]
19. Kumar, S.; Kumar, N.; Vivekadhish, S. Millennium Development Goals (MDGS) to Sustainable Development Goals (SDGS): Addressing Unfinished Agenda and Strengthening Sustainable Development and Partnership. *Indian J. Community Med.* **2016**, *41*, 1. [CrossRef]
20. Nilsson, M.; Griggs, D.; Visbeck, M. Policy: Map the Interactions between Sustainable Development Goals. *Nature* **2016**, *534*, 320–322. [CrossRef]
21. Galabada, J.K. Towards the Sustainable Development Goal of Zero Hunger: What Role Do Institutions Play? *Sustainability* **2022**, *14*, 4598. [CrossRef]
22. Lencucha, R.; Kulenova, A.; Thow, A.M. Framing Policy Objectives in the Sustainable Development Goals: Hierarchy, Balance, or Transformation? *Glob. Health* **2023**, *19*, 5. [CrossRef] [PubMed]
23. Grosso, G.; Mateo, A.; Rangelov, N.; Buzeti, T.; Birt, C. Nutrition in the Context of the Sustainable Development Goals. *Eur. J. Public Health* **2020**, *30*, i19–i23. [CrossRef] [PubMed]
24. Meybeck, A.; Gitz, V. Sustainable Diets within Sustainable Food Systems. *Proc. Nutr. Soc.* **2017**, *76*, 1–11. [CrossRef] [PubMed]
25. The Economist Intelligence Unit and Barilla Foundation Fixing Food 2021: An Opportunity for G20 Countries to Lead the Way. Available online: https://foodsustainability.eiu.com (accessed on 15 January 2022).
26. Haines, A.; Scheelbeek, P. European Green Deal: A Major Opportunity for Health Improvement. *Lancet* **2020**, *395*, 1327–1329. [CrossRef] [PubMed]
27. Donini, L.M.; Dernini, S.; Lairon, D.; Serra-Majem, L.; Amiot, M.-J.; del Balzo, V.; Giusti, A.-M.; Burlingame, B.; Belahsen, R.; Maiani, G.; et al. A Consensus Proposal for Nutritional Indicators to Assess the Sustainability of a Healthy Diet: The Mediterranean Diet as a Case Study. *Front. Nutr.* **2016**, *3*, 37. [CrossRef] [PubMed]

28. Dernini, S.; Berry, E.; Serra-Majem, L.; La Vecchia, C.; Capone, R.; Medina, F.; Aranceta-Bartrina, J.; Belahsen, R.; Burlingame, B.; Calabrese, G.; et al. Med Diet 4.0: The Mediterranean Diet with Four Sustainable Benefits. *Public Health Nutr.* **2017**, *20*, 1322–1330. [CrossRef]

29. Dernini, S.; Meybeck, A.; Burlingame, B.; Gitz, V.; Lacirignola, C.; Debs, P.; Capone, R.; Bilali, H.E. Developing a Methodological Approach for Assessing the Sustainability of Diets: The Mediterranean Diet as a Case Study. *New Medit* **2013**, *12*, 28–36.

30. Rousseau, R. Forgotten Founder of Bibliometrics. *Nature* **2014**, *510*, 218. [CrossRef]

31. Zupic, I.; Čater, T. Bibliometric Methods in Management and Organization. *Organ. Res. Methods* **2015**, *18*, 429–472. [CrossRef]

32. Aria, M.; Cuccurullo, C. Bibliometrix: An R-Tool for Comprehensive Science Mapping Analysis. *J. Informetr.* **2017**, *11*, 959–975. [CrossRef]

33. Falagas, M.E.; Pitsouni, E.I.; Malietzis, G.A.; Pappas, G. Comparison of PubMed, Scopus, Web of Science, and Google Scholar: Strengths and Weaknesses. *FASEB J.* **2008**, *22*, 338–342. [CrossRef]

34. Capobianco-Uriarte, M.d.l.M.; Casado-Belmonte, M.d.P.; Marín-Carrillo, G.M.; Terán-Yépez, E. A Bibliometric Analysis of International Competitiveness (1983–2017). *Sustainability* **2019**, *11*, 1877. [CrossRef]

35. Fox, S.; Lynch, J.; D'Alton, P.; Carr, A. Psycho-Oncology: A Bibliometric Review of the 100 Most-Cited Articles. *Healthcare* **2021**, *9*, 1008. [CrossRef]

36. Tricco, A.C.; Lillie, E.; Zarin, W.; O'Brien, K.K.; Colquhoun, H.; Levac, D.; Moher, D.; Peters, M.D.J.; Horsley, T.; Weeks, L.; et al. PRISMA Extension for Scoping Reviews (PRISMA-ScR): Checklist and Explanation. *Ann. Intern. Med.* **2018**, *169*, 467–473. [CrossRef] [PubMed]

37. Page, M.J.; Moher, D.; Bossuyt, P.M.; Boutron, I.; Hoffmann, T.C.; Mulrow, C.D.; Shamseer, L.; Tetzlaff, J.M.; Akl, E.A.; Brennan, S.E.; et al. PRISMA 2020 Explanation and Elaboration: Updated Guidance and Exemplars for Reporting Systematic Reviews. *BMJ* **2021**, *372*, n160. [CrossRef] [PubMed]

38. Liberati, A.; Altman, D.G.; Tetzlaff, J.; Mulrow, C.; Gøtzsche, P.C.; Ioannidis, J.P.A.; Clarke, M.; Devereaux, P.J.; Kleijnen, J.; Moher, D. The PRISMA Statement for Reporting Systematic Reviews and Meta-Analyses of Studies That Evaluate Health Care Interventions: Explanation and Elaboration. *PLoS Med.* **2009**, *6*, e1000100. [CrossRef] [PubMed]

39. Wulff, D.U.; Meier, D.S. Text2sdg: Detecting UN Sustainable Development Goals in Text. 2021. Available online: https://zenodo.org/record/5553980 (accessed on 2 July 2023).

40. Van Eck, N.J.; Waltman, L. Software Survey: VOSviewer, a Computer Program for Bibliometric Mapping. *Scientometrics* **2010**, *84*, 523–538. [CrossRef] [PubMed]

41. Li, H.; An, H.; Wang, Y.; Huang, J.; Gao, X. Evolutionary Features of Academic Articles Co-Keyword Network and Keywords Co-Occurrence Network: Based on Two-Mode Affiliation Network. *Phys. Stat. Mech. Its Appl.* **2016**, *450*, 657–669. [CrossRef]

42. Bandara, T. Scientific Footprint of Indian Major Carp Research in South Asia: A Scientometric Study between 1955 and 2018. *J. Appl. Aquac.* **2021**, *33*, 267–278. [CrossRef]

43. de Oliveira, O.J.; da Silva, F.F.; Juliani, F.; Barbosa, L.C.F.M.; Nunhes, T.V. *Bibliometric Method for Mapping the State-of-the-Art and Identifying Research Gaps and Trends in Literature: An Essential Instrument to Support the Development of Scientific Projects*; IntechOpen: London, UK, 2019; ISBN 978-1-78984-713-0.

44. Agbo, F.J.; Sanusi, I.T.; Oyelere, S.S.; Suhonen, J. Application of Virtual Reality in Computer Science Education: A Systemic Review Based on Bibliometric and Content Analysis Methods. *Educ. Sci.* **2021**, *11*, 142. [CrossRef]

45. Burlingame, B.; Dernini, S. Sustainable Diets and Biodiversity: Directions and Solutions for Policy, Research and Action. In Proceedings of the Intenational Scientific Symposium Biodiversity and Sustainable Diets United Against Hunger, Rome, Italy, 3–5 November 2010.

46. Dernini, S.; Berry, E.M. Mediterranean Diet: From a Healthy Diet to a Sustainable Dietary Pattern. *Front. Nutr.* **2015**, *2*, 15. [CrossRef] [PubMed]

47. Germani, A.; Vitiello, V.; Giusti, A.M.; Pinto, A.; Donini, L.M.; del Balzo, V. Environmental and Economic Sustainability of the Mediterranean Diet. *Int. J. Food Sci. Nutr.* **2014**, *65*, 1008–1012. [CrossRef] [PubMed]

48. Van Dooren, C.; Marinussen, M.; Blonk, H.; Aiking, H.; Vellinga, P. Exploring Dietary Guidelines Based on Ecological and Nutritional Values: A Comparison of Six Dietary Patterns. *Food Policy* **2014**, *44*, 36–46. [CrossRef]

49. Ruini, L.F.; Ciati, R.; Pratesi, C.A.; Marino, M.; Principato, L.; Vannuzzi, E. Working toward Healthy and Sustainable Diets: The "Double Pyramid Model" Developed by the Barilla Center for Food and Nutrition to Raise Awareness about the Environmental and Nutritional Impact of Foods. *Front. Nutr.* **2015**, *2*, 9. [CrossRef] [PubMed]

50. Sáez-Almendros, S.; Obrador, B.; Bach-Faig, A.; Serra-Majem, L. Environmental Footprints of Mediterranean versus Western Dietary Patterns: Beyond the Health Benefits of the Mediterranean Diet. *Environ. Health* **2013**, *12*, 118. [CrossRef]

51. Gaforio, J.J.; Visioli, F.; Alarcón-de-la-Lastra, C.; Castañer, O.; Delgado-Rodríguez, M.; Fitó, M.; Hernández, A.F.; Huertas, J.R.; Martínez-González, M.A.; Menendez, J.A.; et al. Virgin Olive Oil and Health: Summary of the III International Conference on Virgin Olive Oil and Health Consensus Report, JAEN (Spain) 2018. *Nutrients* **2019**, *11*, 2039. [CrossRef]

52. Serra-Majem, L.; Tomaino, L.; Dernini, S.; Berry, E.M.; Lairon, D.; Ngo de la Cruz, J.; Bach-Faig, A.; Donini, L.M.; Medina, F.-X.; Belahsen, R.; et al. Updating the Mediterranean Diet Pyramid towards Sustainability: Focus on Environmental Concerns. *Int. J. Environ. Res. Public Health* **2020**, *17*, 8758. [CrossRef]

53. Hachem, F.; Vanham, D.; Moreno, L.A. Territorial and Sustainable Healthy Diets. *Food Nutr. Bull.* **2020**, *41*, 87S–103S. [CrossRef]
54. Cambeses-Franco, C.; González-García, S.; Feijoo, G.; Moreira, M.T. Encompassing Health and Nutrition with the Adherence to the Environmentally Sustainable New Nordic Diet in Southern Europe. *J. Clean. Prod.* **2021**, *327*, 129470. [CrossRef]
55. Saxe, H. The New Nordic Diet Is an Effective Tool in Environmental Protection: It Reduces the Associated Socioeconomic Cost of Diets. *Am. J. Clin. Nutr.* **2014**, *99*, 1117–1125. [CrossRef]
56. Jensen, J.D.; Poulsen, S.K. The New Nordic Diet—Consumer Expenditures and Economic Incentives Estimated from a Controlled Intervention. *BMC Public Health* **2013**, *13*, 1114. [CrossRef] [PubMed]
57. Agnihotri, N.; Øverby, N.C.; Bere, E.; Wills, A.K.; Brantsæter, A.L.; Hillesund, E.R. Childhood Adherence to a Potentially Healthy and Sustainable Nordic Diet and Later Overweight: The Norwegian Mother, Father and Child Cohort Study (MoBa). *Matern. Child Nutr.* **2021**, *17*, e13101. [CrossRef]
58. Agnihotri, N.; Rudjord Hillesund, E.; Bere, E.; Wills, A.K.; Brantsæter, A.L.; Øverby, N.C. Development and Description of New Nordic Diet Scores across Infancy and Childhood in the Norwegian Mother, Father and Child Cohort Study (MoBa). *Matern. Child Nutr.* **2021**, *17*, e13150. [CrossRef] [PubMed]
59. Mazzocchi, A.; De Cosmi, V.; Scaglioni, S.; Agostoni, C. Towards a More Sustainable Nutrition: Complementary Feeding and Early Taste Experiences as a Basis for Future Food Choices. *Nutrients* **2021**, *13*, 2695. [CrossRef] [PubMed]
60. Bagnato, C.; Perfetto, C.; Labanca, F.; Negrin, L.C. The Mediterranean Diet: Healthy and Sustainable Dietary Pattern in the Time of SARS-CoV-2. *Mediterr. J. Nutr. Metab.* **2021**, *14*, 365–381. [CrossRef]
61. Ben Hassen, T.; El Bilali, H.; Allahyari, M.S.; Morrar, R. Food Attitudes and Consumer Behavior towards Food in Conflict-Affected Zones during the COVID-19 Pandemic: Case of the Palestinian Territories. *Br. Food J.* **2022**, *124*, 2921–2936. [CrossRef]
62. Castellini, G.; Savarese, M.; Graffigna, G. The Impact of COVID-19 Outbreak in Italy on the Sustainable Food Consumption Intention From a "One Health" Perspective. *Front. Nutr.* **2021**, *8*, 622122. [CrossRef]
63. Docimo, R.; Costacurta, M.; Gualtieri, P.; Pujia, A.; Leggeri, C.; Attinà, A.; Cinelli, G.; Giannattasio, S.; Rampello, T.; Di Renzo, L. Cariogenic Risk and COVID-19 Lockdown in a Paediatric Population. *Int. J. Environ. Res. Public Health* **2021**, *18*, 7558. [CrossRef]
64. Cakmak, I. Plant Nutrition Research: Priorities to Meet Human Needs for Food in Sustainable Ways. *Plant Soil* **2002**, *247*, 3–24. [CrossRef]
65. Frison, E.A.; Cherfas, J.; Hodgkin, T. Agricultural Biodiversity Is Essential for a Sustainable Improvement in Food and Nutrition Security. *Sustainability* **2011**, *3*, 238–253. [CrossRef]
66. Godfray, H.C.J.; Crute, I.R.; Haddad, L.; Lawrence, D.; Muir, J.F.; Nisbett, N.; Pretty, J.; Robinson, S.; Toulmin, C.; Whiteley, R. The Future of the Global Food System. *Philos. Trans. R. Soc. B Biol. Sci.* **2010**, *365*, 2769–2777. [CrossRef] [PubMed]
67. Ritchie, H.; Reay, D.S.; Higgins, P. The Impact of Global Dietary Guidelines on Climate Change. *Glob. Environ. Chang.* **2018**, *49*, 46–55. [CrossRef]
68. Mabhaudhi, T.; Chimonyo, V.G.P.; Hlahla, S.; Massawe, F.; Mayes, S.; Nhamo, L.; Modi, A.T. Prospects of Orphan Crops in Climate Change. *Planta* **2019**, *250*, 695–708. [CrossRef] [PubMed]
69. Mugambiwa, S.S.; Tirivangasi, H.M. Climate Change: A Threat towards Achieving 'Sustainable Development Goal Number Two' (End Hunger, Achieve Food Security and Improved Nutrition and Promote Sustainable Agriculture) in South Africa. *Jàmbá J. Disaster Risk Stud.* **2017**, *9*, 350. [CrossRef]
70. Sinha, S.K.; Swaminathan, M.S. Deforestation, Climate Change and Sustainable Nutrition Security: A Case Study of India. *Clim. Chang.* **1991**, *19*, 201–209. [CrossRef]
71. Springmann, M.; Godfray, H.C.J.; Rayner, M.; Scarborough, P. Analysis and Valuation of the Health and Climate Change Cobenefits of Dietary Change. *Proc. Natl. Acad. Sci. USA* **2016**, *113*, 4146–4151. [CrossRef]
72. Kumanyika, S.; Afshin, A.; Arimond, M.; Lawrence, M.; McNaughton, S.A.; Nishida, C. Approaches to Defining Healthy Diets: A Background Paper for the International Expert Consultation on Sustainable Healthy Diets. *Food Nutr. Bull.* **2020**, *41*, 7S–30S. [CrossRef]
73. Henchion, M.; Hayes, M.; Mullen, A.; Fenelon, M.; Tiwari, B. Future Protein Supply and Demand: Strategies and Factors Influencing a Sustainable Equilibrium. *Foods* **2017**, *6*, 53. [CrossRef]
74. Maya, C.; Cunha, L.M.; de Almeida Costa, A.I.; Veldkamp, T.; Roos, N. Introducing Insect- or Plant-Based Dinner Meals to Families in Denmark: Study Protocol for a Randomized Intervention Trial. *Trials* **2022**, *23*, 1028. [CrossRef]
75. Schösler, H.; Boer, J.D.; Boersema, J.J. Can We Cut out the Meat of the Dish? Constructing Consumer-Oriented Pathways towards Meat Substitution. *Appetite* **2012**, *58*, 39–47. [CrossRef]
76. Kim, T.-K.; Cha, J.Y.; Yong, H.I.; Jang, H.W.; Jung, S.; Choi, Y.-S. Application of Edible Insects as Novel Protein Sources and Strategies for Improving Their Processing. *Food Sci. Anim. Resour.* **2022**, *42*, 372–388. [CrossRef] [PubMed]
77. Mason, J.B.; Black, R.; Booth, S.L.; Brentano, A.; Broadbent, B.; Connolly, P.; Finley, J.; Goldin, J.; Griffin, T.; Hagen, K.; et al. Fostering Strategies to Expand the Consumption of Edible Insects: The Value of a Tripartite Coalition between Academia, Industry, and Government. *Curr. Dev. Nutr.* **2018**, *2*, nzy056. [CrossRef] [PubMed]
78. Melina, V.; Craig, W.; Levin, S. Position of the Academy of Nutrition and Dietetics: Vegetarian Diets. *J. Acad. Nutr. Diet.* **2016**, *116*, 1970–1980. [CrossRef] [PubMed]

79. Eker, S.; Reese, G.; Obersteiner, M. Modelling the Drivers of a Widespread Shift to Sustainable Diets. *Nat. Sustain.* **2019**, *2*, 725–735. [CrossRef]
80. Wolfson, J.A.; Willits-Smith, A.M.; Leung, C.W.; Heller, M.C.; Rose, D. Cooking at Home, Fast Food, Meat Consumption, and Dietary Carbon Footprint among US Adults. *Int. J. Environ. Res. Public Health* **2022**, *19*, 853. [CrossRef] [PubMed]

MDPI AG
Grosspeteranlage 5
4052 Basel
Switzerland
Tel.: +41 61 683 77 34

Sustainability Editorial Office
E-mail: sustainability@mdpi.com
www.mdpi.com/journal/sustainability